Instant Notes

Microbiology

Second Edition

The INSTANT NOTES series

Series editor
B.D. Hames
Department of Biochemistry and Molecular Biology, University of Leeds, Leeds, UK

Animal Biology
Genetics
Chemistry for Biologists
Immunology
Biochemistry 2nd edition
Molecular Biology 2nd edition
Neuroscience
Psychology
Developmental Biology
Plant Biology
Ecology 2nd edition
Microbiology 2nd edition

Forthcoming titles
Bioinformatics

The INSTANT NOTES Chemistry series
Consulting editor: Howard Stanbury

Organic Chemistry
Inorganic Chemistry
Physical Chemistry
Medicinal Chemistry
Analytical Chemistry

Instant Notes

Microbiology

Second Edition

J. Nicklin
Department of Biology,
Birkbeck College, London, UK

K. Graeme-Cook
Department of Biosciences,
University of Hertfordshire, Hatfield, UK

and

R. Killington
Department of Microbiology,
University of Leeds, Leeds, UK

First published 1999 (ISBN 1 85996 156 8)
Second Edition 2002 (ISBN 1 85996 267 X)

A CIP catalogue record for this book is available from the British Library.

ISBN 1 85996 267 X

BIOS Scientific Publishers Ltd
9 Newtec Place, Magdalen Road, Oxford OX4 1RE, UK
Tel. +44 (0)1865 726286. Fax +44 (0)1865 246823
World Wide Web home page: http://www.bios.co.uk/

Distributed exclusively in the United States, its dependent territories, Canada, Mexico, Central and South America, and the Caribbean by Springer-Verlag New York Inc., 175 Fifth Avenue, New York, USA, by arrangement with BIOS Scientific Publishers Ltd., 9 Newtec Place, Magdalen Road, Oxford, OX4 1RE, UK

Production Editor: Andrea Bosher
Typeset by Phoenix Photosetting, Chatham, Kent, UK
Printed by Biddles Ltd, Guildford, UK

CONTENTS

* Contributed by Dr Simon Baker, Department of Biology, Birkbeck College, London, UK

ABBREVIATIONS

A	adenine
ABC	ATP-binding cassette
ACP	acyl carrier protein
ADP	adenosine 5′-diphosphate
Ala	alanine
AMP	adenosine 5′-monophosphate
A-site	amino-acyl site (ribosome)
ATP	adenosine 5′-triphosphate
ATPase	ATP synthase
BHK	baby hamster kidney
Bp	base pair
C	cytosine
C-phase	Chromosome replication phase (bacterial cell cycle)
cAMP	cyclic adenosine 5′-monophosphate
CAP	catabolite activator protein
CAT	chloramphenicol acetyl transferase
CFU	colony-forming unit
CMV	cytomegalovirus
CNS	central nervous system
CoA	coenzyme A
CPE	cytopathic effect
CRP	cAMP receptor protein
CTL	cytotoxic T lymphocyte
Da	Dalton
D-Ala	D-alanine
DAP	*meso*-diaminopimelic acid
D-Glu	D-glutamic acid
DHA	dihydroxyacetone
DNA	deoxyribonucleic acid
dNTP	deoxyribonucleoside triphosphate
DOM	dissolved organic matter
D-phase	division phase (bacterial cell cycle)
Ds	double-stranded
EF	elongation factor
EM	electron microscopy
ER	endoplasmic reticulum
FAD	flavin adenine dinucleotide (oxidized)
$FADH_2$	flavin adenine dinucleotide (reduced)
FMN	flavin mononucleolides
G	guanine
G-phase	gap phase (bacterial cell cycle)

GTP	guanosine 5′-triphosphate
HA	hemagglutination
Hfr	high frequency recombination
HMP	hexose monophosphate pathway
HSV	herpes simplex virus
I	inosine
ICNV	International Committee on Nomenclature of Viruses
Ig	immunoglobulin
IHF	integration host factor
Inc group	incompatible group (of plasmids)
IS	insertion sequence
Kb	kilobase
KDO	2-keto-2-deoxyoctonate
KDPE	2-keto-2-deoxy-6-phosphogluconate
Lac	lactose
LBP	luciferin-binding protein
LPS	lipopolysaccharide
MAC	membrane-attack complex
MCP	methyl-accepting chemotaxis protein
MEM	minimal essential medium
MHC	major histocompatibility complex
m.o.i.	multiplicity of infection
mRNA	messenger ribonucleic acid
MTOC	microtubule organizing centre
NAD^+	nicotinamide adenine dinucleotide (oxidized form)
NADH	nicotinamide adenine dinucleotide (reduced form)
$NADP^+$	nicotinamide adenine dinucleotide phosphate (oxidized form)
NADPH	nicotinamide adenine dinucleotide phosphate (reduced form)
NAG	*N*-acetyl glucosamine
NAM	*N*-acetyl muramic acid
NB	nutrient broth
NTP	ribonucleoside triphosphate
O	operator
OD	optical density
Omp	outer membrane protein
P	promoter

Abbreviation	Meaning
PCBs	polychlorinated biphenyls
PCR	polymerase chain reaction
PEP	phosphoenol pyruvate
Pfu	plaque-forming unit
PHB	poly-β-hydroxybutyrate
Phe	phenylalanine
P_i	inorganic phosphate
PMF	proton motive force
PMN	polymorphonucleocyte
PP_i	inorganic pyrophosphate
PPP	pentose phosphate pathway
PS	photosystem
PSI and II	photosystems I and II
P-site	peptidyl site (ribosome)
R	resistance (plasmid)
r	rho factor
RBC	red blood cell
redox	reduction-oxidation
RER	rough endoplasmic reticulum
RNA	ribonucleic acid
rRNA	ribosomal RNA
rubisco	ribulose bisphosphate carboxylase
S	Svedberg coefficient
snRNA	small nuclear ribonucleic acid
SPB	spindle pole bodies
ss	single-stranded
T	thymine
TCA	tricarboxylic acid
TCID	tissue culture infective dose
tRNA	transfer RNA
Trp	tryptophan
TSB	tryptone soya broth
U	uracil
U_L,U_S	unique long, unique short
UDP	uridine diphosphate
UDPG	uridine disphosphate glucose
UV	ultraviolet light

PREFACE

The second edition of *Instant Notes in Microbiology* has been updated throughout the sections, including suggestions from readers of the first edition, new developments in the taxonomy of microbes, and new insights in molecular biology.

The section on Biochemistry has been completely rewritten (Section B) reflecting a change in authorship. Recent changes in the taxonomy of the Prokarya have necessitated the inclusion of a new topic of the Archea (D5) and the bacteriology and molecular biology sections have been updated to reflect the latest understanding of these rapidly evolving subjects.

The first edition sections on Algae and Protozoa have been combined into a new section, the Protista, reflecting the newest evidence and ideas on the evolution of this group of micro-organisms. Current taxonomic terms have been adopted throughout the sections on the fungi and protista.

The virology text remains a basic introduction to the topic. However, our knowledge of viruses, their replication mechanisms and interactions with their hosts is forever increasing as molecular and immunological techniques become more rapid and sophisticated. The second edition makes such revisions in our knowledge base. Virus classification has been updated and account has been made of trends in emerging viruses e.g. hepatitis C. A chapter on prions (whilst not viruses) and transmissible spongiform encephalopathies, has been introduced in the virology section.

We would like to thank the readers for their feedback, they are much appreciated as reviewers and we hope that this new edition has included as many of their suggestions as possible.

A1 The microbial world

Key Notes

What is a microbe?

The word microbe (microorganism) is used to describe an organism that is so small that, normally, it cannot be seen without the aid of a microscope. Viruses, Bacteria, Archaea, fungi, and protista are all included in this category.

Prokaryotes and eukaryotes

Microbes are found in all three major kingdoms of life: the Bacteria, the Archaea and the Eukarya. The Bacteria and Archaea are prokaryotes, while all other microbes are **eukaryotes.** There are many differences between prokaryote and eukaryote cells, the major distinction being the presence of a nucleus and other membrane-bound organelles in eukaryotes.

The importance of microbiology

Microbes are essential to life. Among their many roles, they are necessary for geochemical cycling and soil fertility. They are used to produce food as well as pharmaceutical and industrial compounds. On the negative side, they are the cause of many diseases of plants and animals and are responsible for the spoilage of food. Finally, microbes are used extensively in research laboratories to investigate cellular processes.

What is a microbe?

A **microbe** or **microorganism** is a member of a large, extremely diverse, group of organisms that are lumped together on the basis of one property – the fact that, normally, they are so small that they cannot be seen without the use of a microscope. The word microbe is therefore used to describe **viruses**, **Bacteria, Archaea**, **fungi** and **protista**: the relative sizes and nature of these are shown in *Table 1*. However, there are a few macroscopic microbes that can be seen by the naked eye including the fruiting bodies of many fungi; and a recently isolated bacterium, *Thiomargarita namibiensis*, whose cells grow up to 0.75 mm in width.

Microbes generally do not have complex multicellular structures. Most of the Bacteria, Archaea, protista and fungi are single-celled microorganisms. Microbes that are multicellular tend to have a limited range of cell types. Viruses are not cells, just genetic material surrounded by a protein coat, and are incapable of independent existence.

Table 1. Types of microbes, their sizes and cell type

Microbe	Approximate range of sizes	Nature of cell	Section of book
Viruses	0.01–0.25 μm	Acellular	J
Bacteria	0.1–750 μm	Prokaryote	D,E,F
Fungi	2 μm–>1 m	Eukaryote	G,H
Protista	2–1000 μm	Eukaryote	I

The science of microbiology did not start until the invention of the microscope in the mid 16th century and it was not until the late 17th century that Robert Hooke and Antoine van Leeuwenhoek made their first records of fungi, bacteria and protists. The late 19th century was the time when the first real breakthroughs on the role of microbes in the environment and medicine were made. Louis Pasteur disproved the theory of **spontaneous generation** (that living organisms spontaneously arose from inorganic material) and Robert Koch's development of **pure culture** techniques (see Topic D8) allowed him to show unequivocally that a bacterium was responsible for a particular disease. Since then the science has grown dramatically as microbiology impinges on all aspects of life and the environment.

Prokaryotes and eukaryotes

Within the microbial world can be found examples of the three distinct cell lineages that have evolved from the first original cell (*Fig.* 1). These lineages (called **kingdoms** or **domains**) have been established using DNA sequencing technology which has shown that these groups called the **Bacteria** (previously known as the Eubacteria), the **Archaea** and the **Eukarya** diverged very early in history. All the Bacteria and the Archaea are microbes but the Eukarya contain higher plants and animals as well as those fungi and protista considered to be microbes. Bacteria and Archaea have a **prokaryotic** cell structure. Their cells lack a distinct nuclear membrane, and they do not have complex internal organelles, such as mitochondria or chloroplasts which are associated with energy generation in eukaryotes. Prokaryotes have neither endoplasmic reticulum nor Golgi apparatus membranes. The Eukarya are **eukaryote** meaning they have a nucleus but there are many other differences between the two cell types. A comparison of the main features of these two categories of cell is shown in *Table* 2, but other differences do occur which will be examined in the individual sections. It is also now recognized that the organelles found in eukaryotes arose as a result of endosymbiotic events early in their evolution, as mitochondria and chloroplasts show considerable similarities to some prokaryotic cells.

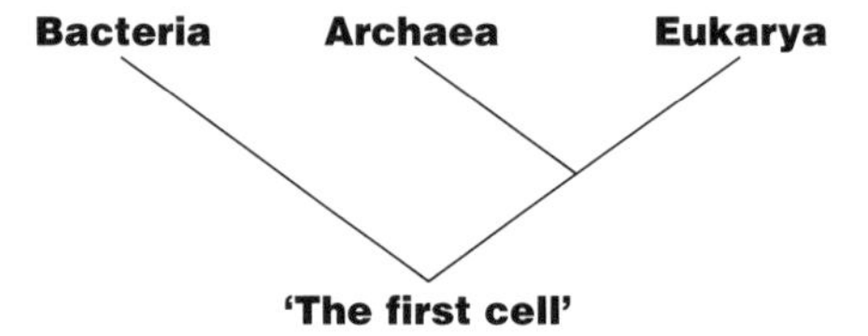

Fig. 1. The three cell lineages evolved from a common ancestor.

The importance of microbiology

Microbes impinge on all aspects of life; just a few of these are listed below.

- **The environment.** Microbes are responsible for the cycling of carbon, nitrogen and phosphorus (geochemical cycles), all essential components of living organisms (Topic F1). They are found in association with plants in symbiotic relationships, maintain soil fertility and may also be used to clean up the environment of toxic compounds (bio-remediation; Topic H4). Some microbes are devastating plant pathogens (Topic H5), which destroy important food crops, but others may act as biological control agents against these diseases.

Table 2. The major differences between prokaryote and eukaryote genetic and cellular organization

Prokaryotes	Eukaryotes
Organization of the genetic material and replication	
DNA free in the cytoplasm	DNA is contained within a membrane bound nucleus. A nucleolus is also present
Generally only one chromosome present but there are exceptions	>1 chromosome. Two copies of each chromosome may be present (diploid)
DNA associated with histone-like proteins	DNA complexed with histone proteins
May contain extrachromosomal elements called plasmids	Plasmids rarely found
Introns very rarely found in mRNA (except Archaea)	Introns found in all genes
Cell division by binary fission – asexual replication only	Cells divide by mitosis
Transfer of genetic information occurs by conjugation, transduction and transformation	Exchange of genetic information occurs during sexual reproduction. Meiosis leads to the production of haploid cells (gametes) which can fuse
Cellular organization	
Cytoplasmic membrane contains hopanoids (except Archaea). Lipopolysaccharides and teichoic acids found	Cytoplasmic membrane contains sterols
Energy metabolism associated with the cytoplasmic membrane	Mitochondria present in most cases (not present in some anaerobic microbes)
Photosynthesis associated with membrane systems and vesicles in cytoplasm	Chloroplasts present in algal and plant cells
	Internal membranes, endoplasmic reticulum and Golgi apparatus present associated with protein synthesis and targetting
	Membrane vesicles such as lysosomes and peroxisomes present
	Cytoskeleton of microtubules present
Flagella consist of one protein, flagellin	Flagella have a complex structure with 9+2 microtubular arrangement
Ribosomes – 70S	Ribosomes – 80S (mitochondrial and chloroplast ribosomes are 70S)
Peptidoglycan cell walls (Bacteria only: different polymers in archaebacteria)	Polysaccharide cell walls, where present, are generally either cellulose or chitin

- **Medicine.** The disease causing ability of some microbes is well known. Human pathogens include viruses (e.g. Variola virus causes smallpox, Topic J8), protista (e.g. *Plasmodium* causes malaria Topic I5) and bacteria (e.g. *Vibrio cholera* causes cholera, Topic F3). To date there are no known instances of the Archaea acting as human pathogens. Microorganisms have also provided us with the means to control some non-viral infections in the form of antibiotics (Topic F7). They also provide us with many other medicinally important drugs.
- **Food.** Microbes have been used for thousands of years, in many different processes, to produce foods such as cheese and bread, alcoholic drinks including beer and wine, and condiments like soy sauce (Topic F2). At the other end of the scale, microbes are responsible for food spoilage, and disease-causing microbes are frequently carried on food (Topic F5).

- **Biotechnology.** Traditionally, microbes have been used to synthesize many important chemicals including acetone, butanol and acetic acid (Topic F2). More recently, the advent of genetic engineering techniques has led to the cloning of pharmaceutically important polypeptides (for example, insulin) into microbes, which may then be produced on a large scale.
- **Research.** Microbes have been used extensively as model organisms for the investigation of biochemical and genetical processes as they are much easier to work with than more complex animals and plants. Millions of copies of the same single cell can be produced in large numbers very quickly and at low cost to give plenty of homogeneous experimental material. An additional advantage is that most people have no ethical objections to experiments with these microorganisms.

B1 Heterotrophic Pathways

Key Notes

High-energy compounds

Heterotrophy refers to the breaking down of organic molecules to obtain energy. This energy is generally stored in the form of high-energy compounds, such as ATP and NAD^+. The formation of such compounds relies on balanced redox reactions that generate organic molecules containing oxygen and phosphate groups.

Glycolysis

Glycolysis is a cytoplasmic pathway that is used by most microorganisms to break down sugars (such as glucose and fructose) to pyruvate, yielding two molecules of ATP. Pyruvate then enters the citric acid cycle, and its utilization through this pathway yields energy-rich compounds including ATP and NADH.

Alternatives to glycolysis

There are a number of hexose monophosphate pathways (including the Entner-Douderoff pathway, the phosphoketolase pathway and the pentose phosphate pathway) that can be used as alternatives to glycolysis for the oxidation of glucose. These pathways yield less ATP per molecule of glucose than glycolysis, but they generate important metabolic intermediates including NADPH and pentose sugars for nucleic acid synthesis.

Citric acid cycle and respiration

The citric acid cycle occurs in the cytoplasm of aerobic bacteria and in the mitochondria of aerobic eukaryotes. Respiration is the complete oxidation of an organic substrate to carbon dioxide and water. It requires an external electron acceptor, usually oxygen, and results in the formation of large amounts of ATP. For each glucose molecule oxidized by the citric acid cycle, 12 molecules of ATP are generated. Important intermediates for fatty acid synthesis, nucleotide synthesis and amino acid synthesis are also generated by the citric acid cycle.

Fermentation

Fermentation is the incomplete oxidation of an organic substrate and it occurs under anaerobic conditions. Energy yields from fermentation are lower than comparative yields from respiration. The products of incomplete oxidation can include pyruvate, lactate, formate and ethanol.

High-energy compounds

The ability to produce high-energy compounds for metabolism and storage is a prerequisite for cell survival. Energy is acquired by cells through a series of balanced **oxidation-reduction** (**redox**) reactions from organic or inorganic substrates. The simplest redox reaction can be seen in the reaction below

$$H_2 + \frac{1}{2} O_2 \rightarrow H_2O$$

H_2 = reductant (electron donor) that becomes oxidized
O_2 = oxidant (electron acceptor) that becomes reduced

The energy that is released in redox reactions is stored in a variety of organic molecules that contain oxygen atoms and phosphate groups. **ATP**, adenosine triphosphate, is a **high-energy compound** found in almost all living organisms. It is synthesized in catabolic reactions, where substrates are oxidized, and utilized in anabolic, biosynthetic reactions. Intermediates called **carriers** participate in the flow of energy from the electron donor to the terminal electron acceptor. The co-enzyme **nicotinamide adenine nucleotide** (NAD^+) is a freely diffusable carrier that transfers two electrons and a proton, and a second proton from water, to the next carrier in the chain.

$$NAD^+ + 2H^+ + 2e^- \rightleftarrows NADH + H^+$$

The reactions for the phosphorylated derivative ($NADP^+$) are similar. NAD^+ is usually used in energy-generating reactions and $NADP^+$ in biosynthetic reactions.

All protozoa, all fungi and most bacteria synthesize ATP by oxidizing organic molecules. This can be either via **respiration** or by **fermentation**. Respiration requires a terminal electron acceptor. This is usually oxygen, but nitrate or sulfate are among the compounds used in anoxic conditions. Fermentation requires an organic terminal oxygen acceptor.

Microorganisms can be grouped according to the source of energy they use, and by the source of carbon which may either be an organic molecule or from CO_2 (carbon dioxide fixation) (*Table 1*).

Table 1. Classification of microorganisms by energy and carbon source utilized

	Type	Electron donor	Energy source	Carbon source	Examples
Organotrophs	Chemo-organotroph	Organic compounds	Redox reactions of organic compounds	Organic compounds	All fungi, all protists, most terrestrial bacteria
	Photo-organotroph	Organic compounds	Light	Carbon dioxide and organic compounds	Nonsulphur bacteria
Lithotrophs	Chemo-lithotrophs	Inorganic compounds	Redox reactions of inorganic compounds	CO_2	*Thiobacillus, Nitrosomonas, Nitrobacter, Hydrogenomonas, Beggiotia*
	Photolithotrophs	Inorganic compounds	Light	CO_2	Photosynthetic green and purple bacteria, photosynthetic protista

Glycolysis (Embden-Meyerhof-Parnas)

The reactions termed **glycolysis** take place in the cytoplasm of all prokaryotes and eukaryotes. The pathway generates two ATP molecules per molecule of glucose degraded, and feeds substrates into subsequent metabolic pathways.

The steps in glycolysis are shown in *Fig. 1* but the overall net reaction can be summarized as follows:

$$\text{Glucose} + 2\text{ADP} + 2\text{P}_i + 2\text{NAD}^+ \rightarrow 2\text{ pyruvate} + 2\text{ATP} + 2\text{NADH} + 2\text{H}^+$$

The reactions at the beginning of the pathway require two ATP molecules, but the gross yield of ATP per glucose molecule is four, giving a net gain of two ATP per glucose.

The initial reactions at the beginning of the pathway transform the 6 carbon sugar glucose into glucose 1,6-bisphosphate, via two phosphorylation reactions.

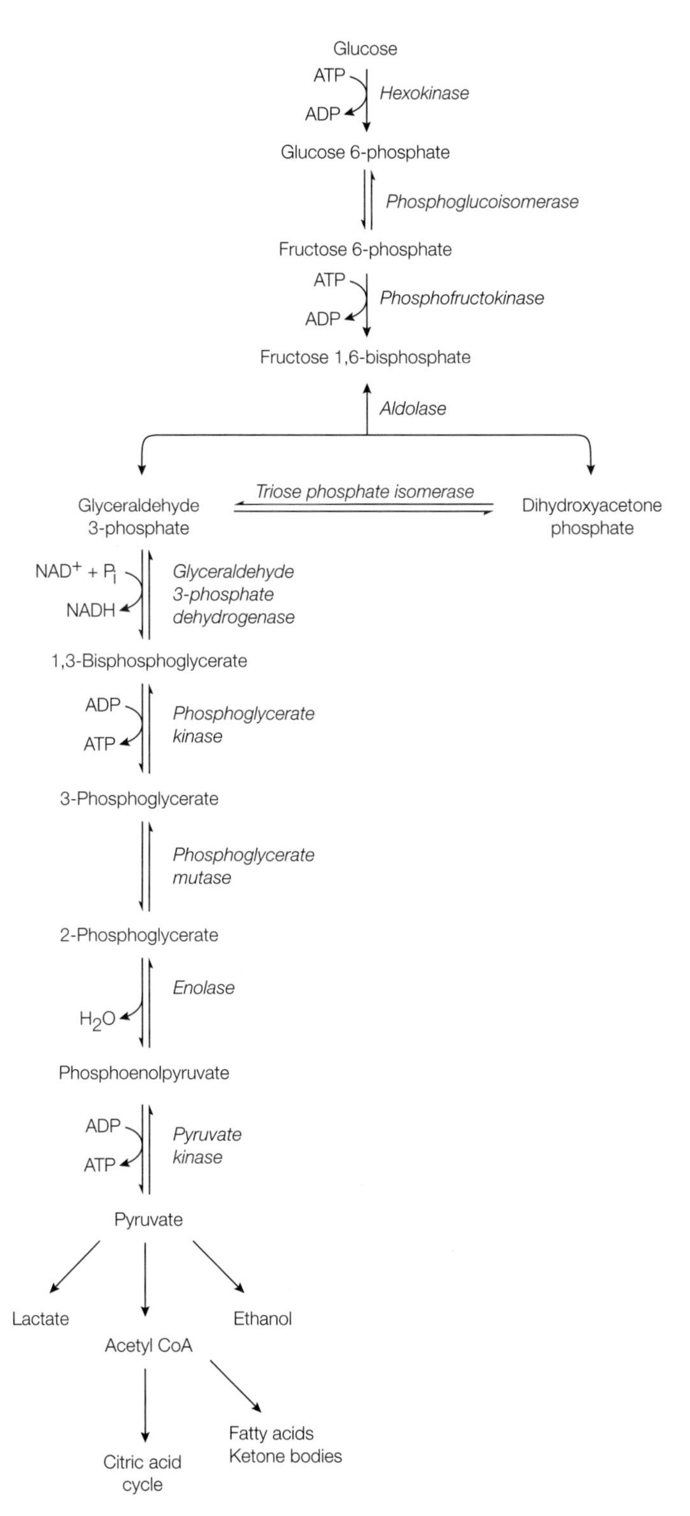

Fig. 1. Glycolysis.

There follows a near symmetrical split into two 3C phosphorylated compounds (glyceraldehyde-3-phosphate and dihydroxyacetone phosphate (DHA)). These compounds will interconvert as an equilibrium reaction via the enzyme triose phosphate isomerase. Glyceraldehyde-3-phosphate is the substrate for subsequent reactions of glycolysis.

Further energy is added to the glyceraldehyde-3-phosphate by the addition of a second high-energy phosphate group, from NADPH to the aldehyde group. There then follow two reactions where the high-energy phosphate groups of 1,3-bisphosphoglycerate are used to form ATP from ADP, mediated by two kinase enzymes, phosphoglycerate kinase and pyruvate kinase. These reactions are termed **substrate level phosphorylations**.

The final product of glycolysis is pyruvate, which feeds into respiration in aerobic conditions.

Alternatives to glycolysis

Some important groups of bacteria, for example some Gram-negative rods, do not use glycolysis to oxidize glucose. They use a different mechanism, the **Entner-Douderoff** *(Fig. 2)*, which yields one mole of ATP, NADPH and NADH from every mole of glucose. This is a **hexose monophosphate pathway** (HMP), and in this pathway only one molecule of ATP is produced per molecule of glucose metabolized.

Another HMP is the **phosphoketolase** pathway, which is another method for glucose breakdown found in *Lactobacillus* and *Leuconostoc* spp. when grown on 5-carbon sugars (pentoses). The pathway produces lactic acid, CO_2 and either ethanol or acetate (*Fig. 3*).

An important HMP is the **pentose phosphate pathway** (PPP), which often operates in conjunction with glycolysis or other HMP pathways. The PPP is an important provider of intermediates that serve as substrates for other biosynthetic

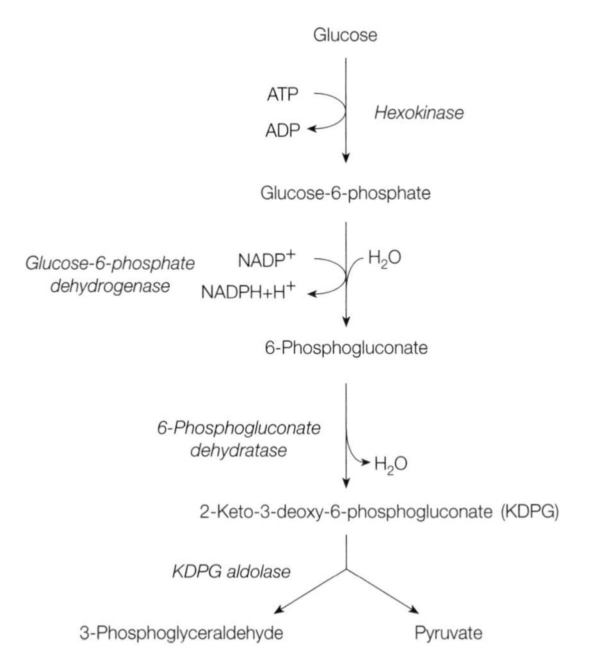

Fig. 2. Entner-Douderoff pathway.

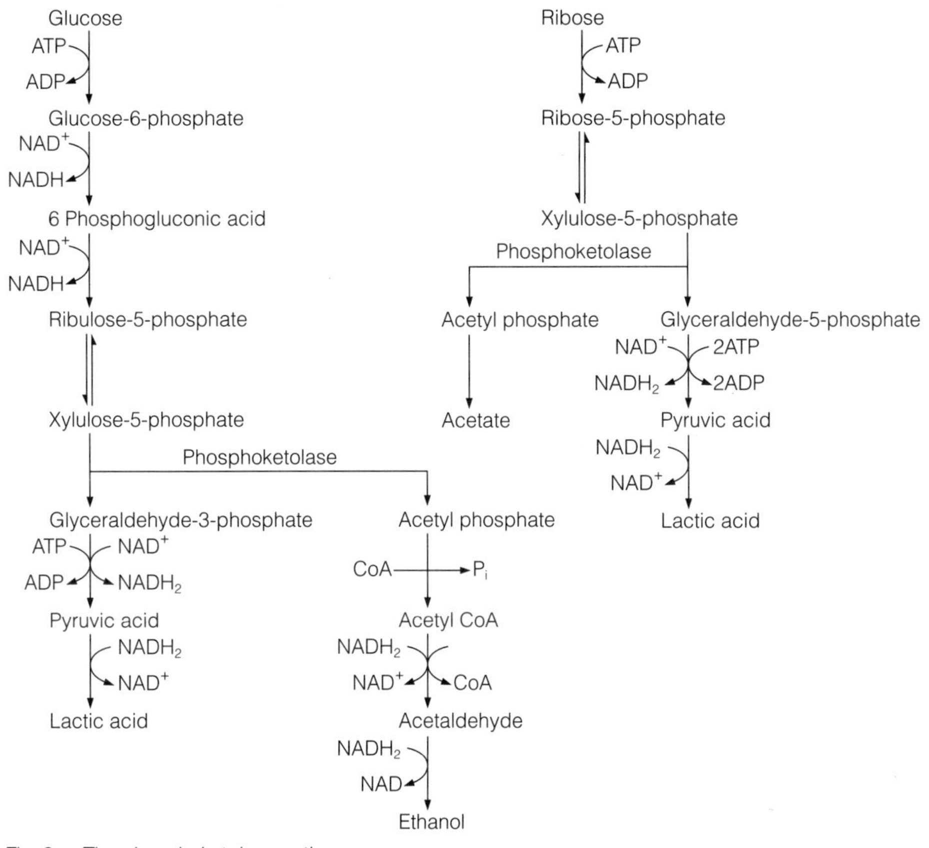

Fig. 3. The phosphoketolase pathway.

pathways. This pathway yields NADPH/+H^+ and pentoses which are used in the synthesis of nucleotides including, FAD, ATP and coenzyme A (CoA).

The reactions can be summarized as

$$\text{Glucose-6-phosphate} + 2\,\text{NADP}^+ + \text{water} \rightarrow \text{Ribose-5-phosphate} + 2\,\text{NADPH} + 2\text{H}^+ + \text{CO}_2$$

There are three important stages to this pathway:

1. Glucose-6-phosphate is converted to ribulose-5-phosphate, generating two NADPH + $2H^+$

NADP+ NADPH + H+ | H₂O H+ | NADP+ NADPH CO₂

Glucose 6-phosphate — 6-Phosphoglucono-δ-lactone — 6-Phosphogluconate — Ribulose 5-phosphate

2. Ribulose-5-phosphate isomerises to ribose-5-phosphate

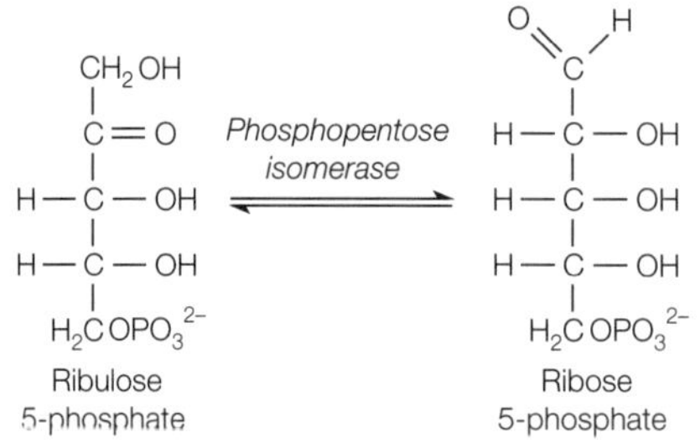

$$\underset{\text{Ribulose 5-phosphate}}{CH_2OH-C(=O)-CH(OH)-CH(OH)-H_2COPO_3^{2-}} \underset{}{\overset{\textit{Phosphopentose isomerase}}{\rightleftharpoons}} \underset{\text{Ribose 5-phosphate}}{OHC-CH(OH)-CH(OH)-CH(OH)-H_2COPO_3^{2-}}$$

3. Excess ribose-5-phosphate is converted to fructose-6-phosphate and glyceraldehyde, via a series of reactions, to enter glycolysis.

The citric acid cycle and respiration

The **citric acid cycle** is found in the cytosol of aerobic Prokaryotes, and the mitochondria of eukaryotes. Anaerobic organisms have incomplete cycles whilst facultative aerobic organisms only have a functional citric acid cycle in the presence of O_2.

Complete oxidation of organic substrates to CO_2 and water via the citric acid cycle requires an external electron acceptor; the best studied are oxygen, nitrate or sulfate. This process yields large amounts of energy stored as ATP. The product of glycolysis, pyruvate, can be completely oxidized using enzymes of the citric acid cycle.

In summary, during the operation of the citric acid cycle three carbon atoms of pyruvate are completely oxidized to CO_2, and four hydrogen atoms reduce NAD^+ and FAD (*Fig. 4*, reactions 1–10).

The cycle begins with an oxidative decarboxylation (reaction 1), where CO_2 is released and NADH is formed. The resulting 2C (acetyl) unit is linked to CoA (reaction 2), and this high energy compound couples with oxaloacetate (4C) to form a 6C unit, citric acid. The acetyl group of the citric acid is then further metabolized, with 2 decarboxylation reactions releasing CO_2 (reactions 15 and 16) and 4 more coenzyme molecules are reduced. At the end of the cycle oxaloacetate is regenerated as the acceptor of further acetyl units (reaction 10).

The reduced co-enzymes are then oxidized by a respiratory electron transport chain which may use oxygen, nitrate or sulfate as terminal electron acceptors. This allows for NAD^+ regeneration and the synthesis of ATP, a process known as oxidative phosphorylation. NAD^+ regeneration is essential as levels are limited in cells.

Each turn of the citric acid cycle yields three NADH molecules and one $FADH_2$ molecule which via oxidative phosphorylation generates ATP molecules. Including the single ATP molecule that is formed in the conversion of succinyl CoA to succinate, each molecule of glucose oxidized by the citric acid cycle produces 12 ATP molecules.

The cycle produces intermediates for many other biosynthetic pathways, including fatty acid biosynthesis (citrate), amino acid synthesis (α-ketoglutarate), and nucleotide synthesis (α-ketoglutarate and oxaloacetate).

Fermentation

Fermentation is an **incomplete oxidation** of an organic substrate. During fermentations an electron donor becomes reduced, and energy is trapped by **substrate level phosphorylation**. Fermentation products include pyruvate if the

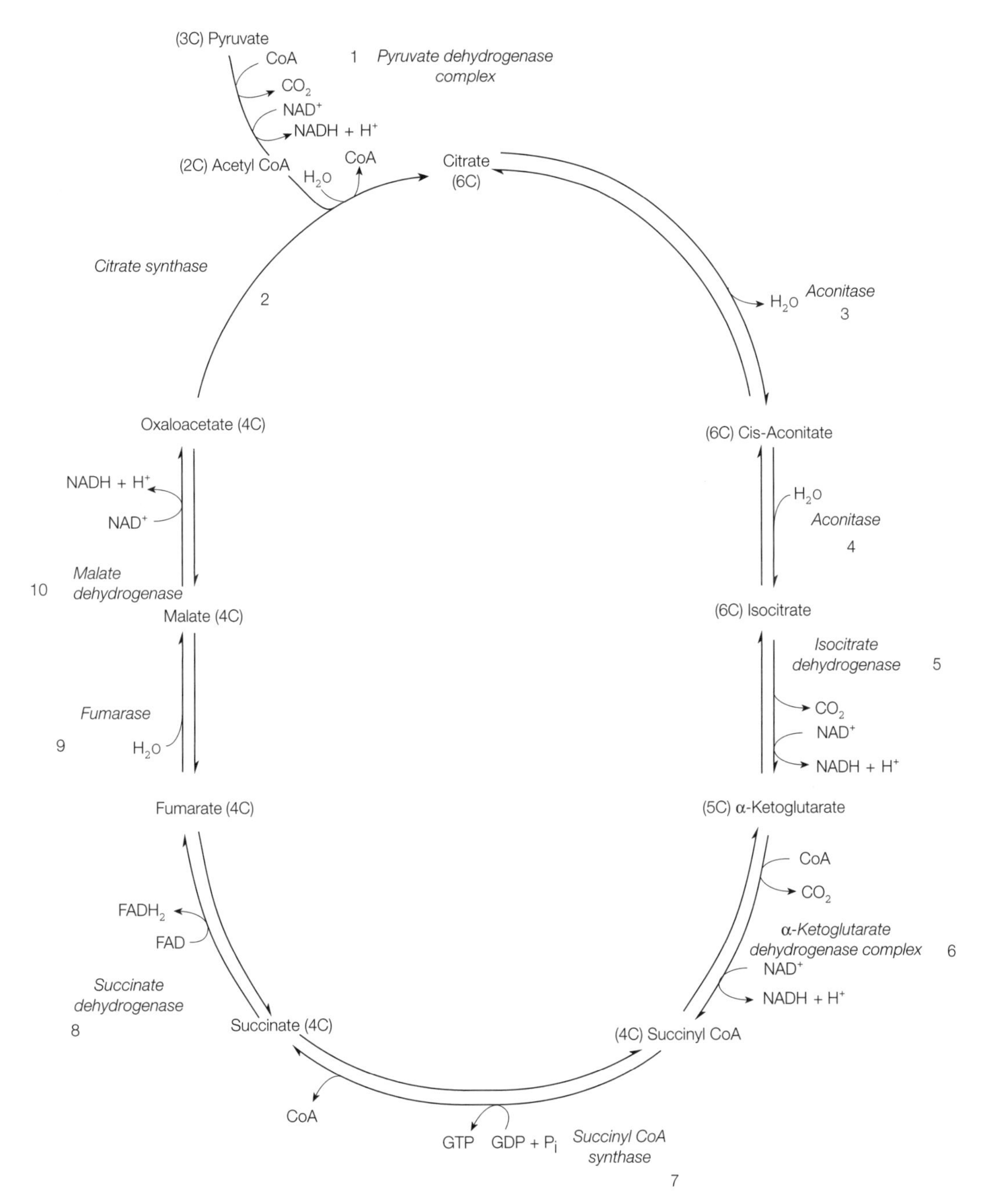

Fig. 4. The citric acid cycle.

glycolytic pathway is used, or lactate, formate, 2,3-butanediol, and ethanol from **the butanediol pathway** (used by *Klebsiella, Erwinia, Enterobacter* and *Serratia* spp.) or succinate, ethanol, acetate and formate from a **mixed acid fermentation** (found in *Escherichia, Salmonella* and *Shigella* spp.). See *Fig. 5* for details.

(a)

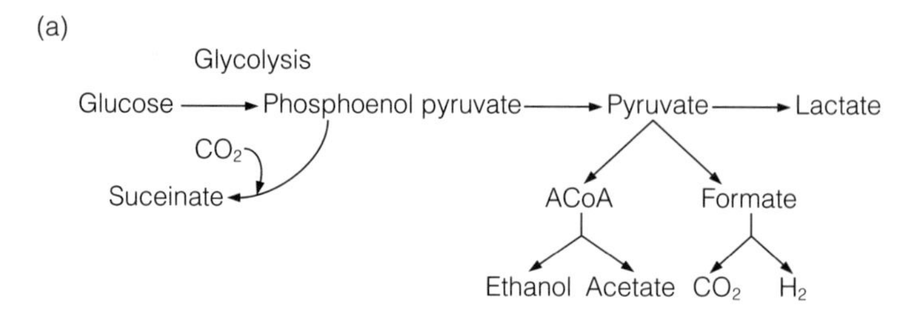

(b)

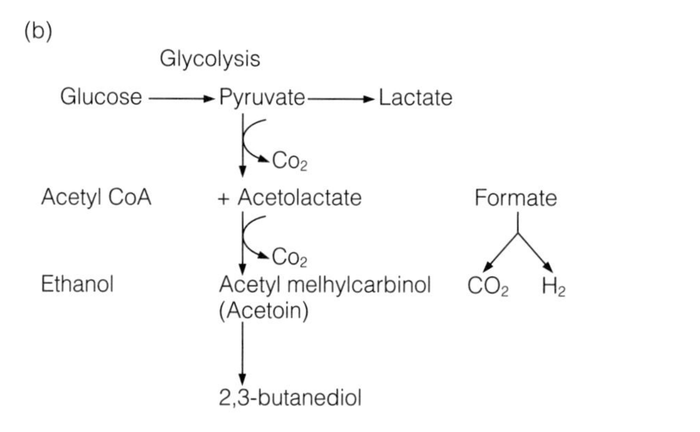

Fig. 5. Products of fermentation. (a) Mixed acid. (b) Butanediol.

B2 ELECTRON TRANSPORT, OXIDATIVE PHOSPHORYLATION AND β-OXIDATION OF FATTY ACIDS

Key notes

Electron transport

Electron transport is used to create a proton motive force (PMF) across membranes. This PMF is used by all microorganisms to generate ATP via a membrane bound ATPase. The Archaea and Bacteria also use PMF to drive the movement of flagellae, allow transport of charged substrates across membranes and maintain their osmotic potential. In eukaryotic microbes, PMF is established across the inner membrane of the mitochondrion. In the electron transport chain, a series of balanced oxidation and reduction reactions drives the movement of electrons through the carrier series from NADH to oxygen. During this process energy is released and ATP is synthesized.

Anapleurotic pathways

Lost intermediates from glycolysis and the citric acid cycle are replenished by anapleurotic reactions, where carbon dioxide is fixed into three-carbon compounds by carboxylation reactions.

Glyoxalate cycle

Some substrates that microbes can utilize as carbon sources, for example the two-carbon compound acetate, can lead to the depletion of citric acid cycle intermediates. Reactions that result in the loss of CO_2 during the cycle can be avoided by using the glyoxalate cycle.

Fatty acid oxidation

Fatty acids can be used as substrates by microorganisms through the fatty acid or beta oxidation pathway. This is located in the mitochondria of eukaryotes and the cytoplasm of prokaryotes.

Anaerobic respiration

Many microbes live in low or no oxygen environments. Alternative electron acceptors, such as nitrate and sulfate, can be utilized instead of oxygen by these organisms to complete the electron transport chain.

Electron transport

Peter Mitchell theorized that the generation of ATP only occurred because mitochondria and bacteria could pump protons across a membrane. These primary pumps lead to a generation of a charge across the membrane, known as the **proton motive force** (PMF). As the protons try to move back across the membrane, the energy of their movement can be harnessed in a number of ways. The most important use of the PMF in many aerobic organisms is the **generation of ATP**, normally using the enzyme f_1f_0 ATPase. This ATP-generating enzyme is found in the cytoplasmic membrane of bacteria and the inner membrane of the mitochondria, and in the absence of the PMF actually cleaves ATP into ADP and phosphate. However when PMF is applied, the ATPase works essentially in reverse, generating ATP from ADP and phosphate. It is thus known as a **secondary pump** (*Fig. 1*). For all this to happen efficiently, the organism or organelle must conform in several ways to the Mitchell hypotheses which are:

1. Protons are pumped across mitochondrial and bacterial membranes in such a way as to generate an **electric potential** across the membrane. A membrane bound-enzyme (ATPase) couples synthesis of ATP to the flow of protons down the electric potential gradient.
2. Solutes can accumulate against a concentration gradient by the coupling of proton flow to the movement of the solute by a **transmembrane protein**. These cotransporters may act as **symports, antiports or uniports.**
3. The flow of protons through the flagellar transmembrane proteins rotates components of the flagellum and allows a bacterium to move.

There are five types of component molecule:

1. Enzymes that catalyse transfer of hydrogen atoms from reduced NAD^+ to flavoproteins (NADH dehydrogenases).
2. Flavoproteins. Flavin mononucleolides (FMN) and flavin adenine dinucleotide (FAD). The flavins are reduced by accepting a hydrogen atom from NADH and oxidized by losing an electron.
3. Electron carriers, cytochromes. Cytochromes are porphyrin containing proteins each of which can be reduced or oxidized by the loss of a single electron:

$$\text{Cytochrome-Fe}^{2+} \rightleftarrows \text{Cytochrome-Fe}^{3+} + \varepsilon\text{-}$$

4. Iron sulfur proteins. These are carriers of electrons with a range of reduction potentials
5. Quinones. These are lipid-soluble carriers that can diffuse through membranes carrying electrons from iron-sulfur proteins to cytochromes.

The current view of electron transport is summarized in *Fig. 1*.

Protons for the final reduction in the transfer chain are supplied by the disassociation of water, providing the build up of hydroxyl ions on the inside of the membrane.

Protons flow back into the mitochondrial matrix or bacterial cell through the enzyme ATP synthase, driving ATP synthesis. This enzyme is in two parts, one localized on the bacterial cytoplasmic or mitochondrial matrical side and the other which spans the membrane to the outside of the bacterial cell or the intermembrane space of the mitochondrion.

The rate of **oxidative phosphorylation** is set by the availability of ADP, electrons only flow down the chain when ATP is needed. When there are high levels of TAP and energy-rich compounds like $NADH^+$ and $FADH_2$ accumulation of citric acid inhibits the citric acid cycle and glycolysis.

However, an alternative theory, the **conformational change hypothesis**, proposes ATP synthesis occurs because of conformational changes created in the ATP synthase enzyme caused by electron transport. This theory is currently being intensively researched.

Anapleurotic pathways

The intermediates of glycolysis and the citric acid cycle are used as precursors of biosynthetic pathways. To maintain the energy-yielding processes of glycolysis

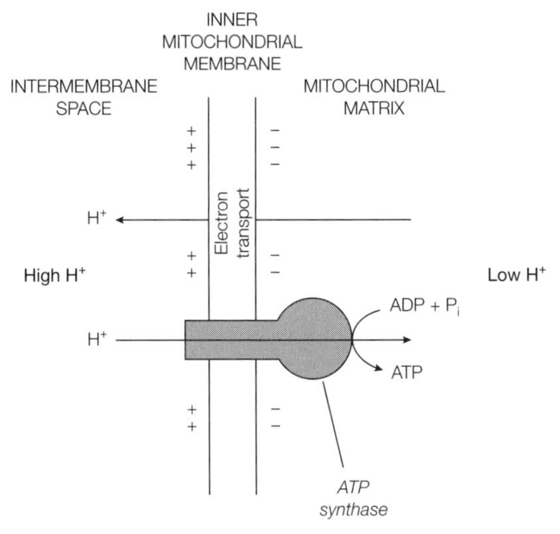

Fig. 1. The role of ATPase as a secondary pump. From Hames and Hooper, Instant Notes in Biochemistry *2nd edn. © BIOS Scientific Publishers 2000.*

and the citric acid cycle these lost intermediates must be replenished **by anapleurotic reactions**. Three carbon compounds are carboxylated to form oxaloacetate. Both pyruvate and phosphoenol pyruvate (PEP) can be used in these reactions, e.g.

$$\text{PEP} + \text{HCO}_3^- \xrightarrow{\text{PEP carboxylase}} \text{oxaloacetate} + \text{P}_i$$

Glyoxalate cycle

A number of organic acids can be used by microorganisms as electron donors and carbon sources. Those that are common to the citric acid cycle, citrate, malate, fumarate and succinate for example, can be metabolized using the enzymes of the citric acid cycle. However, utilization of acetate via the citric acid cycle will cause the depletion of oxaloacetate. If this occurs the citric acid cycle could not operate. To compensate for the loss of oxaloacetate the **glyoxalate shunt occurs** (*Fig. 2*).

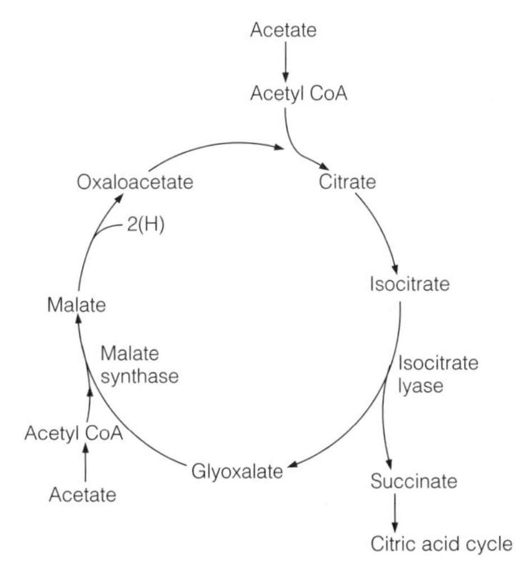

Fig. 2. Glyoxalate shunt.

In this pathway, which shares many of the reactions of the citric acid cycle, reactions that give rise to CO_2 evolution are bypassed, and isocitrate is split into succinate and glyoxalate by the enzyme isocitrate lyase. Succinate can be used in biosynthetic reactions, whilst glyoxalate is combined with Acetyl CoA via malate synthase to form malate, which enters the citric acid cycle.

Fatty acid oxidation

Fatty acids can be used as substrates for microbial metabolism. The metabolic process is called **beta oxidation**, and it occurs in the mitochondria of eukaryotes and the cytoplasm of prokaryotes. Two carbon units are removed from the fatty acid to yield their acyl CoA. The pathway is outlined in *Fig. 3*.

The pathway begins with the activation of the fatty acid by CoA. There then follow two separate dehydrogenation reactions where electrons are transferred

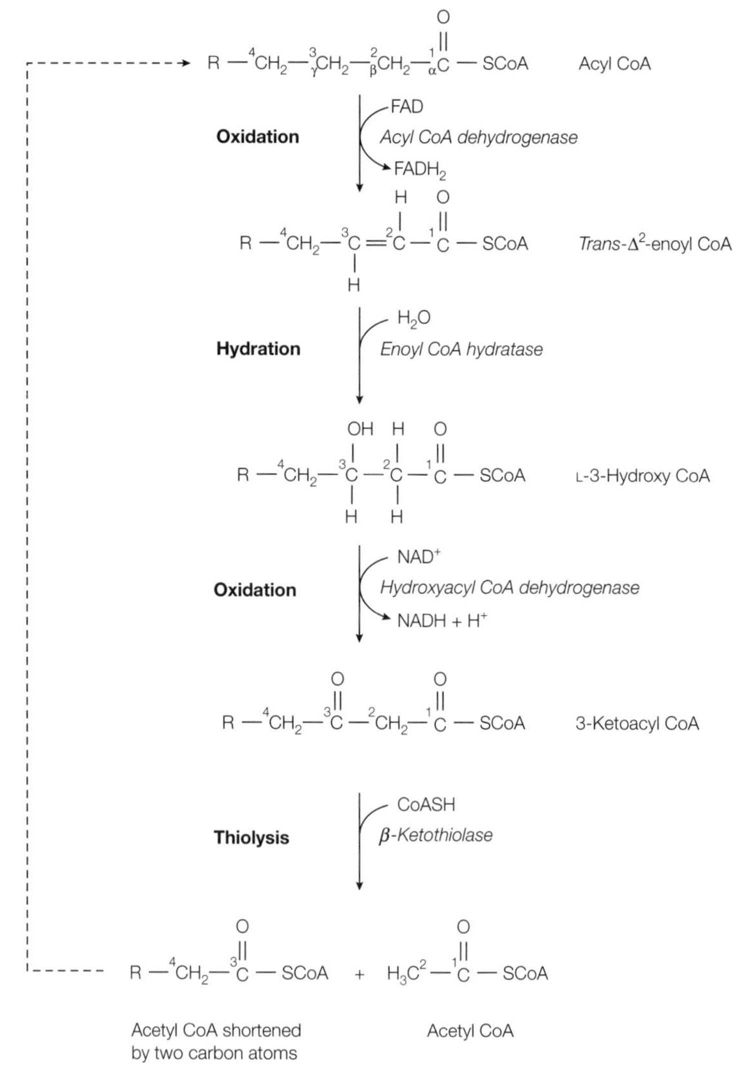

Fig. 3. Fatty acid oxidation. From Hames and Hooper, Instant Notes in Biochemistry *2nd edn. © BIOS Scientific Publishers 2000.*

to FAD and NAD^+, finally yielding CoA and an activated fatty acid to restart the cycle.

Anaerobic respiration

Anaerobic respiration occurs in prokaryotes that are unable to use oxygen as a terminal electron acceptor. These organisms are termed **obligate anaerobes**. Other prokaryotes can use anaerobic respiration facultatively, if oxygen happens to be unavailable. Less energy is generated during anaerobic respiration than in aerobic respiration. Nitrate, sulfate and carbon dioxide can be used as alternative electron acceptors.

Nitrate respiration uses the most common alternative electron acceptor, nitrate (see *Fig. 4*). The first step of the reaction is catalyzed by the enzyme nitrate reductase, an enzyme which is only synthesized under anaerobic conditions. The product is nitrite, which is excreted by most staphylococci and enterobacteriaceae, but other Bacteria will reduce nitrite further to ammonia or nitrogen gas. This reaction, and the enzyme that catalyzes it, is termed **dissimilatory** because nitrogen is reduced during the biological breakdown of organic compounds. This type of respiration leads to **denitrification.**

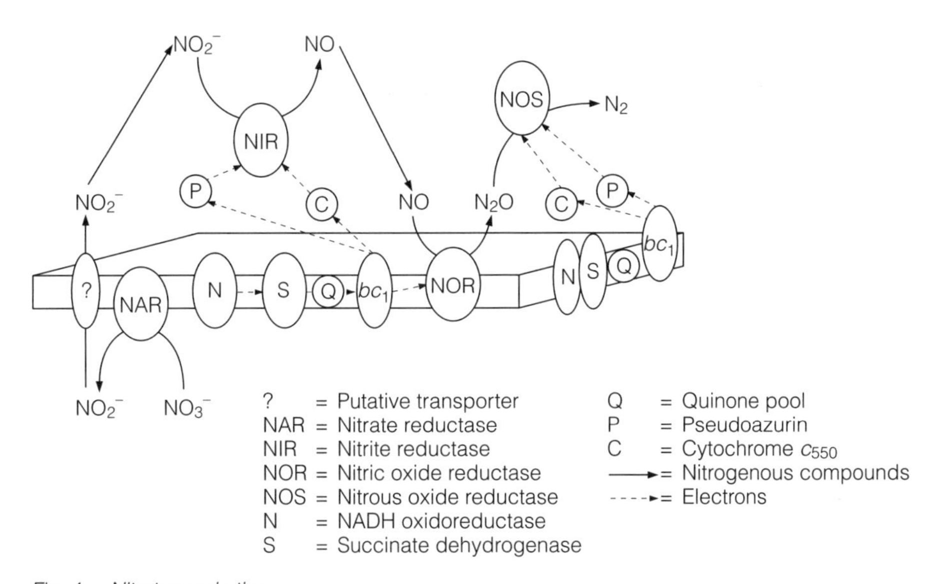

Fig. 4. Nitrate respiration.

B3 Autotrophic reactions

Key Notes

Chemolithotrophy

Autotrophic microorganisms can survive in the absence of organic carbon sources by fixing atmospheric or dissolved CO_2 to form carbohydrates. Chemolithotrophs have the ability to fix CO_2 using the Calvin cycle, and the energy required to drive the reactions comes from the oxidation of inorganic substrates such as ammonia.

Photosynthesis

Photosynthesis can be divided into two sets of reactions, those that are light-dependent (light reactions) and those that are light-independent (dark reactions). The light reactions convert light into chemical energy through the synthesis of ATP, which is then used to drive the Calvin cycle (dark reactions). Photosynthesis may be described as oxygenic if oxygen is generated (as in the cyanobacteria and the photosynthetic eukaryotes) or as anoxygenic if it is not (as in the green and purple bacteria). The light reaction can be driven by photosystem I and II in eukaryotes, but may only be driven by photosystem I in some Prokaryotes.

Light reactions in bacterial photosynthesis

Photosynthetic green and purple bacteria contain chlorophyll A and B, and carry out anoxygenic photosynthesis that utilizes only photosystem I.

Light reactions in eukaryotic photosynthesis

In eukaryotes, photosynthesis occurs in the chloroplasts, and involves photosystems I and II. The light-dependent reactions generate NADPH+ H^+, and the resulting proton gradient is used to generate ATP by non-cyclic phosphorylation.

Dark reactions in eukaryotic photosynthesis

The dark (light-independent) reactions of photosynthesis are called the Calvin cycle and use the energy generated from light-dependent reactions to synthesize carbohydrates from CO_2 and H_2O.

Chemolithotrophy

Chemolithotrophy is found in a limited number of microorganisms. Chemolithotrophs obtain their energy by the oxidation of inorganic substrates and their carbon from CO_2. However, these reactions yield less energy than oxidation of glucose to CO_2, so large quantities of substrates have to be oxidized to generate enough energy for sufficient ATP and NADH generation. The process whereby ammonia is oxidized to nitrate is termed nitrification, and ATP can be generated via this reaction. However, electrons cannot be donated directly for NADH production from ammonia or nitrate because they have a more positive redox potential than NAD^+, a process termed reversed electron flow allows ATP to be used to generate small but sufficient amounts of NADH (*Fig. 1*).

Photosynthesis

A large number of microorganisms have the ability to use sunlight to generate ATP by photophosphorylation. This process may not generate oxygen, a reac-

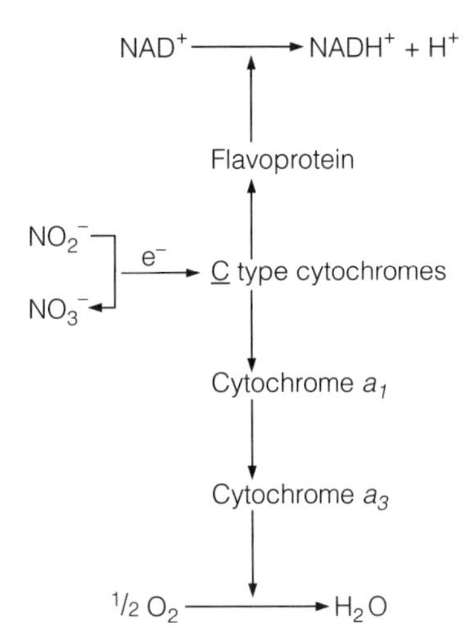

Fig. 1. Reversed electron flow.

tion termed anoxygenic, as found in the green and purple bacteria, or it may generate oxygen, termed oxygenic, by the photolysis of water, as found in the blue green bacteria and algae (*Table 1*). The reactions are complex but can be divided into two sets of reactions, the light reactions where light energy is converted into chemical energy (ATP), and the dark reactions where ATP is used to synthesize glucose.

Table 1. Classification of photosynthetic microorganisms according to hydrogen (reductant) and carbon source.

Nutritional classification	Examples	Carbon source	Hydrogen source	Oxygen evolution
Primarily photolithotrophs	Green sulfur bacteria (Chlorobiaceae)	CO_2, acetate, butyrate	H_2, H_2S, $S_2O_3^{2+}$	Negative
	Purple sulfur bacteria, (Chromatiaceae)	CO_2, acetate, butyrate	H_2, H_2S, $S_2O_3^{2+}$	Negative
Photo-organotrophs	*Purple non-sulfur bacteria (Rhodospirrillaceae)	Organic (CO_2)	H_2, organic	Negative
	*Green gliding bacteria (Chloroflexaceae)	Organic (CO_2)	H_2, organic	Negative
	Halobacteria (Archaea)	Organic	Organic	–
Photolithotrophs	Blue green bacteria (Cyanobacteria)	CO_2	H_2O	Positive
Photolithotrophs	Photosynthetic protista	CO_2	H_2O	Positive

* Can grow as chemoorganotrophs aerobically in the dark.

The light reaction in bacterial photosynthesis

Photosynthetic bacteria contain **bacteriochlorophylls** *a* and *b*, with absorption maxima of 775 and 790 nm respectively. These pigments are contained within sac-like extensions of the plasma membrane called **chlorosomes** in green sulfur and non-sulfur bacteria and **intracytoplasmic** vesicles in purple bacteria. Bacterial photosynthesis is an **anoxygenic** (non-oxygen producing) photosynthesis which relies on photosystem I only, and is termed a cyclic phosphorylation (see *Fig. 2*).

There is no net change in the numbers of electrons in the system. ATP synthesis occurs during the generation of a protein motive force during photosynthesis, which allows ATP synthase to synthesize ATP. The electrons expelled from the reaction center return to the bacteriochlorophyll via the electron transport chain. The photosynthetic apparatus consists of four membrane-bound pigment–protein complexes, plus an ATP synthase. For NADPH synthesis, bacteria must use electron donors like hydrogen, H_2S, sulphur and organic compounds with a more negative reduction potential than water. In this case direct transfer can occur via ferrodoxin.

The purple sulfur bacteria cannot synthesize NADPH directly by photosynthetic electron transport. This is because their acceptor molecules are more positive than the $NADP^+$/NADPH couple (–0.32 volts). In this case electrons enter at the cytochromes from the electron donors, and ATP from the light reactions is used to reduce $NADP^+$ to $NADPH^+$, a process termed energy-dependent reverse electron flow (*Fig. 3*).

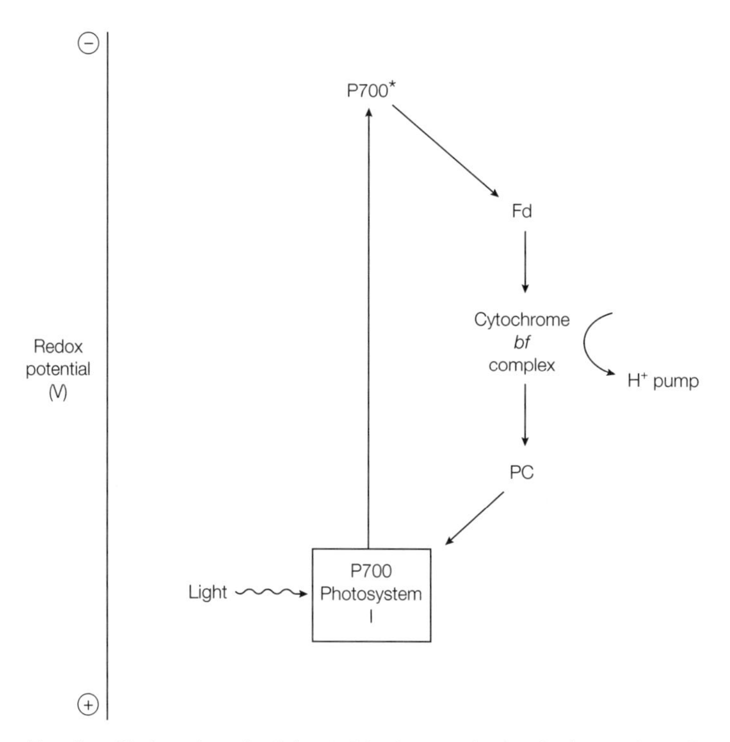

Fig. 2. Photosystem I of bacterial photosynthesis. Redrawn from Brock, Biology of Microorganisms. *Madigan* et al., *Prentice Hall Inc, USA, 1997.*

The light reaction in eukaryotic photosynthesis

Photosynthesis occurs within **antenna complexes** and **reaction centers** in the thylakoid membrane of chloroplasts in eukaryotic microorganisms. Antenna complexes are formed from several hundred chlorophyll molecules plus accessory pigments. Light excitation of the chlorophyll molecule results in an electron in a chlorophyll molecule being excited to a higher orbit, and this energy is transferred between chlorophylls until it is channeled into the chlorophyll molecules of the reaction center.

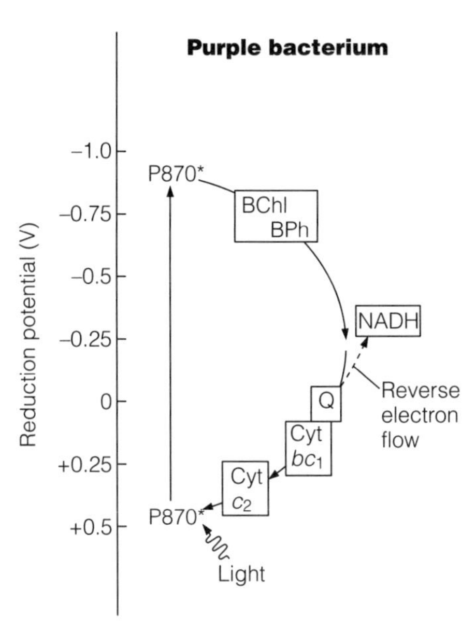

Fig. 3. Energy-dependent reverse electron flow.

The reaction center contains two photosystems, called **photosystem I (PS I)** and **photosystem II (PS II)**, with different light-energy absorption maxima. PS I absorbs at 700 nm, and PS II at 680 nm. The reaction centers are linked by other electron carriers, and if the components are arranged by their redox potentials they assume a Z shape, so the scheme is called the **Z scheme** (*Fig. 4*).

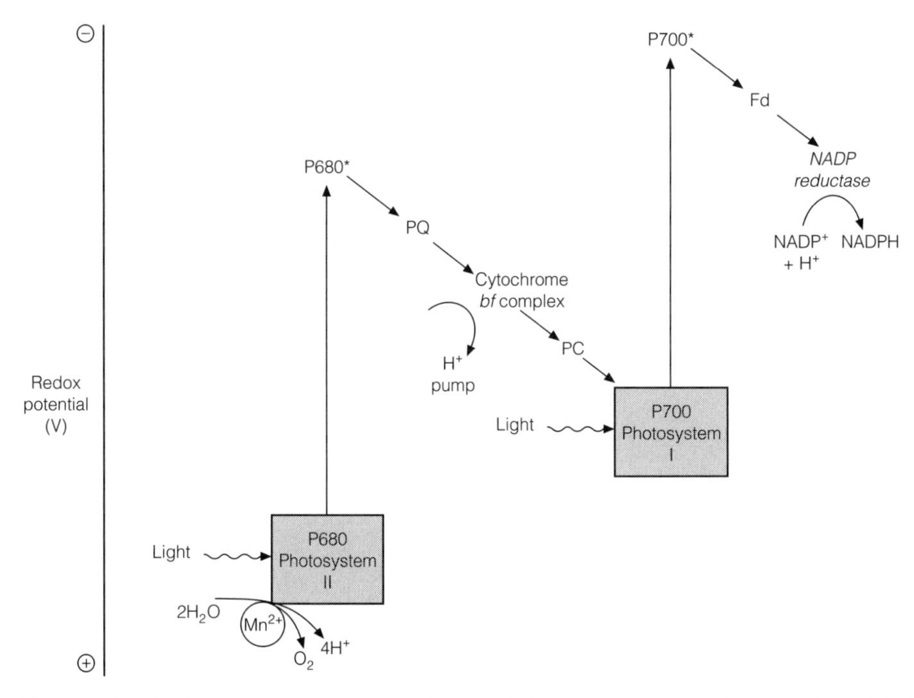

Fig. 4. The Z scheme for noncyclic photophosphorylation. From Hames and Hooper, Instant Notes in Biochemistry, *2nd edn. © BIOS Scientific Publishers 2000.*

The reactions of the Z scheme generate NADPH from NADP. ATP is generated by non-cyclic phosphorylation reactions because of the creation of a **proton gradient** between the thylakoid space and the stroma by the reactions of PS I and PS II. An ATP synthase is present in the thylakoid membrane, and H^+ is pumped from the stroma into the thylakoid space, generating ATP (*Fig. 5*). PS I may operate without PS II in some circumstances, and in this reaction no O_2 is produced; only ATP is produced via the proton gradient.

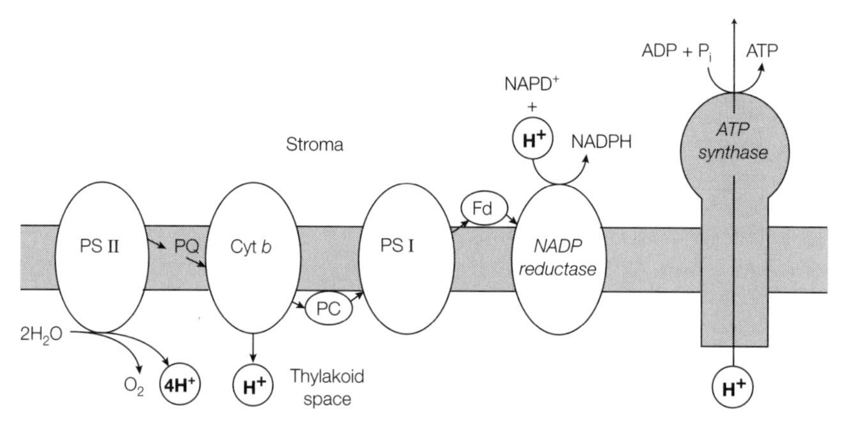

Fig. 5. Formation of the proton gradient and ATP synthesis. From Hames and Hooper, Instant Notes in Biochemistry, *2nd edn. © BIOS Scientific Publishers 2000.*

The dark reaction of photosynthesis

The light-independent reactions of photosynthesis use the NADPH and ATP generated from the light reactions to synthesize carbohydrates from CO_2 and water. This is called the **Calvin cycle**. A key enzyme in this cycle is **rubisco** (*Fig. 6*) (ribulose bisphosphate carboxylase), a large, multi-component enzyme. This enzyme incorporates CO_2 into ribulose 1,5-bisphosphate to form first a six-carbon compound, which then splits to form two three-carbon molecules (3-phosphoglycerate). Subsequent reactions regenerate ribulose 1,5-bisphosphate from one of the 3-phosphoglycerate molecules, to continue the cycle (see Topic B2). The other molecule of 3-phosphoglycerate is transported to the cytosol and used in respiration and to produce storage sugars.

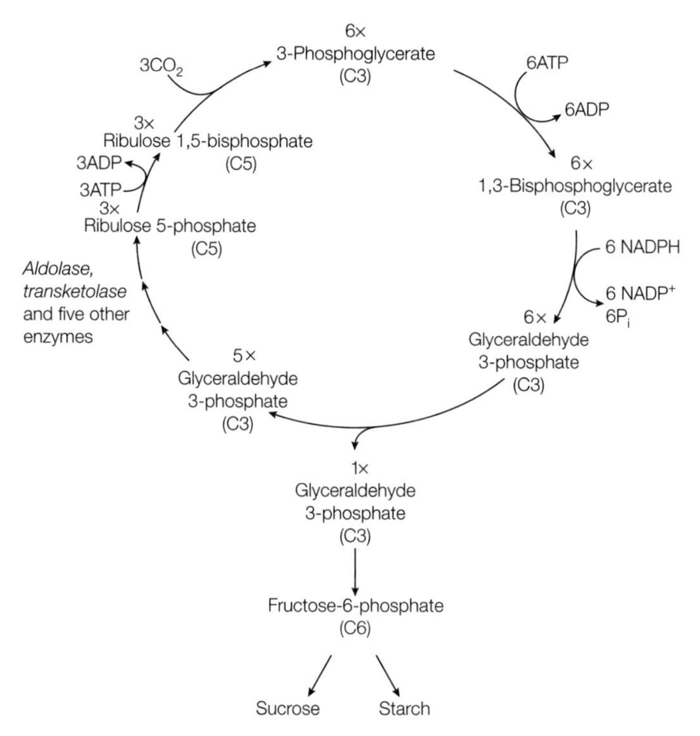

Fig. 6. The Calvin cycle. From Hames and Hooper, Instant Notes in Biochemistry, *2nd edn.* *© BIOS Scientific Publishers 2000.*

B4 BIOSYNTHETIC PATHWAYS

Key Notes

Carbohydrates

Carbohydrates are synthesized from precursors by a process termed gluconeogenesis. This pathway is almost the reverse of glycolysis, except that three irreversible glycolytic enzymes are replaced by three synthetic enzymes specific to this pathway.

Amino acids

Some bacteria can fix atmospheric nitrogen when fixed nitrogen sources (nitrate, nitrite or ammonia) are not available. All other microorganisms must use fixed nitrogen. Only ammonia can be incorporated directly to form amino acids; nitrate and nitrite must be reduced to ammonia first. There are five amino acid families from which 20 different amino acids are synthesized.

Nucleic acids

Nucleic acids are made of nucleotides, which are cyclic nitrogen-containing compounds. Purines are bicylic, while pyrimidines have a single ring. When linked to a phosphorylated pentose sugar they are termed nucleosides. They are the building blocks of DNA and RNA.

Lipids

Lipids are synthesized from fatty acids. They are long chain molecules of around 18 carbon atoms, and they may be saturated (contain no double bonds) or unsaturated (contain one or more double bonds).

Carbohydrates

New cellular material has to be produced by microorganisms in order for them to grow and reproduce. This process is called **biosynthesis**. Small molecules are produced initially from building blocks obtained from the environment, larger molecules being synthesized from these basic units.

Small molecules include glucose, amino acids and nucleic acids. Larger macromolecules include complex carbohydrates, cellulose, lipids and proteins.

All heterotrophic species of organisms must synthesize glucose from sources other than carbon dioxide. The process is termed **gluconeogenesis** (*Fig. 1*) and essentially it is a pathway that reverses the process of glycolysis.

There are several points at which irreversible glycolysis reactions are substituted with gluconeogenic reactions.

- Phosphoenolpyruvate kinase converts pyruvate to phophoenolpyruvate.
- Fructose-6-phosphate is synthesized by fructose bisphosphatase from fructose 1-6 phosphate.
- Glucose synthesis by glucose-6-phosphatase from glucose-6-phosphate.

Amino acids

Assimilation of nitrogen in microbes is very variable. Only one group can utilize atmospheric nitrogen, in a process called **nitrogen fixation**. This reaction is only seen in proteobacteria such as *Azotobacter* or *Rhizobium* as well as in Gram positive bacteria such as some species of *Clostridium*.

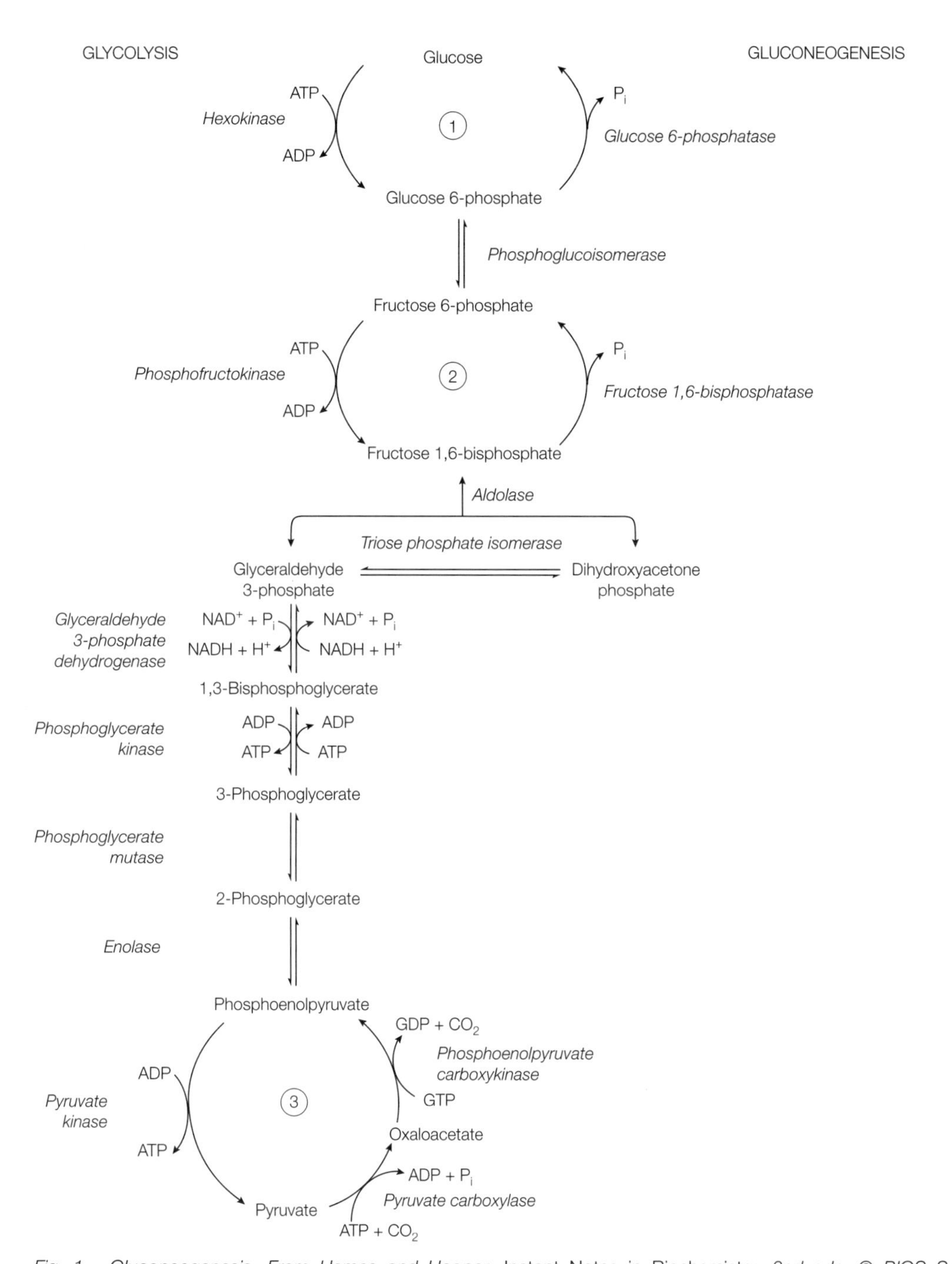

Fig. 1. Gluconeogenesis. From Hames and Hooper, Instant Notes in Biochemistry, *2nd edn. © BIOS Scientific Publishers 2000.*

Nitrogen fixation is mediated by an oxygen-sensitive enzyme called **nitrogenase**

$$N_2 + 6H^+ + 12\ ATP + 12\ H_2O \rightarrow 2NH_3 +\_\ 12\ ADP + 12\ P_i$$

This reaction is an extremely energy expensive one.

All other members of the microbial world utilize nitrate, nitrite or ammonia. Nitrate and nitrite must be reduced to ammonia before assimilation via nitrate and nitrite reductase.

There are 20 different amino acids found in **proteins**. However there are not 20 different biosynthetic pathways because there are five amino acid families that share parts of their biosynthesis. The five families are based on glutamate, aspartate, pyruvate, serine or chorismate (*Fig. 2*).

NADPH glutamate dehydrogenase will synthesize glutamate from α-ketoglutarate by the amination of the 2 carbon organic acid. A series of **transaminations** to other organic acid intermediates of the TCA cycle synthesizes aspartate. Glutamine is synthesized from glutamate by the enzyme glutamine synthetase. Alanine, aspartate and asparagine are formed by transamination of glutamate to pyruvate or oxaloacetate respectively.

Nucleic acids

Nucleic acids are formed of **purines** and **pyrimidines**, cyclic nitrogen containing compounds called **nucleotides**. The purines, adenine and guanine are two-ringed structures synthesized from the precursor ribose-5-phosphate, but six other molecules contribute to its formation (*Fig. 3a*). Pyrimidines, uracil, cytosine

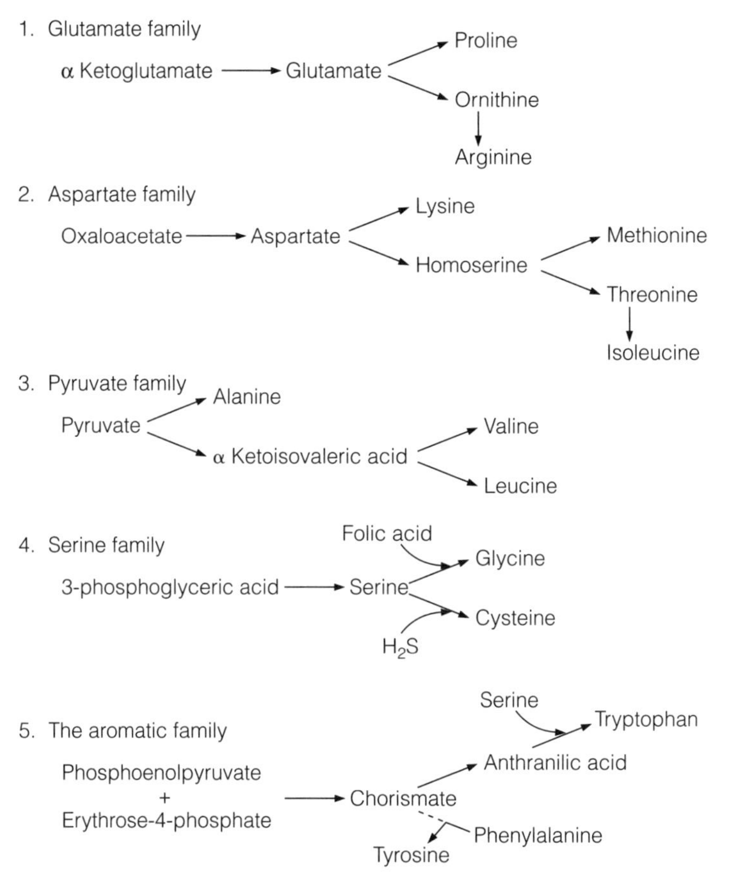

Fig. 2. Amino acid families share common pathways.

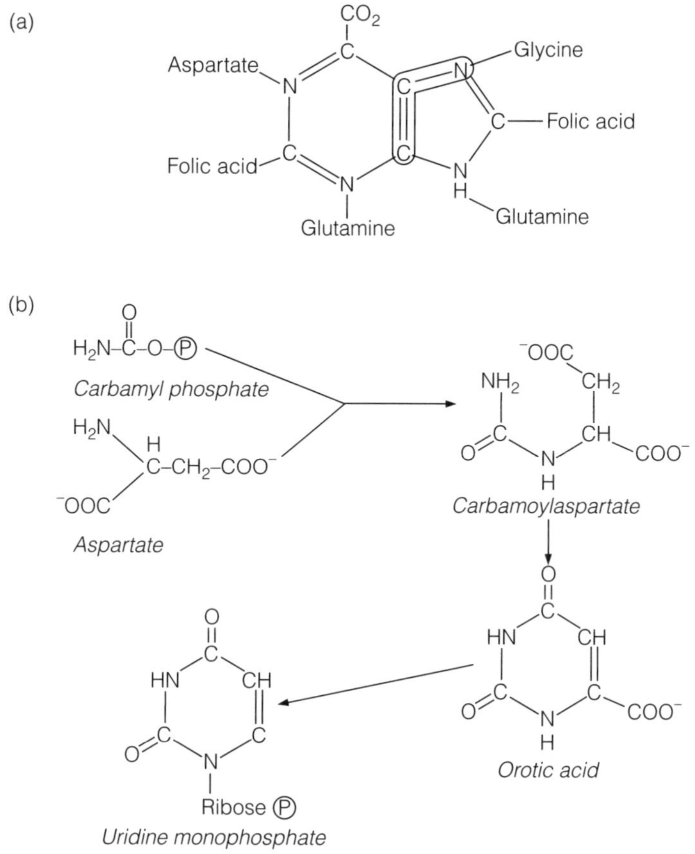

(a)

CO_2 Aspartate Glycine Folic acid Folic acid Glutamine Glutamine

(b)

$H_2N-C(=O)-O-$Ⓟ
Carbamyl phosphate

Aspartate

Carbamoylaspartate

Orotic acid

Ribose Ⓟ
Uridine monophosphate

Fig. 3. (a) Purine structure. (b) Pyrimidine structure.

and thymine are synthesized by the carbamation of aspartic acid via the enzyme aspartate carbamyl transferase, cyclization forming the single ring structure (*Fig. 3b*).

Nucleosides are formed from a purine or pyrimidines linked to a pentaose sugar and if the sugar is phosphorylated it is termed a nucleotide.

Lipids

Fatty acid synthesis occurs in the cytosol using NADPH as the reductant (*Fig. 4*). An **acyl carrier protein** (ACP) is central to the synthesis of fatty acids. Acetyl CoA is carboxylated to malonyl CoA and then both acetyl CoA and malonyl CoA are formed by the transacylase enzymes. AACP and MACP condense to form acetoacetyl ACP. The acetoacetyl ACP undergoes a reduction, dehydration and a further reduction to produce 4 carbon butytryl ACP.

This cycle repeats to synthesize the long chain fatty acids needed for **membrane synthesis**. Lipids are synthesized from long chain saturated fatty acids, containing on average 18 fatty acids with a single unsaturated bond. Some may be branched. In aerobic bacteria and eukaryotic microbes double bonds are introduced in the saturated fatty acid chain by the action of desaturase enzymes. Facultative and anaerobic bacteria form double bonds during the synthesis of the fatty acid.

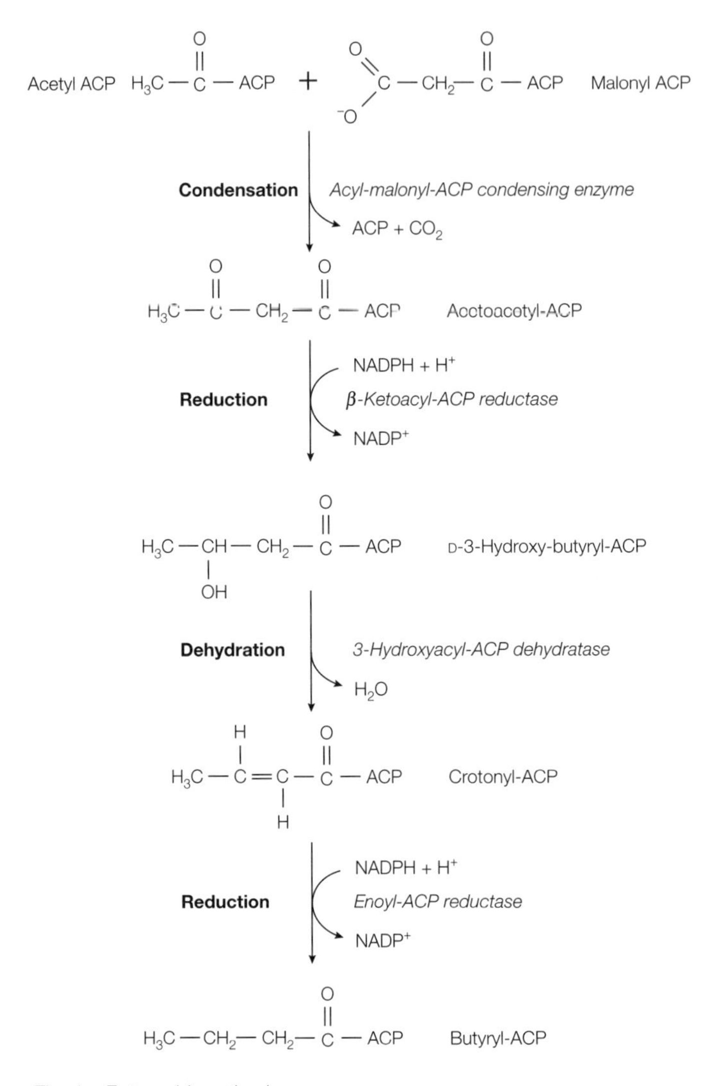

Fig. 4. Fatty acid synthesis.

C1 STRUCTURE AND ORGANIZATION OF DNA

Key Notes

DNA structure

DNA is made up of deoxyribonucleotides joined by 3′,5′-phosphodiester bonds between the deoxyribose sugars. One of four nitrogenous bases, adenine (A), guanine (G), cytosine (C) or thymine (T), is attached to the sugar. Two strands of DNA are held together by specific hydrogen bonding between A and T and G and C to form a right-handed DNA helix. The specific pairing of A:T and G:C is called complementary base-pairing.

DNA conformation

Most DNA in the cell is in the B-form, which is a right-handed helix with 10 base pairs per turn. Other conformations include the A-form and the Z-form.

DNA measurement and description

DNA can be measured by weight, in Daltons; or by length, in micrometers; or number of base pairs. The sequence of the DNA may be described as a sequence of the bases in a 5′→3′ direction; for example, 5′-AGCTTATTCCG-3′.

DNA packaging

DNA is negatively supercoiled, which places the molecule under torsion and reduces its volume. In prokaryotes, DNA is packaged with proteins and polyamines to form a nucleoid. In eukaryotes, DNA is wound round histone proteins to form nucleosomes which are further twisted to form 30 nm chromatin fibers. At cell division the chromatin becomes condensely packed to form the metaphase chromosome.

Chromosomes

Most Bacteria and Archaea have only one chromosome but there are notable exceptions. A cell with only one copy of each chromosome is called haploid. Eukaryotes are typically diploid, having two copies of each chromosome, one from each parent; however many eukaryotic micro-organisms may exist for most of their life cycle in a haploid state.

Related topics

Biosynthetic pathways (B4)
DNA replication (C2)
Transcription (C4)
Plasmids (E5)
Bacteriophage (E7)
Cell division and ploidy (G3)

DNA structure

Deoxyribonucleic acid (DNA) is the genetic material in all living cells but some viruses of prokaryotes or eukaryotes may contain RNA as their genetic material (Topics E7 and J4). The basic repeating subunit of a DNA molecule is a **deoxyribonucleotide**, which consists of a five-carbon deoxyribose sugar, phosphate and one of four nitrogenous bases. The bases are **cytosine (C)** and **thymine (T)** which are monocyclic **pyrimidines**, and **adenine (A)** and **guanine (G)** which are **purines** having two rings (*Fig. 1*). The nucleotides are held together by

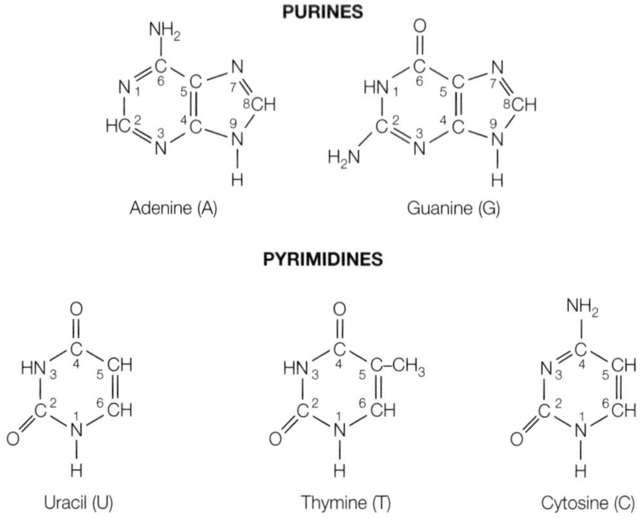

PURINES

Adenine (A) Guanine (G)

PYRIMIDINES

Uracil (U) Thymine (T) Cytosine (C)

Fig. 1. The structure of bases found in DNA and RNA.

phosphodiester bonds between the 3′-carbon on one deoxyribose sugar and the 5′-carbon on the next (*Fig. 2*) to form the backbone of the molecule. The nitrogenous bases are attached to the 1′-position of the sugar by a glycosidic bond. It is the order of the bases on the DNA strands that carries the genetic information. Two strands of DNA, running in opposite directions (**anti-parallel strands**), are held together in a **double helix** by hydrogen bonding between the nitrogenous bases in a highly specific manner (*Fig. 2*). C is always opposite G and A is always opposite T because these pairs of bases are capable of forming hydrogen bonds between them. C forms three hydrogen bonds with G and A forms two hydrogen bonds with T. These A:T and G:C base-pair interactions are called **complementary base-pairs**. DNA replication (Topic C2) and the transfer of genetic information in the DNA into the functioning components of the cell (Topic C4) are possible because of this specific complementary base-pairing. Another point to note is that the base-pairing is such that a purine is always opposite a pyrimidine, thus ensuring that the DNA helix is a constant width of approximately 2 nm. Finally, although hydrogen bonds are non-covalent and comparatively weak they can form a very strong interaction if there are large numbers of them, as is the case in DNA molecules; however, the base pairs can be broken apart when required.

DNA conformation

Most of the DNA in the cell is in what is called the B-form. This is a right-handed helix with 10 base pairs per turn (*Fig. 3*). Other forms of structure that may be found are the A-form, which has 11 base pairs per turn and is associated with DNA–RNA structures, and Z-DNA, which is a left-handed helix associated with particular sequences of alternating G:C and C:G pairs. The two distinctive features of the DNA molecule are the **major** and **minor grooves** which play an important role in the binding of proteins to the DNA.

DNA measurement and description

DNA is measured in different ways depending on the circumstances. Weight is measured in Daltons and length in micrometers or the number of base pairs (bp). The *E. coli* chromosome, for example, is circular, 4.6×10^6 bp long equiva-

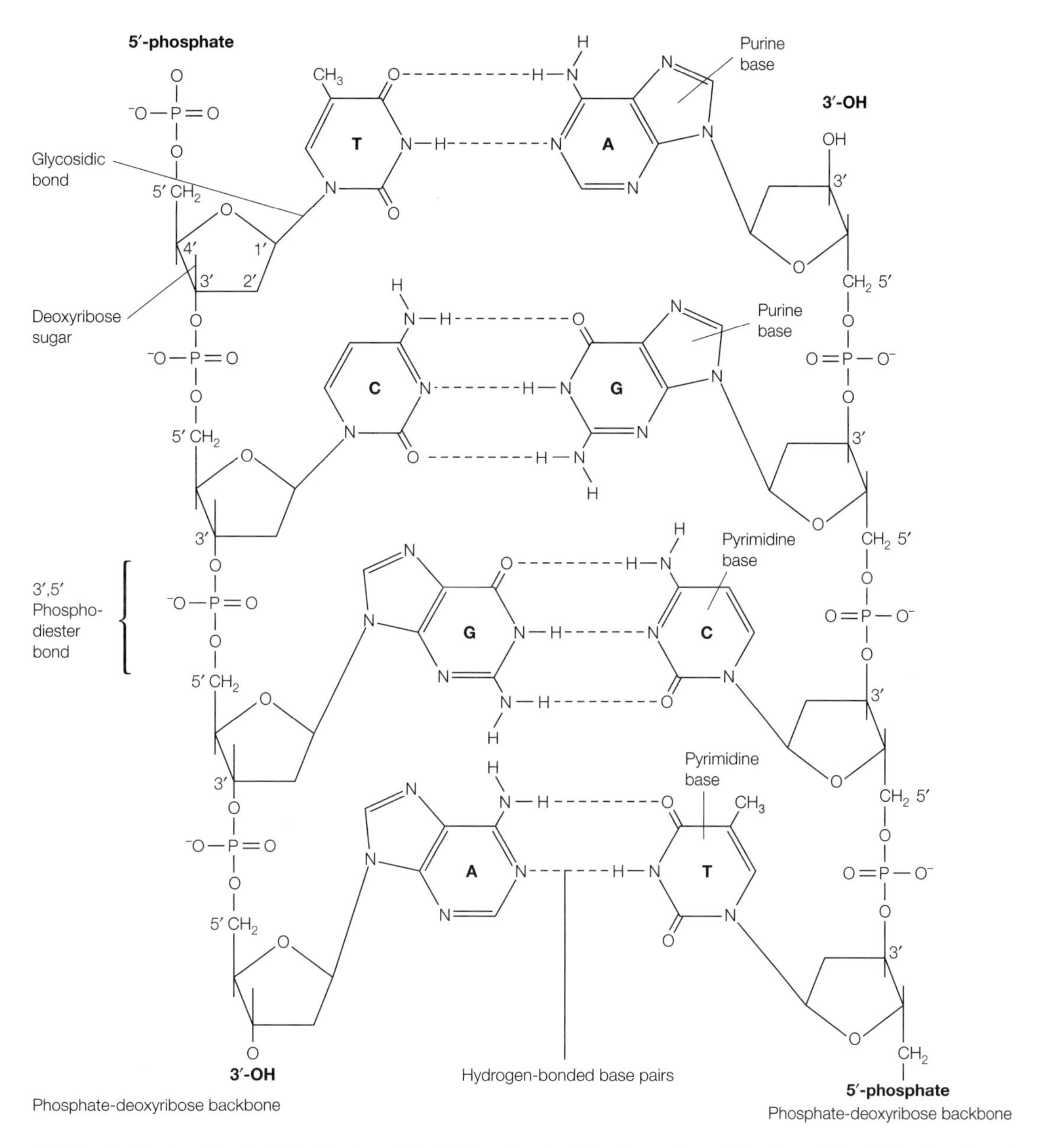

Fig. 2. A diagram showing the structure of a DNA molecule. Note the anti-parallel nature of the DNA strands and hydrogen-bonded base pairs.

lent to 1.2 mm. This holds the information for 4290 genes. Another way of describing a DNA molecule is the sequence of the bases in the molecule in a 5′ → 3′ direction; for example 5′-ATTCGACGT-3′ may describe a short sequence. Only the sequence of one strand is required as the other can be worked out from complementary base-pairing. Currently the complete sequences of 46 Bacteria, 10 Archaebacteria and the yeast *Saccharomyces cerevisiae* are known and this number is increasing all the time as DNA sequencing becomes more efficient.

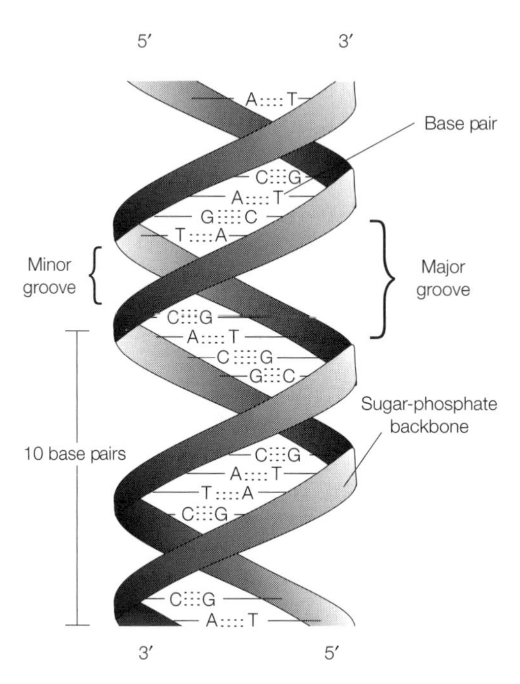

Fig. 3. A schematic view of the DNA double helix.

DNA packaging

The length of DNA in an *E. coli* cell is over 1 mm when the cell itself is only a 1.7 μm × 0.65 μm cylinder. The density of the DNA is therefore approximately 10 mg ml^{-1}. In order to achieve this density the DNA is organized into a much tighter molecule by the introduction of **supercoiling** (twists) into the DNA molecule which puts it under torsion. The enzymes that alter supercoiling are called **topoisomerases** of which **DNA gyrase** is the enzyme responsible for introducing supercoils. Natural DNA is described as negatively supercoiled in that the coiling is in the opposite direction to the DNA right-handed helix, resulting in an underwound DNA helix.

The further packaging of the DNA is dependent on the nature of the cell. In Bacteria, the DNA is associated with positively charged polyamine molecules and proteins, which hold it into about 40–50, independently super-coiled, loops to form a structure sometimes referred to as the **nucleoid**. In Archaea the DNA is associated with histone proteins similar to those in eukaryotes (see below). In both cases, the nucleoid is free in the cytoplasm of the cell.

In eukaryotes, the linear chromosomes are made up of **chromatin** which consists of DNA packaged with specialist proteins called **histones**. At the first level of chromatin organization the DNA is wrapped round the histones to form a chain of 11 nm beads called **nucleosomes** (*Fig. 4*). This is then further wound into a helical coil called a solenoid, which contains six nucleosomes per turn, to give a **30 nm chromatin fiber** (*Fig. 4*). A third level of organization is seen in eukaryotic cells that are about to divide. Here, the DNA is associated with non-histone scaffold proteins to form a tightly packed condensed structure called the **metaphase chromosome** (*Fig. 4*). For most of the eukaryotic cell cycle the chromosomes are located in the nucleus, separated from the cytoplasm by the nuclear membrane. The exception is during cell division when this membrane

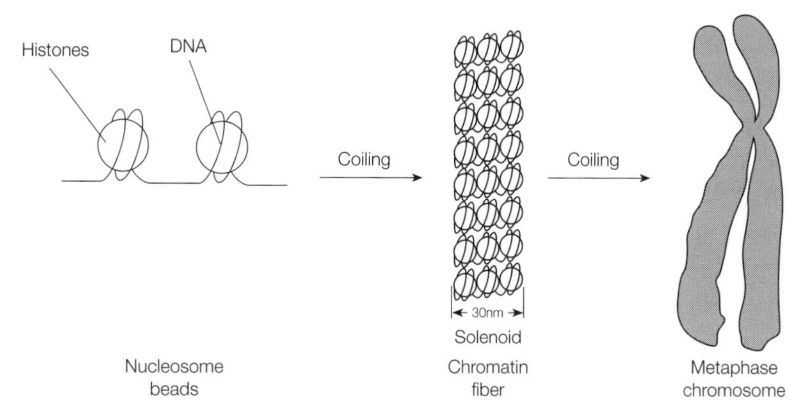

Fig. 4. Levels of chromatin structure in eukaryotic chromosomes.

breaks down allowing the replicated chromosomes to segregate into progeny cells (see Topic G3).

Chromosomes

Until recently it was thought that all bacterial chromosomes were circular and present as only one copy per cell. However, exceptions to this are being found as more genomes are studied and sequenced. To date it has been shown that *Borrelia burgdorferi* and *Streptomyces lividans* contain linear chromosomes, and *Rhodobacter sphaeroides* and members of the *Brucella* genus have two chromosomes. However most Bacteria and Archaea studied in laboratories are genetically **haploid** (monoploid) carrying only one copy of each chromosome in the cell. Typically, eukaryotic cells are diploid as each chromosome is present as two copies due to the combination of genomes from two parents during sexual reproduction. This has led to the requirement of a very complex mechanism for cell division called mitosis to ensure that each offspring receives a copy of each chromosome and a special mechanism of cell division (meiosis) for the production of haploid gametes for reproduction. However it should be noted that a significant proportion of microbial eukaryotes spend most of their life cycle as haploid cells and can reproduce aesexually (Topic H3). The other main feature of many eukaryotes is the presence of a separate DNA genome in the mitochondria and chloroplasts of the cell, which tend to have the characteristics of prokaryotic cells rather than eukaryotic ones (Topic G2). Mitochondrial genomes are inherited via a mechanism separate from chromosome replication.

C2 DNA REPLICATION

Key Notes

Overview

DNA replication is the synthesis of new strands of DNA using the original DNA strands as templates. Complementary base-pairing between the bases on the precursor deoxyribonucleotide triphosphates and the exposed bases on the DNA molecule is responsible for the selection of the correct nucleotide to insert into the growing strand. DNA polymerase is the enzyme that forms the phosphodiester bonds.

The replication bubble

Initiation of DNA replication occurs at one site, *oriC*, on the bacterial chromosome. The helix is opened by a DNA helicase to expose the bases, and DNA synthesis starts on both strands. Replication is bidirectional forming a pair of replication forks that move around the chromosome until they meet at a termination site. Topoisomerases remove the positive supercoiling created by the unwinding of the helix for replication.

DNA polymerase

DNA polymerases synthesize DNA in a $5' \rightarrow 3'$ direction. They also have exonuclease activity which can remove mismatched bases (proof-reading). Because DNA polymerase can only add a nucleotide to an existing 3′-OH group two problems are created: (i) a primer is required to start DNA synthesis, and (ii) one strand must be synthesized in small fragments rather than continuously.

Primers

The primers to initiate DNA synthesis are short stretches of RNA synthesized by a RNA polymerase called primase.

Leading and lagging strands

The strand of DNA synthesized continuously is called the leading strand. The other strand (the lagging strand) is synthesized as a series of short fragments called Okazaki fragments. A complex of proteins called a primosome is responsible for the repeated initiation of DNA synthesis on this strand.

Eukaryote replication

The main differences between prokaryote and eukaryote replication are that in eukaryotes: (i) there are multiple origins of replication on one chromosome; (ii) different polymerases are used for the leading and lagging strand; (iii) a telomerase is required to replenish the telomers at the ends of the chromosome.

Rolling circle replication

Replication of DNA by the rolling circle mechanism produces concatemers of DNA that are essential for the life-cycle of some bacteriophage and viruses.

Related topics

Prokaryote growth and cell cycle (D7)
DNA repair mechanisms (E4)
F plasmids and conjugation (E6)
Replication of bacteriophage (E8)
Cell division and ploidy (G3)

Overview

DNA replication is carried out by enzymes called DNA polymerases using deoxyribonucleoside triphosphates (dNTPs) as precursors (Topic B4). The process starts with the separation of the two strands of the DNA helix to expose the bases. New strands of DNA are synthesized using the old strands as templates. Complementary base-pairing between the bases on the incoming nucleotides and the template DNA strand (A:T; G:C) dictates which nucleotide will be inserted into the growing DNA chain (*Fig. 1*). This is called **semi-conservative replication** as, after the whole DNA molecule has been copied, each new DNA helix consists of one original DNA strand and one newly synthesized DNA strand. The role of the DNA polymerase is to check that the hydrogen bonding beween the bases is correct, then, to form a phosphodiester bond between the two adjacent nucleotides. As the basic mechanism for DNA replication is similar for all organisms, this will be described first in relation to bacterial DNA replication and in particular that of *Escherichia coli*. Differences found in eukaryotes will be discussed subsequently.

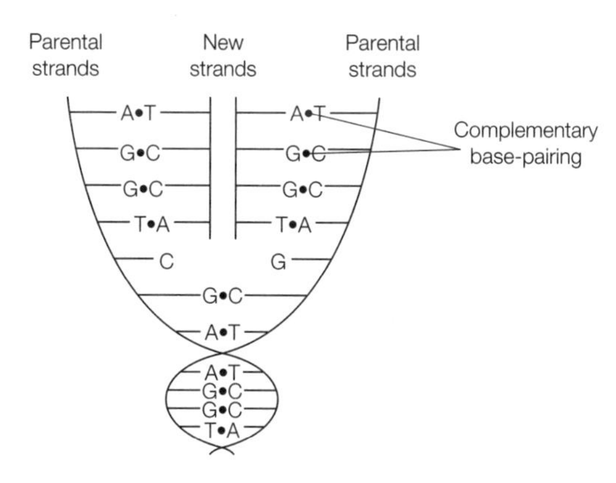

Fig. 1. Semi-conservative replication of DNA showing complementary base-pairing.

The replication bubble

The first stage of replication involves the identification of the origin of replication. There is only one on the chromosome of *E. coli* called *oriC*. A complex of proteins binds to this site, opens up the helix and initiates replication. DNA synthesis occurs in both directions (bidirectional) and on both strands, creating a **replication bubble** which appears as a θ (theta) structure in electron micrographs (*Fig. 2*). The two sites at which DNA synthesis occurs are called the **replication forks**. As replication proceeds the replication forks move round the molecule opening up the DNA strands, synthesizing two new complementary strands and rewinding up the DNA behind, until the two replication forks meet at a termination site. The two completed circles of DNA are still linked and have to be separated by an enzyme **topoisomerase IV** which makes a transient break in both strands of one of the molecules, allowing them to disconnect.

Other enzymes/proteins associated with DNA replication are:

- **DNA helicase** which unwinds and opens the DNA helix to allow access to the bases.
- **Single-stranded binding proteins** that bind to the DNA strands, preventing them from base-pairing before they have been copied.

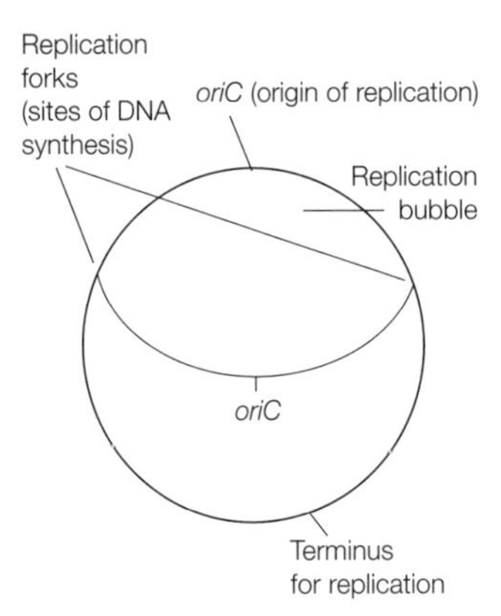

Fig. 2. Diagram showing θ structure as a result of bidirectional replication of a circular bacterial chromosome.

- There is a problem in that as the helix is opened up the DNA ahead of the replication fork would become overwound because the DNA ends are fixed (the chromosome is circular) and therefore cannot rotate relative to each other. This is solved by **topoisomerase I** which makes a break in one of the DNA strands which allows the DNA to rotate round the intact strand thus releasing the torsion.

The complex of proteins that initiate DNA synthesis is called the **primosome**.

DNA polymerase

The enzyme responsible for normal DNA synthesis in bacteria is **DNA polymerase III**. DNA polymerase I also has a role in DNA replication and in DNA repair (Topic E4). These enzymes function by adding a new nucleotide to the 3′-OH end of a growing polynucleotide chain (*Fig. 3*) with the release of diphosphate thus synthesizing a new strand of DNA in a 5′ → 3′ direction. DNA polymerases also have exonuclease activity which allows the sequential removal of nucleotides in both the 5′ → 3′ and 3′ → 5′ directions. This ability to remove incorrectly inserted nucleotides confers proof-reading properties on DNA polymerase III. Immediately after the enzyme has inserted a nucleotide into the growing chain the enzyme can detect if the base-pairing is correct or if a mismatch has occurred. If there is a mismatch, the 3′ → 5′ exonuclease activity removes the nucleotide and the polymerase can then replace it with the correct one. This proof-reading ability ensures that the error rate (mutation rate) in DNA synthesis is less than 10^{-8} per base pair.

There are several features of DNA replication that are common between all DNA polymerase enzymes.

1. DNA polymerases cannot begin from a single-stranded template alone – they must have a 3′-OH group to add a nucleotide on to. These enzymes therefore require some sort of **primer** to start DNA synthesis.
2. DNA can only be synthesized in one direction, 5′ → 3′. As the two strands of DNA are antiparallel and are both copied at the same time at the replication fork, only one strand can be synthesized continuously. The other strand must be synthesized in short fragments.

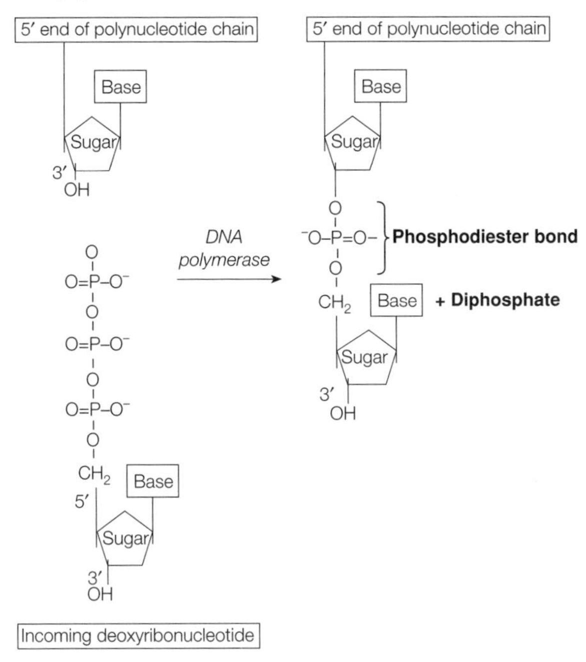

Fig. 3. Diagram showing the joining of a deoxyribonucleotide to a growing DNA strand by DNA polymerase. Base-pairing to the complementary DNA strand is not shown.

Primers

The problem of how to start DNA synthesis is solved by using a specialist RNA polymerase called a **primase** to lay down a short stretch of ribonucleotides (primer) complementary to the DNA template at the start point for DNA synthesis. RNA polymerases do not require a 3′-OH group to start RNA synthesis so can start from scratch. The primer provides the necessary 3′-OH group to which DNA polymerase III can join a deoxyribonucleotide and start to synthesize the DNA strand (*Fig. 4*). Synthesis continues until the DNA polymerase III reaches a terminator sequence or another primer (see below).

Leading and lagging strands

As DNA can only be synthesized in the 5′ → 3′ direction, only one strand can be synthesized continuously. This is called the **leading strand** and DNA synthesis need only be initiated once. The other strand, called the **lagging strand**, is

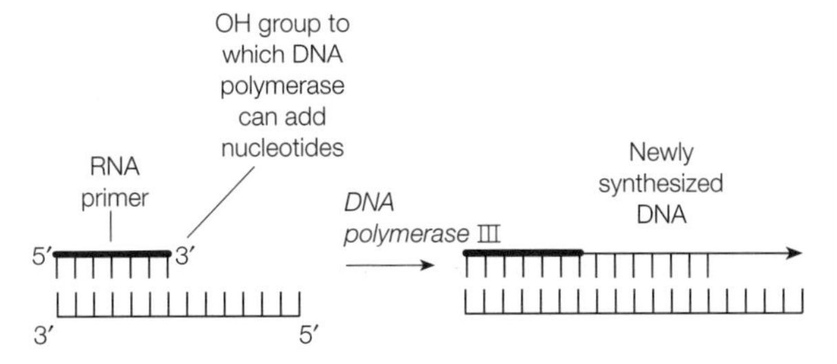

Fig. 4. An RNA primer is required to initiate DNA synthesis.

synthesized as short pieces, approximately 1000 bases long, called **Okazaki fragments**. To do this the primosome initiates DNA synthesis approximately every 1000 bases. DNA polymerase III then copies the template strand until it reaches the RNA primer of a previous fragment. DNA polymerase III is released and DNA polymerase I binds in its place. DNA polymerase I has 3′ → 5′ exonuclease activity which removes the ribonucleotides one by one and its polymerase activity replaces these with deoxyribonucleotides. The final gap in the deoxyribonucleotide chain is sealed by a specialist enzyme, **DNA ligase**, which can join adjacent 5′-phosphate and 3′-OH groups as described in the section above, with the consequent joining of the fragments by the removal of the RNA primers by DNA polymerase I and sealing of the gaps by DNA ligase (*Fig. 5*).

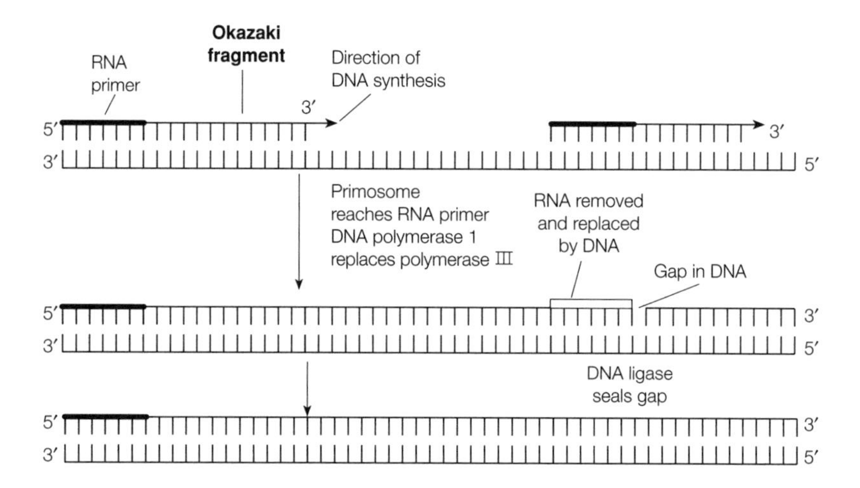

Fig. 5. The joining together of Okazaki fragments by DNA polymerase I and ligase on the lagging strand of DNA synthesis.

Eukaryote replication

The major differences between prokaryotic and eukaryotic replication are generally as a result of the larger size of the genomes and the fact that eukaryotic genomes are linear. Some of the main differences in eukaryotic replication are listed below:

- There are many origins of replication on each chromosome: approximately one every 3–300 kb of DNA.
- Leading and lagging strands are made by different DNA polymerases. The Okazaki fragments are shorter and the method of removing the primers is different.
- Because of the need to reinitiate DNA synthesis constantly on the lagging strands there are short segments at the ends (**telomeres**) which do not get copied. The chromosomes would therefore gradually shorten with time. This problem is remedied by a specialist RNA-containing enzyme, a **telomerase**, which extends the ends of chromosomes by adding numerous hexanucleotide repeat sequences. These then act as templates for DNA replication.

Rolling circle replication

An alternative mechanism for DNA replication is used by some bacteriophage (Topic E8) and viruses (Topic J7) and in the transfer of plasmid DNA between cells by conjugation (Topic E6). Called rolling-circle replication, one strand of the DNA is nicked to provide a 3′-OH group. Helicase and single-stranded-binding proteins bind to create a replication fork, and the nicked strand is displaced. Deoxyribonucleotides are added to the 3′-OH group to create the leading strand and the displaced strand acts as a template for lagging strand synthesis (*Fig. 6*). In some cases of phage, viral and plasmid replication the strand may remain single. This process can continue for longer than one round of replication, if required, producing long linear molecules of repeated genome lengths called **concatamers**. These are essential for the life-cycle of some bacteriophage and viruses where the cutting of the concatamers into genome lengths is essential for the correct assembly of the virus particles.

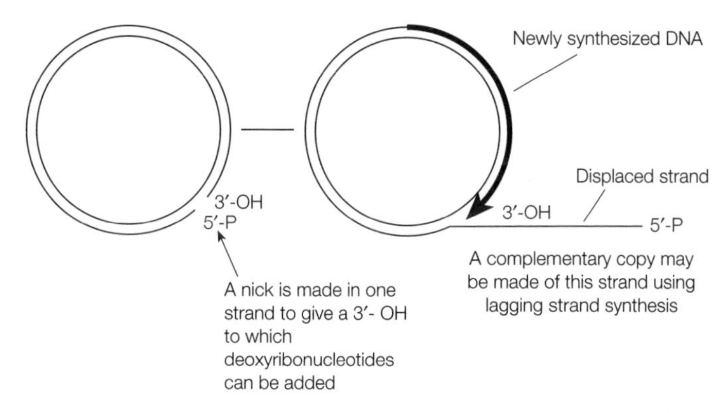

Fig. 6. Diagram showing the mechanism of circle replication.

C3 RNA MOLECULES IN THE CELL

Key Notes

Structure of RNA — RNA is a polymer of ribonucleotides with a similar structure to DNA except that it contains a ribose sugar instead of deoxyribose and uracil instead of thymine. RNA is normally single-stranded but can form secondary structures by base-pairing within the molecule.

RNA molecules in the cell — RNA molecules in the cell can be split into two groups. Comparatively long-lived stable RNAs have a number of different structural and functional roles in the cell including ribosomal (rRNAs) and transfer RNAs (tRNAs). Messenger RNA (mRNA) molecules that act as templates for protein synthesis have a much shorter life.

Catalytic RNA molecules — Some RNA molecules have enzymatic activity, these are known as ribosomes. The most common of these are self-splicing introns found in a range of microbes.

Related topics — Transcription (C4) Translation (C7)

Structure of RNA

RNA is similar in structure to DNA (Topic C1) with the following differences:

- The sugar is ribose instead of deoxyribose, the difference being the presence of an OH group on the 2′-carbon. This gives the RNA molecule the capacity to have enzymic activity (*Fig. 1*).
- The pyrimidine thymine is replaced with uracil which also base-pairs with adenine (see Topic C1, *Fig. 1*).
- RNA is single-stranded but it is capable of forming double-stranded secondary structures by base-pairing within the same molecule. Some of these structures may be extremely complex (see the structure of tRNA, Topic C7, *Fig. 3*).

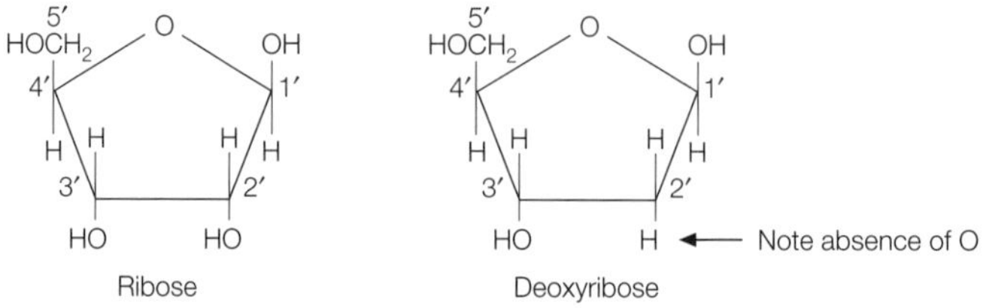

Fig. 1. Structures of sugars in nucleic acids.

- Unusual nucleosides such as inosine and pseudouridine are found in stable RNA molecules formed by the post-transcriptional modification of nucleosides.
- The precursors for the synthesis of RNA are ribonucleoside triphosphates (Topic B4).
- An RNA sequence can be written as the base sequence in the 5′ → 3′ direction: for example, ACGUCCAUG.

RNA molecules in the cell

There are a number of different RNA molecules in the cell, the most abundant of which are stable RNAs associated with the protein synthetic apparatus. **Ribosomal RNAs (rRNAs)** help to maintain the structure of the ribosomes and play a role in their function (see Shine–Dalgarno sequences, Topic C4). Prokaryote and eukaryote ribosomes contain rRNA molecules that differ in size and number. These various classes of rRNA are distinguished by their size and sedimentation behavior, measured as a Svedberg coefficient (S), which reflects the molecules' density, mass and shape. This is the rate at which the particles sediment in a gradient, under centrifugal force, and takes into account the size, the shape and density of the particle. The sizes of rRNA molecules in prokaryote and eukaryote cells are shown in *Table 1*. The sequences of rRNA molecules in organisms of the same strain are highly conserved, and some regions of rRNA are conserved between the Bacteria, Archaea or eukaryote. Consequently, sequences of regions within the 16S RNA (prokaryotes) and the 18S RNA (eukaryotes) have been used to measure evolutionary relationships between different organisms (Topics A1 and D1).

Table 1. Classes of rRNA molecules in prokaryotic and eukaryotic cells

Organism	Ribosomal subunit	Svedberg coefficient	Size (approximate number of nucleotides)
Prokaryotes	Large	23S	2900
		5S	120
	Small	16S	1500
Eukaryotes	Large	28S	4200
		5.8S	160
		5S	120
	Small	18S	1900

Transfer RNAs (tRNAs) are the adapter molecules that convert a sequence of bases into an amino acid sequence (Topic C7). Other RNA molecules are found associated with a number of large molecules including spliceosomes and telomerases in eukaryotes. Finally, the major group of RNA molecules are the **messenger RNAs** (mRNAs) produced by **transcription** from DNA, which act as the templates for protein synthesis. These RNAs have comparatively short half-lives compared with the stable RNA molecules.

Catalytic RNA molecules

In the early 1980s it was shown that some RNA molecules (sometimes called **ribozymes**) have enzymatic activity. The most common of these are self-splicing introns which have the ability to remove themselves from a piece of transcribed RNA joining the two adjacent exons together (see Topic C4). Ribozymes have been found in a wide range of microorganisms including archaea, mitochondria of fungi, chloroplasts of photosynthetic protista and some bacteriophage.

C4 Transcription

Key Notes

Overview

Transcription is the synthesis of messenger RNA (mRNA) and stable RNA molecules from a DNA template by RNA polymerase, using ribonucleoside triphosphates as precursors. Each mRNA carries the sequence of one gene or in the case of prokaryote operons, a number of genes. Regulation of gene expression may occur at the level of transcription.

DNA-dependent RNA polymerases

Bacterial cells contain only one RNA polymerase whereas eukaryotes have at least three. In Bacteria a holoenzyme consisting of five subunits (α_2, β, β', σ) is responsible for accurate initiation of transcription. The sigma (σ) subunit (factor) is released to give the core enzyme (α_2, β, β') which is responsible for elongation.

Stages of transcription

There are three stages of transcription: initiation, which involves the binding of RNA polymerase to promoter sequences in the DNA that signal the start site for transcription; elongation, which is the synthesis by RNA polymerase of a sequence of ribonucleotides that is complementary to the sequence of one of the strands of the DNA; and termination of RNA synthesis at the end of the gene or the operon.

Promoters

Promoters consist of conserved sequences necessary for the initiation of transcription. There are many different types of promoters. Many *E. coli* promoters contain two conserved regions at –35 and –10 bases before the start site for transcription. The first transcribed base is normally a purine.

Termination

There are several types of terminator sequence in *E. coli*. Some require a protein factor, rho (ρ), for accurate termination of transcription; the others rely on the formation of hairpin structures in the transcribed RNA causing termination.

Transcription in eukaryotes

The promoter sites for RNA polymerases I and II are located before the start site for transcription. RNA polymerase III recognizes sites within the gene itself. Eukaryotic RNA polymerases require the presence of additional transcriptional factors for DNA binding and the initiation of transcription.

RNA processing

Generally in Bacteria only structural RNA molecules are processed post-transcriptionally to give a functional molecule. In contrast, eukaryotic RNAs undergo a number of processes to produce a mature molecule. Introns, non-coding transcribed sequences, are removed by splicing, a methylated guanine cap is added to the 5′ end of the RNA and a poly (A) tail to the 3′ end. Introns are also found in the Archaea.

Related topics

DNA replication (C2)
Control of gene expression (C5)

Overview

The sequence of bases in the DNA that codes for a single polypeptide chain or stable RNA molecule is called a **gene**. Proteins cannot be synthesized directly from the DNA but require the production of an RNA intermediate called messenger RNA (mRNA). The synthesis of RNA is called **transcription** and the enzyme responsible is **RNA polymerase**. The mechanism of RNA synthesis is similar to DNA replication (Topic C2) in that a sequence of ribonucleotides is polymerized which is complementary to the sequence of bases on the DNA molecule. However there are a number of important differences:

- Only one strand of the DNA is copied.
- The precursors for RNA synthesis are ribonucleoside triphosphates (NTPs).
- In eukaryotes the mRNA molecule carries the information for one gene product, but, in prokayotes, more than one gene may be found on one mRNA molecule. These are **polycistronic mRNAs** and are a feature of the organization of the bacterial genome. Gene products that are required together are often clustered into one transcriptional unit, allowing the opportunity to coordinately regulate their synthesis. These units are called **operons** (see Topic C5).
- RNA polymerases bind to specific sites that signal the start of a gene or operon called a **promoter**.
- RNA polymerases do not require a primer.
- Transcription provides an opportunity to regulate the production of a gene product in that additional protein factors may activate or inhibit RNA synthesis of a particular gene (see Topic C5).

DNA-dependent RNA polymerases

The complete *E. coli* RNA polymerase, called the **holoenzyme**, consists of five subunits, two α subunits (α_2), one each of β, β' and σ. The σ subunit (**sigma factor**) is required for the accurate initiation of transcription by binding to the promoter, but is released immediately after the start of transcription to give a **core** enzyme consisting of the α_2, β, β' subunits which is capable of polymerizing the RNA chain. Only one DNA-dependent core RNA polymerase is found in Bacteria but at least three are present in eukaryotic cells, numbered I, II and III, which are responsible for different types of transcription in different regions of the nucleus (*Table 1*). Archaea contain only one RNA polymerase but it can be one of a number of different types depending on the species which together show greater similarity to eukaryote RNA polymerase II than Bacterial RNA polymerases.

Table 1. Distribution and function of RNA polymerases

	Prokaryote	Eukaryote		
	RNA Polymerase	RNA Polymerase I	RNA Polymerase II	RNA Polymerase III
Location	Cytoplasm	Nucleolus	Nucleus	Nucleus
Products	All RNAs	28S, 18S, 5.8S rRNAs	mRNAs and most small nuclear RNAs (snRNAs)	tRNAs, 5S rRNA, one snRNA, other structural RNAs

A bacterium can have more than one σ factor that recognizes different promoter sequences. This provides a useful mechanism for the regulation of gene expression, a clear example of which is the regulation of *Bacillus subtilis* sporulation. All the genes required for sporulation need a set of specific sigma factors for their transcription. These sigma factors are only synthesized when the conditions are right for sporulation. Response to heat shock and some nitrogen assimilation genes are controlled in a similar way.

Unlike prokaryotic RNA polymerases, eukaryotic RNA polymerases require additional transcription factors to facilitate the initiation of RNA synthesis.

Stages of transcription

Transcription consists of three stages:

- **Initiation.** RNA polymerase holoenzyme identifies and binds to the promoter. The RNA polymerase opens up the DNA helix, exposing the bases, to create a single-stranded template. This structure is called a **transcription bubble** (*Fig. 1*). The first two ribonucleotides can then enter the complex and base-pair to the template strand. RNA polymerase forms the phosphodiester bond between them with the release of diphosphate.
- **Elongation**. The sigma factor is released from the enzyme and the core RNA polymerase moves down the DNA template, unwinding a small section of the DNA at a time and polymerizing a complementary copy of RNA. RNA is synthesized in the 5′ → 3′ direction like DNA. As the transcription bubble moves down the DNA molecule the newly synthesized RNA is dissociated from the DNA and the DNA helix is reformed behind the transcription bubble.
- **Termination**. Specific sequences in the DNA signal the end of the gene. Transcription ceases and RNA polymerase and the completed mRNA chain are released from the DNA.

The strand of DNA that acts as the template for RNA synthesis is called the **anti-sense** strand as the mRNA produced has the same sequence as the non-template strand called the **sense** strand.

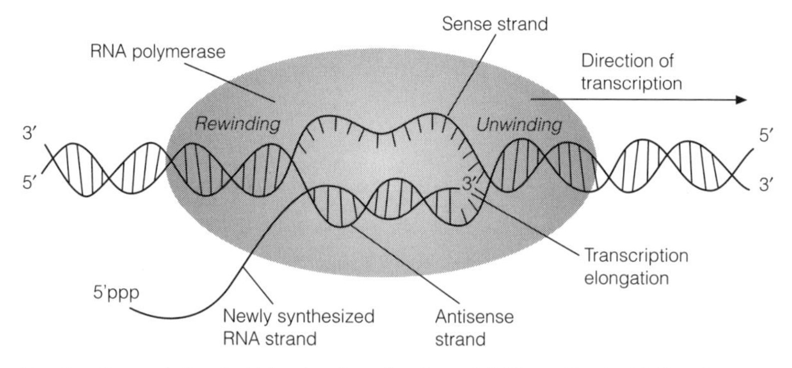

Fig. 1. Transcription bubble showing direction of RNA synthesis. RNA polymerase moves along the DNA molecule inserting ribonucleotides into the growing RNA chain. The enzyme unwinds the DNA ahead and rewinds it behind. From Hames and Hooper, Instant Notes in Biochemistry *2nd edn. © BIOS Scientific Publishers 2000.*

Promoters

There are many different promoter sequences, depending on the organism and how the gene is regulated (see Topic C5). Promoters consist of conserved sequences which are required for the binding of RNA polymerase and the initiation of transcription. Some of the best studied promoters are those of *E. coli*. In many of these promoters, two sets of conserved DNA sequences have been identified before the start sites of transcription. They are called the –10 and –35 sequences as the convention is to number the first transcribed base as +1; therefore, regions before the start site which are not transcribed are given a negative number. The consensus sequences are shown in *Fig. 2*. Promoters which are frequently transcribed, called strong promoters, have sequences at the –10 and –35 sites which are very close to these consensus sequences. Weak promoters show greater differences to the consensus. The first base in *E. coli* mRNAs is usually a purine (>90%). Regulatory regions that control transcription are also found in this upstream region before the start site for transcription.

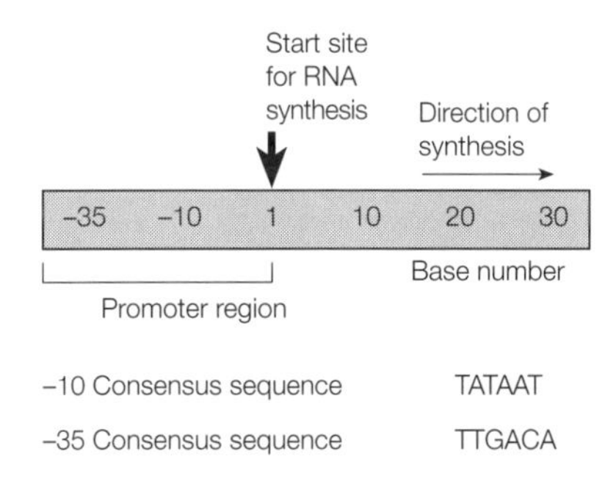

Fig. 2. Consensus sequences of E. coli *promoters.*

Termination

Elongation continues until the RNA polymerase reaches a signal that indicates the end of the RNA molecule. A number of termination signals have been found in Bacteria. One group requires the presence of an additional protein factor, the rho factor (ρ). Other termination signals are due to the presence of certain sequences in the RNA molecule which once synthesized form hairpin structures by base-pairing. These secondary structures in the RNA cause the RNA polymerase to pause. This hairpin is usually followed by a run of Us (uracil) in the RNA which will be comparatively weakly base-paired with the DNA strand (A–U base pairs have only two hydrogen bonds) and consequently the mRNA–DNA–enzyme complex dissociates (*Fig. 3*). A third group of terminators consisting of GC-rich regions followed by an AU rich region have also been identified.

Transcription in eukaryotes

In eukaryotes the promoter sites for RNA polymerases I and II are located at the 5′ end of the gene but the polymerase III promoter lies within the gene. Most RNA polymerase II promoters have a region at –25 called the TATA box which has a consensus sequence of TATAAA but, RNA polymerase II can also transcribe a number of genes that lack this sequence. In contrast to prokaryotes, RNA polymerase II requires the assistance of a number of additional general transcription factors to aid binding and initiation of transcription.

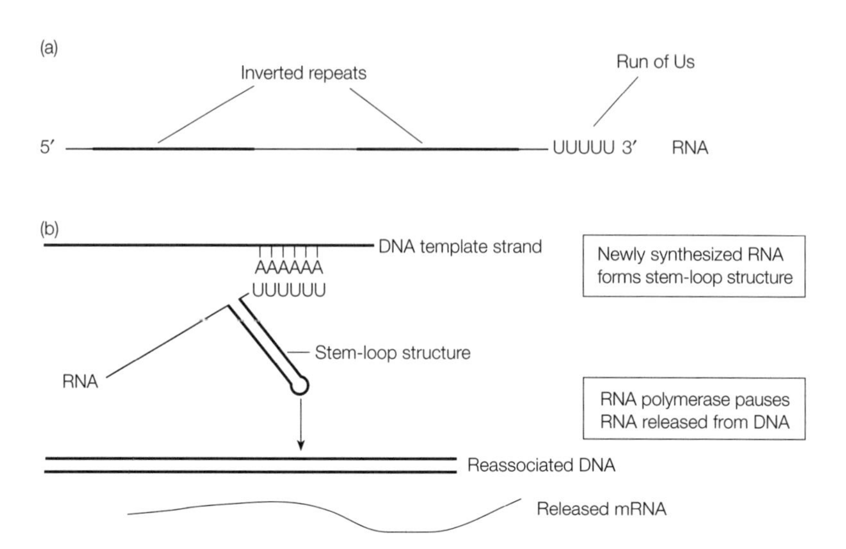

Fig. 3. Model for rho-independent termination of transcription. (a) Inverted repeats and run of U's in RNA required for terminator formation. (b) Formation of stem–loop structure and termination during transcription.

RNA processing

In Bacteria, protein-encoding mRNAs generally do not require any further processing prior to translation. In fact, protein synthesis normally starts as soon as the ribosome-binding sites (the start signals for translation) have been transcribed and are free of RNA polymerase. It is not unusual to see multiple ribosomes attached to the mRNA while it is still being transcribed. However a number of mRNAs in bacteria and bacteriophage have been identified recently that do require removal of internal sequences. In contrast, many structural rRNA and tRNA molecules are synthesized as precursor molecules containing a number of these RNAs, which require cleavage by specific ribonucleases to produce the finished molecules.

In eukaryotes, the vast majority of the protein-encoding genes do contain sequences, called **introns**, which are not found in the translated proteins. After transcription, these sequences are spliced out of the transcript to leave just the protein-encoding regions, the **exons**, in the mRNA (*Fig. 4*). Eukaryote mRNAs are also further modified by the addition of at least a **methylated guanine cap** to the 5′ end (this occurs immediately after initiation) and a string of adenylate residues to the 3′ end, the **poly (A) tail** during termination. Additional modifications may also occur. These modifications help to stabilize the mRNAs and prevent their degradation by ribonucleases. Unlike prokaryotes, transcription and translation take place in two separate compartments of the cell, and the complete mRNA molecule must leave the nucleus before translation starts.

Transcription in the Archaea appears to have a number of features common with both the Eukarya and the Bacteria. Many of the promoters studied so far contain eukaryote-like TATA boxes rather than the –10 regions of Bacteria. However those termination sequences that have been identified appear to be more Bacterial, consisting of inverted repeats followed by strings of Us.

Introns are also found in rRNA and tRNA genes (Topic D5).

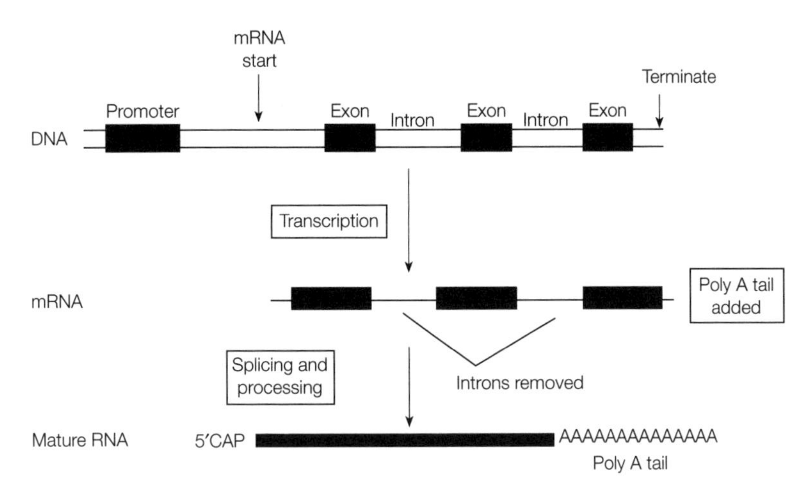

Fig. 4. The processing of mRNA in eukaryotes.

C5 CONTROL OF GENE EXPRESSION

Key Notes

Why control gene expression?

Not all genes in the cell need be expressed at all times. Cellular regulatory systems allow the control of gene expression so that their products are only synthesized when required. Regulation can occur at all stages in the synthesis of proteins in the cell, but much of it is at the level of transcription which gives rapid and sensitive control.

General features of regulation

Constitutive genes are expressed at all times, whereas inducible genes are expressed when required. Regulation can be specific, controlling just one gene, or operon, or global, affecting a large number of genes. Regulation requires some way of detecting the current situation and a way of responding by controlling transcription. The control may be either negative, where the gene is normally switched on unless a repressor is present, or positive, in which an activator is required to allow transcription to occur. Genes or operons may be controlled by more than one regulatory system.

Lactose operon

The *lacZ*, *lacY* and *lacI* gene products are required for the use of lactose as a carbon source. Gene expression is controlled by two regulatory systems. A negative control system, which switches off the genes in the absence of lactose, and a positive control system, called catabolite repression, which switches on the genes when glucose is not present in the surrounding medium.

Negative control

In the absence of lactose, the lactose repressor protein binds to an operator (O) site and inhibits transcription of the *lac* genes. If lactose is present, a product of lactose, allolactose, acts as the inducer molecule. It binds to the *lac* repressor and prevents it binding to the operator site, thereby allowing transcription to occur.

Positive control

A number of operons associated with carbon utilization cannot be transcribed unless a catabolite activator protein (CAP) is bound to the promoter. The CAP requires the presence of cAMP in order to bind to the DNA and activate transcription. cAMP acts as the signal that glucose levels are low in the cell and therefore alternative carbon sources are required.

The tryptophan operon

The *trp* operon is negatively controlled by a repressor protein which is only active if it is complexed with tryptophan (a corepressor). Gene expression is also controlled by attenuation (premature termination of *trp* RNA synthesis in the presence of tryptophan). The combined effect of the two regulatory systems is to reduce transcription by about 700-fold.

Two-factor control systems	These systems consist of a sensor component, which detects the environment around or within the cell, and an effector protein that regulates gene expression in response to that environment. The sensor protein modulates the activity of the effector protein by phosphorylation.
Other regulatory mechanisms in prokaryotes	There is a range of other regulatory mechanisms found in prokaryotes.
Regulation in eukaryotes	Regulation in eukaryotes involves *trans*-acting transcriptional factors that interact with control sequences near the promoter regions called *cis*-acting elements. These regulate gene expression in response to nutritional and environmental signals but are also important during development and differentiation.
Related topic	Transcription (C4)

Why control gene expression?

Many of the genes in the cell encode functions that are not always required. For example, if bacteria are growing in a glucose-containing medium, with all the amino acids provided, they do not need to produce the enzymes that allow them to use alternative carbon sources such as lactose or cellulose: neither do they need to express the genes involved in the synthesis of amino acids. Regulatory systems therefore exist in the cell to control the expression of genes in response to their nutritional or physical environment, preventing the wastage of vital energy. This regulation of **gene expression** can occur at virtually any level in the flow of genetic information in the cell: transcription, mRNA processing, mRNA turnover, translation or enzyme function. However, in many cases, especially in Bacteria, this regulation is at the level of transcription, where regulatory proteins modulate the binding of RNA polymerase to promoters and consequently the amount of RNA synthesis, giving rapid and sensitive control.

General features of regulation

- Some genes, such as those that encode essential RNAs and enzymes for key metabolic processes, are expressed all the time. Such genes are described as **constitutive**. Genes that are expressed only when required are called **inducible**.
- Regulation can be specific, controlling just one gene or operon, or **global**, in which a wide range of genes or operons are regulated together in response to some nutritional or physical signal.
- Regulatory systems consist of a means to detect the current situation of the cell and a way to cause a response by switching transcription on or off.
- Regulation may be either negative or positive. In negative control, genes are normally switched on unless switched off by a **repressor**. This normally occurs in the case of biosynthetic genes where the presence of the final product of a metabolic pathway switches off the expression of the genes encoding that pathway. In the case of positive control, an **activator** protein is required for transcription to occur.
- Genes may be controlled by more than one regulatory system which responds to different signals.

It should be pointed out that there are many different mechanisms of gene regulation acting on a wide range of genes; examples of some of the gene regulatory systems in *E. coli* are shown in *Table 1*. In this topic the regulation of only three different sets of genes will be described: the lactose (*lac*) operon; the tryptophan operon and osmotic regulation of two outer membrane porins, OmpF and OmpC. These provide examples of the major types of regulation in the bacterial cell, showing how the regulatory proteins can respond to the environmental or nutritional signals and interact with the promoter regions of the controlled genes.

Table 1. Some of the global regulatory systems found in E. coli

Global control system	Signal
Anaerobic respiration	Lack of oxygen
Catabolite repression	cAMP concentrations
Heat shock	Increase in temperature
Nitrogen utilization	NH_3 limitation
Osmotic stress	High osmolytes
Oxidative stress	Oxidizing agents
SOS response	DNA damage
Stringent response (starvation)	ppGpp; pppGpp

Lactose operon

The *lac* operon consists of three structural genes, *lacZ*, *lacY* and *lacA*, which are transcribed together from a single promoter, P_{lac}. The genes code for β-galactosidase, the enzyme that splits lactose into glucose and galactose; lactose permease, the lactose transport protein; and lactose transacetylase, whose role is unknown, respectively (*Fig. 1*). These proteins are only required if lactose is present in the medium and glucose is absent. Consequently, the *lac* operon is controlled at two different levels.

- A negative control system, requiring a repressor (LacI), switches off the genes in the absence of lactose.
- A positive, global control system, called **catabolite repression**, allows gene expression but only when the level of glucose drops below a certain concentration.

Negative control

The DNA sequence in front of the *lac* operon contains two regulatory regions, the promoter (P_{lac}) where RNA polymerase binds (see Topic C4) and an operator site (O_{lac}), the site where regulation occurs. There is also a gene, *lac I*, which encodes a regulatory protein, the *lac* repressor. In the absence of lactose, the repressor protein binds to the operator site, preventing transcription (*Fig. 2a*). In the presence of an inducer molecule (a signal that lactose is present), the

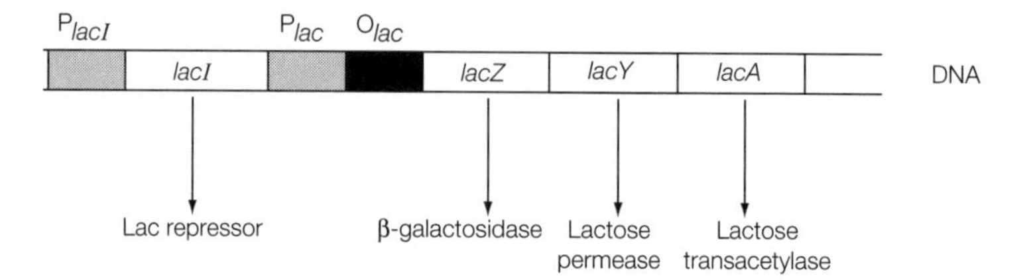

Fig. 1. Organization of the lac *operon.*

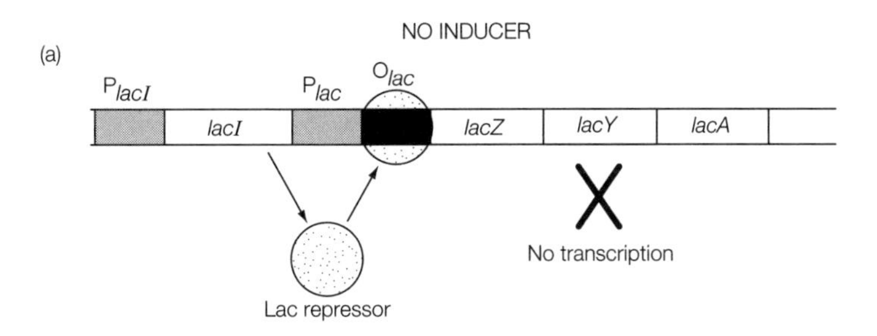

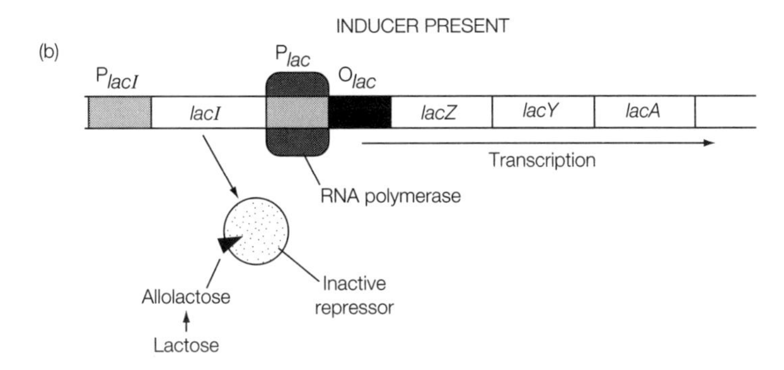

Fig. 2. Negative control of the lac *operon. (a) In the absence of lactose the Lac repressor binds to the operator and prevents RNA polymerase binding. (b) The presence of inducer (allolactose) inactivates the Lac repressor allowing RNA polymerase to bind and transcribe the* lac *genes.*

repressor becomes inactivated so that it no longer can bind to the operator and transcription can now proceed (*Fig. 2b*). The natural inducer of the *lac* operon is allolactose which is produced by β-galactosidase from lactose. This indicates that a small amount of transcription of the *lac* genes does occur even in repressing conditions, which allows the transport of lactose into the cell and its conversion into an inducing molecule.

Positive control

Catabolite repression controls a wide range of operons involved in using carbon sources other than glucose. Transcription of these operons cannot occur unless an activator protein, the **catabolite activator protein** (**CAP**; sometimes called the cAMP receptor protein, CRP) is bound to the operator. However, the CAP protein cannot bind to DNA unless it is associated with an **effector** molecule, cyclic AMP (**cAMP**). Glucose reduces the level of cAMP in the cell, so cAMP concentrations only rise when glucose levels drop; hence, cAMP is a signal that the glucose concentration is low. The presence of CAP–cAMP bound to a specific site in the *lac* promoter stimulates RNA polymerase binding and consequently transcription (*Fig. 3*).

Thus, there is maximum transcription from the *lac* operon promoter when the repressor is switched off by the presence of an inducer, allolactose, and if an activator protein, CAP–cAMP, is bound to the DNA.

The tryptophan operon

The *trp* operon encodes enzymes involved in tryptophan biosynthesis so these proteins are not required when tryptophan is present in the medium. This operon is regulated at a number of different levels. First, it is another example

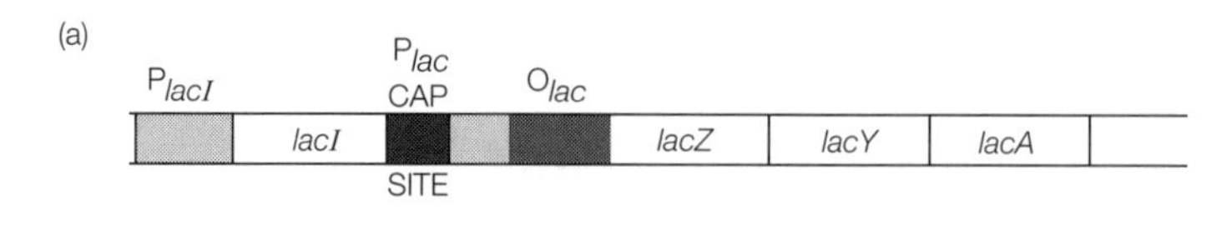

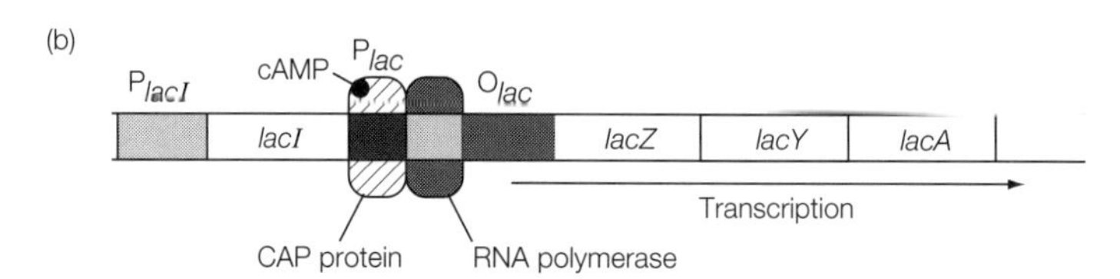

Fig. 3. Positive regulation of the lac *operon. (a) In the absence of CAP–cAMP, RNA polymerase cannot bind to the promoter. (b) When CAP–cAMP is present, RNA polymerase binds to the promoter.*

of negative control but in this case the repressor, sometimes called the **aporepressor,** is unable to prevent transcription unless a **corepressor** is present. The corepressor is tryptophan, a signal that there is no need to produce the amino acid as it is already present. In the absence of tryptophan the repressor is inactive and the genes are expressed. Tryptophan gene expression is also controlled by a mechanism called **attenuation** in which transcription actually starts but is stopped prematurely in the presence of tryptophan. This mechanism of gene control is common to a number of biosynthetic operons and is based on the fact that transcription and translation are coupled in bacteria (Topic C4) and that RNA can fold into secondary structures by base-pairing within the molecule if not prevented by the presence of ribosomes. The mRNA sequence before the translation start site for the first structural gene *trpE* contains a number of interesting features (*Fig. 4a*):

- A short sequence of translatable mRNA, *trpL,* which codes for a short leader peptide that contains two tryptophan residues.
- Four regions of mRNA (1–4) that are capable of forming two different sets of structures by base-pairing; region 1 with 2 and region 3 with 4, or region 2 with 3. The base-pairing between 2 and 3 is more stable.
- If regions 3 and 4 base-pair, they form a hairpin loop followed by a run of Us which can act as a transcription termination signal (see Topic C4).

In the presence of tryptophan, transcription can start, and the leader peptide gene, *trpL,* is transcribed and translated. The ribosome therefore covers part of the region 2 (*Fig. 4b*) preventing it from base-pairing with region 3. Region 3 is therefore free to base-pair with region 4, forming a transcription termination signal. RNA polymerase stops synthesis and transcription is attenuated. In the absence of tryptophan, the ribosome is stalled at the tryptophan codons as there are no tRNAs loaded with the amino acid and translation cannot continue (see Topic C7). In this case the full region 2 is available to base-pair with region 3 (*Fig. 4c*). Region 3 is therefore not available to base-pair with region 4 and the termination signal is not formed. RNA synthesis therefore continues to transcribe the full set of genes.

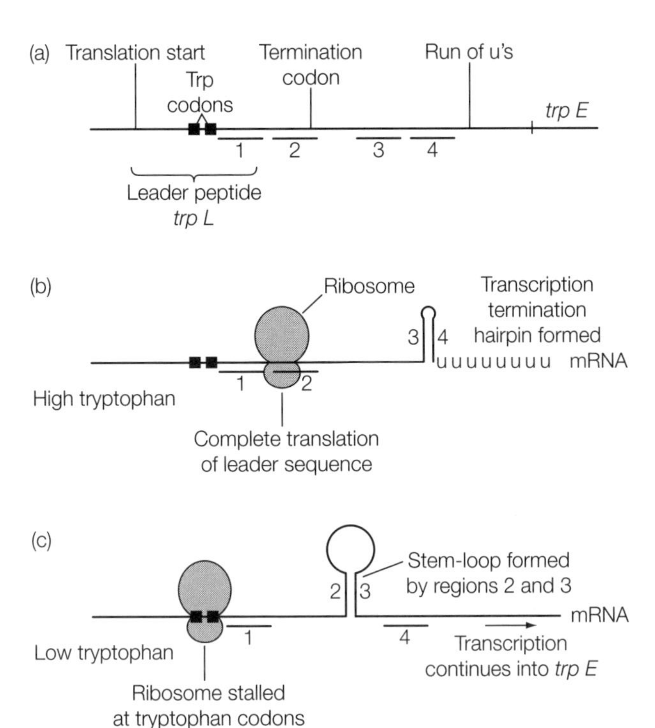

Fig. 4. Attenuation in the trp *operon. (a) Regions 1–4 in the RNA upstream from the start site for* trpE *and the position of two trp codons in the* trpL *leader peptide. (b) Possible RNA structures formed in high tryptophan leading to termination of transcription. (c) Stalling of ribosome on tryptophan codons in low tryptophan prevents the formation of the termination hairpin and transcription can continue.*

The combined effects of the two regulatory systems for tryptophan is to reduce transcription by about 700-fold. Repression accounts for up to a 70-fold decrease in initiation of RNA synthesis, and attenuation reduces the level of full transcripts by another 10-fold.

Two-factor control systems

Many of the regulatory systems in which cells sense and respond to environmental signals consist of two components: a **sensor** protein in the membrane, which detects the situation, and an effector protein (**response regulator protein**) in the cytoplasm. The sensor protein modulates the activity of the effector protein by phosphorylation. In many cases the effector protein is a DNA-binding protein that regulates transcription, and the presence or absence of the phosphate group dictates whether it will have a positive or negative effect on RNA synthesis. One example of this type of control system is the osmotic regulation of the porins, OmpF and OmpC (see Topic D3), in the outer membrane of *E. coli*. OmpF is preferentially expressed in low osmolarity medium and OmpC at high osmolarities. These genes are regulated by EnvZ, a membrane sensor protein, and OmpR, a response regulator protein. At high osmolarity, EnvZ phosphorylates OmpR which causes it to act as an activator of transcription of OmpC and a repressor of OmpF. At low osmolarity, OmpR is not phosphorylated, and the opposite occurs – OmpF is synthesized and OmpC is repressed.

Other regulatory mechanisms in prokaryotes

Other mechanisms of regulation in prokaryotes include anti-termination of transcription used to control λ-lysogeny (Topic E7); antisense RNA, also involved in the osmoregulation of porins; control at the level of translation (regulation of ribosome proteins); and degradation of mRNA.

Finally, it has become evident that Bacteria are capable of cell–cell signaling called **quorum sensing**. Bacteria secrete small molecules (autoinducers) into their environment which act as cell density indicators: low concentrations indicate low cell density; and high concentrations, high cell density. These molecules interact with receptors within cells which in turn induce or repress gene expression. Luminescence in *Vibrio fischeri*, cell division in *E. coli* and the production of virulence factors in a number of Bacterial pathogens have been shown to be regulated in this way.

Regulation in eukaryotes

Gene regulation in eukaryote cells, but especially in multicelled organisms, differs from that of eukaryotes for a number of reasons.

- Transcription and translation happen in separate cellular compartments.
- RNA polymerases do not have strong affinities for promoters and need additional factors for effective initiation.
- mRNA is more stable in eukaryotes having half-lives up to a number of hours.
- Genes need to be regulated during development and differentiation, as well as in response to the nutritional and environmental state of the cell.

Control of gene expression can be at the level of transcription, mRNA processing and translation. Most transcriptional regulation is as a result of additional **transcriptional factors** (called ***trans*-acting factors**) binding to control sequences adjacent to the promoter regions (called ***cis*-acting elements**). The nature of the elements in the promoter region will dictate when that gene will be expressed and what factors can effect its expression. A second group of regulatory elements are **enhancers,** which in association with *trans*-acting factors, greatly increase the rate of transcription from a promoter. These sequences may be located a long distance (10–50 kb) from the gene.

C6 STRUCTURE OF PROTEINS

Key Notes

Proteins in the cell

Proteins in the cell consist of polypeptides of L-amino acids joined by peptide bonds in a sequence dictated by genetic information in the cell. There are 20 amino acids found in proteins, which differ in the nature of their side chains. The sequence of the amino acids in the protein dictates its structure, the nature of any changes made to the protein, its final destination in or outside the cell and its function.

Structure of proteins

There are four levels of structure in proteins: primary structure which is the sequence of amino acids in the polypeptide; secondary structure which is the result of hydrogen bonding between side chains of amino acids to form α-helices and β-sheets; tertiary structure which results from the spontaneous folding of the protein as a result of interactions between the amino acid side-chains, often controlled by chaperone proteins; and quaternary structure which occurs when more than one polypeptide makes up the functional protein.

Related topic

Translation (C7)

Proteins in the cell

Most of the enzymes and many of the structural components of the cell are proteins. These consist of one or more polypeptide chains arranged in a three-dimensional structure. The polypeptide chains are made up of sequences of L-amino acids, joined together by peptide bonds between the amino group of one amino acid and the carboxyl group of another (*Fig. 1*), in an order dictated by the genetic information in the cell. In most organisms, proteins are synthesized from the same 20 amino acids specified by the genetic code (*Fig. 2*; Topic C7) but note that the side-chains of these may be further modified post-translation. In addition, selenocysteine has been found to be inserted during protein synthesis in a number of prokaryotes and eukaryotes.

It is the nature and the order of the amino acids that gives each protein its unique structure and function and there is the capacity to create an enormously varied set of proteins which can carry out the multitude of functions in the cell. The structure of a protein has to be such that it can exist in the correct environment and also carry out its function. For example, proteins may be found either in aqueous environments such as the cytoplasm, or in hydrophobic regions such as the interiors of membranes. Proteins that have to exist in membrane interiors must have hydrophobic amino acids in positions where the protein is in contact with the lipid bilayer (Topic D3).

All proteins are synthesized initially as a sequence of amino acids, but subsequent to this the polypeptides are folded and may undergo many different types of modification to give the final proteins. However, it is important to note that all the information about the final fate of the protein is carried in the amino acid sequence. This includes:

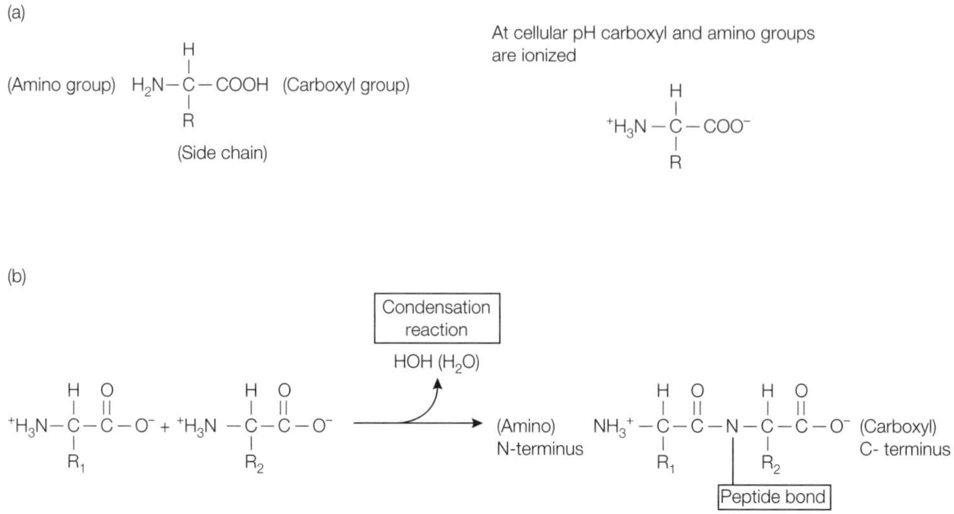

(a)

(Amino group) $H_2N-\overset{H}{\underset{R}{C}}-COOH$ (Carboxyl group)

(Side chain)

At cellular pH carboxyl and amino groups are ionized

$^{+}H_3N-\overset{H}{\underset{R}{C}}-COO^{-}$

(b)

Condensation reaction

HOH (H_2O)

$^{+}H_3N-\overset{H}{\underset{R_1}{C}}-\overset{O}{\overset{\|}{C}}-O^{-} + {}^{+}H_3N-\overset{H}{\underset{R_2}{C}}-\overset{O}{\overset{\|}{C}}-O^{-} \longrightarrow$ (Amino) N-terminus $NH_3^{+}-\overset{H}{\underset{R_1}{C}}-\overset{O}{\overset{\|}{C}}-N-\overset{H}{\underset{R_2}{C}}-\overset{O}{\overset{\|}{C}}-O^{-}$ (Carboxyl) C- terminus

Peptide bond

Fig. 1. (a) The basic structure of L*-amino acids in the cell; (b) the formation of a peptide bond.*

- its final structural conformation;
- whether additional groups are to be added – these include acetylation, hydroxylation, phophorylation, methylation, glycosylation or the addition of nucleotides;
- whether it is to be processed (cut) in any way;
- whether it is associated with cofactors such as NAD^+ or other molecules;
- its final destination, for example, a membrane, secretion out of the cell or into an organelle (eukaryotes);
- its functional activity, for example, as an enzyme, transport protein, structural protein, signalling molecule, regulatory protein or nutritional protein.

Structure of proteins

There are four levels of protein structure; primary, secondary, tertiary and quaternary as shown in *Fig. 3*.

Primary

This is the linear sequence of amino acids joined together by the peptide bonds to form a polypeptide. Each polypeptide has an amino end (N-terminus) and a carboxyl end (C-terminus). The final structure of the protein is determined by the side-groups of these amino acids, for example whether they are polar or non-polar, charged or non-charged, dictates how the side-chains will interact with each other to fold the protein into its subsequent shape.

Secondary

Hydrogen bonds between amino acids in the polypeptide leads to the formation of regular structures in the majority of protein molecules. The two most common ones are **α-helices** (rod-shaped structures) and **β-sheets** (flat, ribbon-like structures). Fibrous proteins tend to have large amounts of these highly structured regions whereas globular proteins have smaller regions interspersed with random coils of amino acids.

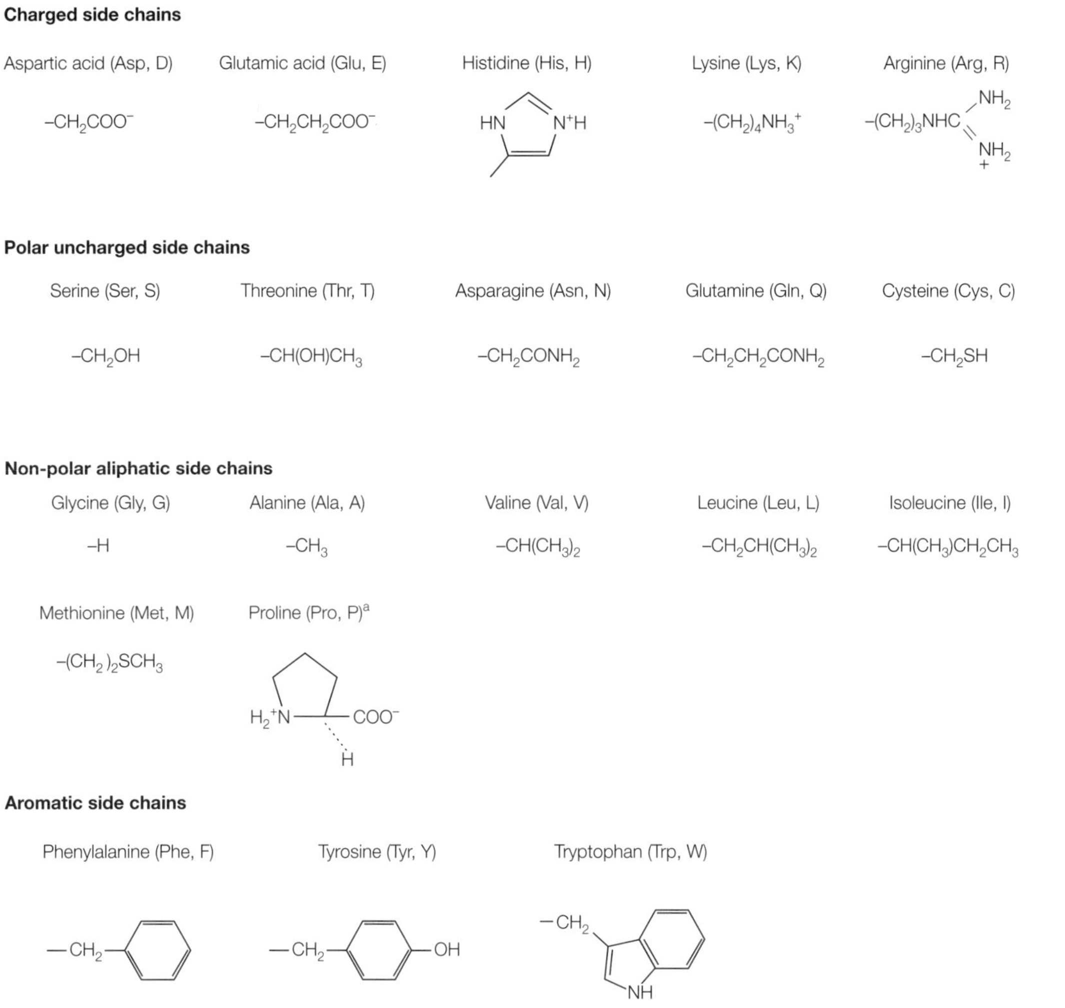

Charged side chains

Aspartic acid (Asp, D) $-CH_2COO^-$

Glutamic acid (Glu, E) $-CH_2CH_2COO^-$

Histidine (His, H) HN N⁺H

Lysine (Lys, K) $-(CH_2)_4NH_3^+$

Arginine (Arg, R) $-(CH_2)_3NHC(NH_2)NH_2^+$

Polar uncharged side chains

Serine (Ser, S) $-CH_2OH$

Threonine (Thr, T) $-CH(OH)CH_3$

Asparagine (Asn, N) $-CH_2CONH_2$

Glutamine (Gln, Q) $-CH_2CH_2CONH_2$

Cysteine (Cys, C) $-CH_2SH$

Non-polar aliphatic side chains

Glycine (Gly, G) –H

Alanine (Ala, A) $-CH_3$

Valine (Val, V) $-CH(CH_3)_2$

Leucine (Leu, L) $-CH_2CH(CH_3)_2$

Isoleucine (Ile, I) $-CH(CH_3)CH_2CH_3$

Methionine (Met, M) $-(CH_2)_2SCH_3$

Proline (Pro, P)[a] H_2^+N — COO^- H

Aromatic side chains

Phenylalanine (Phe, F) $-CH_2-$

Tyrosine (Tyr, Y) $-CH_2-$ OH

Tryptophan (Trp, W) $-CH_2$ NH

Fig. 2. Side chains (R) of the 20 common amino acids found in the cell. The standard three-letter abbreviations and one-letter code are shown in parentheses. [a]The full structure of proline is shown as it is a secondary amino acid. Reproduced from Turner et al., Instant Notes in Molecular Biology, *1997, published by BIOS Scientific Publishers.*

Tertiary

Proteins spontaneously fold into tertiary structures dictated by the amino acid side chains which form a number of different types of weak bond that stabilize the structure:

- hydrophobic interactions between non-polar groups;
- ionic bonds between opposite charges;
- hydrogen bonds;
- Van der Waals forces due to interactions of electrons on clusters of amino acids.

There are also covalent bonds associated with this level of structure. The most common are **disulfide** bonds between cysteine residues, which can hold together two separated parts of the same chain or two different chains (quater-

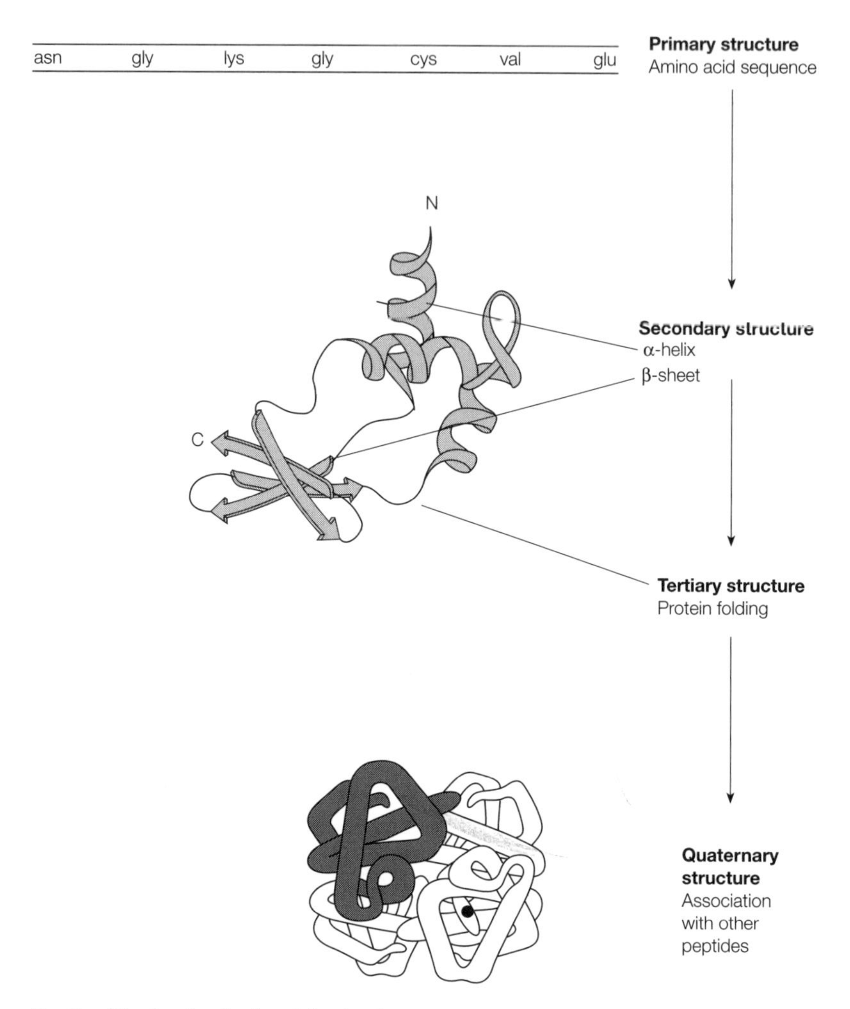

Fig. 3. The four levels of protein structure.

nary structure). The protein may also have other atoms or molecules associated with it that contributes to its final function. The correct folding of proteins is often associated with **chaperone** proteins which prevent them folding into an incorrect shape before their synthesis is completed.

Quaternary

Finally, if a protein is made up of more than one polypeptide chain, it is said to have quaternary structure. These chains may be different polypeptides such as the case of RNA polymerase (Topic C4) or multiple copies of the same one. The bonds that contribute to the quaternary structure are the same as those involved in the tertiary structure.

C7 TRANSLATION

Key Notes

The genetic code

A sequence of three bases, called a codon, in the mRNA codes for one amino acid. tRNA molecules act as the adapter molecules between the mRNA sequence and the amino acid sequence. They carry an anticodon which base-pairs with the codon sequence and a site at which an amino acid is attached. The process of matching the anticodon and codon and the synthesis of the amino acid chain occurs on the ribosome and is called translation.

Ribosomes

The role of the ribosomes is to hold the complex of mRNA and the amino-acid loaded tRNAs together, form the peptide bond between the amino acids and ensure accuracy of protein synthesis. Ribosomes consist of a large and a small subunit made up of proteins and RNA. The tRNA molecules are held at the aminoacyl (A) site and the peptidyl (P) site within the ribosome.

Loading of tRNA molecules

Amino acids are attached to the correct tRNA molecule, called the cognate tRNA, by aminoacyl-tRNA synthetases to form aminoacyl tRNAs.

Protein synthesis

There are three stages of protein synthesis: initiation; elongation and termination. Initiation involves the identification of the first codon in the protein sequence, normally AUG, which codes for methionine. Various protein factors are also required and energy is provided by guanosine 5′-triphosphate (GTP) hydrolysis.

Prokaryotic initiation of translation

The small, 30S, ribosomal subunit binds to the mRNA so that the start codon (AUG) is located in the P site. The Shine–Dalgarno sequence on the mRNA is responsible for the correct positioning. The initiator tRNA carrying formyl methionine base-pairs with the AUG codon; the large, 50S, subunit then binds to form the complete ribosome, and the next aminoacyl-tRNA enters the A site.

Eukaryotic initiation of translation

The small, 40S, ribosomal subunit binds to the initiator tRNA which carries methionine. This complex then binds to 5′ cap of the mRNA and moves to the first AUG in the sequence. The large, 60S, unit then binds to form the complete ribosome.

Elongation

A peptidyl transferase enzyme forms a peptide bond between the two adjacent amino acids, transferring the new amino acid on to the growing peptide chain. The ribosome then translocates (moves) down the mRNA in a $5' \rightarrow 3'$ direction so that the next codon is at the A site and the growing peptide chain is attached to the tRNA at the P site.

Termination

The codons UGA, UAA and UAG act as termination signals.

Wobble base-pairing Some tRNA molecules can recognize more than one codon in the ribosome due to base-pairing which does not follow the normal rules (wobble base-pairing).

Related topic Structure of proteins (C6)

The genetic code

A sequence of three bases in the mRNA, called a **codon,** codes for one amino acid. Each amino acid is coded for by between one and six codons and there are also codons that act as signals for the start and end of protein synthesis. This genetic code, as it is called, is shown in *Table 1* and is universal for all biological systems. The few exceptions include: a few small differences in mitochondrial DNA and lower eukaryotic genomes, which often effect stop codons; and the context-dependent use of UGA (a stop codon) for the insertion of selenocysteine into selected proteins in a small number of organisms.

Table 1. The 'universal' genetic code; see Figure 2, Topic C6 for abbreviations

First position	Second position				Third position
(5′ end)	U	C	A	G	(3′ end)
U	Phe	Ser	Tyr	Cys	U
	Phe	Ser	Tyr	Cys	C
	Leu	Ser	**Stop**	**Stop**	A
	Leu	Ser	**Stop**	Trp	G
C	Leu	Pro	His	Arg	U
	Leu	Pro	His	Arg	C
	Leu	Pro	Gln	Arg	A
	Leu	Pro	Gln	Arg	G
A	Ile	Thr	Asn	Ser	U
	Ile	Thr	Asn	Ser	C
	Ile	Thr	Lys	Arg	A
	[Met]	Thr	Lys	Arg	G
G	Val	Ala	Asp	Gly	U
	Val	Ala	Asp	Gly	C
	Val	Ala	Glu	Gly	A
	[Val]	Ala	Glu	Gly	G

Note: the boxed codons are used for initiation.

Conversion of the base sequence in the mRNA into an amino acid sequence involves RNA adapter molecules called transfer RNAs (tRNAs). These have a sequence of three bases, called the **anticodon**, which can base-pair with the codon sequence in the mRNA and, attached to the other end of the tRNA molecule, is an amino acid corresponding to that codon. The matching up of tRNA molecules to the mRNA and the subsequent joining together of the amino acids into a polypeptide chain is carried out on the **ribosomes**. The process of protein synthesis in the cell is called **translation**.

Ribosomes

The role of the ribosome is to hold the complex of mRNA and tRNAs together, form the peptide bonds between the amino acids and ensure the accuracy of protein synthesis. Ribosome particles consist of two subunits, one large and one small, made up of stable rRNA molecules and proteins. The size of ribosomes and the RNA molecules within them have traditionally been measured by their sedimentation coefficient in Svedberg units. This is the rate at which the particles sediment in a gradient, under centrifugal force, and takes into account the size, the shape and density of the particle. Although ribosomes from prokaryotes and eukaryotes look similar in structure there are differences in the size of the ribosomes (70S and 80S, respectively), their subunits and the composition of the proteins and rRNA within them (*Table 2*). This makes protein synthesis an ideal target for antibiotics as drugs that affect bacterial ribosomes will not affect eukaryotic cells (Topic F7). Within the ribosome there are two binding sites for tRNAs, the **amino-acyl (A) site** and the **peptidyl (P) site** as well as binding sites for mRNA (*Fig. 1*).

Table 2. The size of prokaryote and eukaryote ribosomes and their components

Whole ribosome	Prokaryotes (70S)		Eukaryotes (80S)	
Ribosomal subunits	30S (small)	50S (large)	40S (small)	60S (large)
Proteins	21	34	~ 33	~ 49
rRNA	16S	23S 5S	18S	5S 5.8S 28S

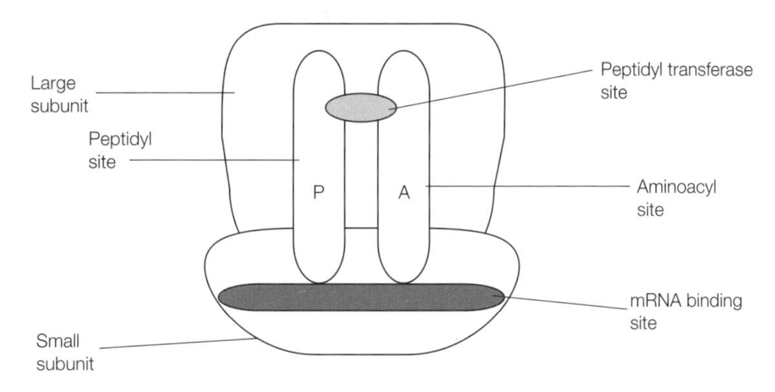

Fig. 1. The structure of a ribosome showing important functional sites.

Loading of tRNA molecules

A tRNA molecule is a single strand of RNA (73–93 nucleotides long) folded into a cloverleaf secondary structure which carries an anticodon and an amino acid attachment site (*Fig.* 2). Transfer RNA molecules contain some unique nucleotides, which are modified from the original A, U, G and C nucleotides post-transcriptionally, including pseudouridine and inosine. There is at least one tRNA molecule for each amino acid in the cell but not necessarily one for each codon (see wobble base-pairing).

The loading of the tRNA molecules with the correct amino acid is a key part of protein synthesis and it is the main point at which accuracy of translation is ensured. If the wrong amino acid is added to the tRNA molecule, it will subsequently be inserted incorrectly into the growing polypeptide chain, as the ribosome would not detect that the wrong amino acid was present. The

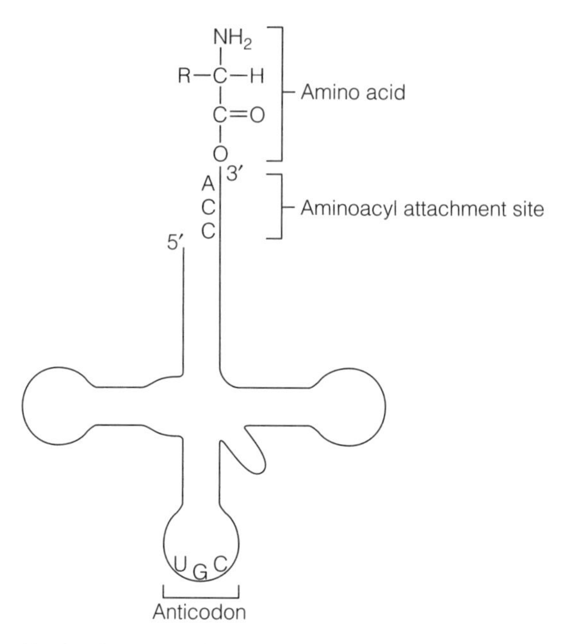

Fig. 2. Structure of an aminoacyl-tRNA.

enzymes responsible for the addition of the amino acid to the tRNA are **aminoacyl-tRNA synthetases**. There is at least one enzyme for each amino acid. The enzyme binds the correct amino acid and the appropriate tRNA molecule. It then catalyzes the addition of the amino acid to the 3′ end of the tRNA molecule in two stages using ATP:

$$\text{Amino acid} + \text{ATP} \rightarrow \text{aminoacyl-AMP} + 2\text{P}_i$$

$$\text{Aminoacyl-AMP} + \text{tRNA} \rightarrow \text{aminoacyl-tRNA} + \text{AMP}$$

This process is sometimes called charging of the tRNA. The tRNA for an amino acid, called the **cognate** tRNA, is designated by a superscript (e.g. tRNA^{Ala}); once loaded with the amino acid it is written as $\text{Ala-tRNA}^{\text{Ala}}$.

Protein synthesis

Protein synthesis can be divided into three stages: initiation, elongation and termination. At initiation the start codon for the protein is recognized: normally it is AUG which codes for methionine, but very occasionally it may be GUG (valine). A complex is formed between the mRNA, the ribosome and the initiating tRNA. During elongation, amino acids are added sequentially to the growing peptide chain in accordance with the codon sequence in the mRNA. At termination, the end of the polypeptide chain is recognized and the complex of mRNA, polypeptide, tRNA and ribosomes breaks apart. Each stage requires a number of different factors to ensure the correct order of events, which are different between prokaryotic and eukaryotic translation. The energy for protein synthesis is provided by the hydrolysis of GTP.

Initiation of translation in bacteria

The first stage of protein synthesis is the binding of the small (30S) ribosomal subunit to the mRNA so that the first AUG codon is positioned in the P site. The correct positioning is achieved by base-pairing between a short purine-rich sequence (consensus sequence: 5′-GGAGG-3′), called the **Shine–Dalgarno** sequence, which is located in the mRNA, 8–13 nucleotides before the start site of translation, and a complementary sequence on the 16S RNA (consensus sequence: 3′-CCUCC-5′) in the small subunit of the ribosome (*Fig. 3a*). This mechanism, which is unique to prokaryotes, allows translation to start in the middle of a mRNA sequence as bacterial mRNAs frequently contain a number of genes which are translated independently (see Topic C4). Once the AUG is located at the P site, the first tRNA, the initiator tRNA, can enter this site and its anticodon can base-pair with the AUG (*Fig. 3b*). The initiator tRNA is distinct from the tRNA which inserts methionine in response to AUG in internal sites in the polypeptide in that:

- it carries *N*-formylmethionine (fMet) instead of methionine (Met);
- it can enter the P site of the ribosome when the large subunit is absent.

Once the fMet–$tRNA^{fMet}$ has bound to the AUG, the large (50S) ribosomal subunit joins to form the complete 70S ribosome (*Fig. 3c*) and the next tRNA can enter the A site to base-pair with the next codon (*Fig. 3d*). Three initiation factors (IFs) and GTP hydrolysis are required for the formation of the initiation complex.

Eukaryotic initiation of translation

In eukaryotes the first AUG codon on the mRNA is the initiation codon, consequently the mechanism to identify the start site for protein synthesis is slightly different. The initiating tRNA carries Met rather than fMet and forms a complex with the small, 40S, ribosomal subunit before it binds to the ribosome. This complex binds to the 5′-methyl cap on the mRNA molecule then moves up the RNA until an AUG codon is located at the P site. The large, 60S, ribosomal subunit then binds and the next tRNA enters the completed A site.

Elongation

The ribosome detects that there is accurate base-pairing between the anti-codon and codon at the A site and then a **peptidyl transferase** enzyme, part of the ribosome, forms a peptide bond between the two adjacent amino acids. In doing this the amino acid on the tRNA at the P site is transferred on to the amino acid attached to the tRNA at the A site. The ribosome then moves in relation to the mRNA by a process called **translocation** so that the tRNA with the dipeptide attached is now at the P site. The empty tRNA is released and the A site is then free to accept a new tRNA molecule to base-pair with the new codon at the A site (*Fig. 3e*). Elongation factors, EF-tu, EF-ts, and EF-G, are required in Bacteria (eEF-1 and eEF-2 in eukaryotes) to ensure that the process occurs in the correct order, and two molecules of GTP are hydrolyzed. The process of checking, synthesis of the peptide bond and translocation is repeated down the length of the mRNA molecule in a $5' \rightarrow 3'$ direction. The polypeptide sequence grows from the amino end (N-terminus) to the carboxyl end (C-terminus).

Once the ribosome-binding site and the AUG codon become available on the mRNA another ribosome can bind and start translation. Consequently, under the electron microscope, mRNA appears as a string of beads owing to the presence of multiple ribosomes. This structure is called a **polysome** or polyribosome.

Termination

There are three codons, UGA, UAA and UAG, which do not have a corresponding tRNA molecule. These act as termination signals indicating the end of an amino acid sequence. When the ribosome reaches such a codon it stalls as

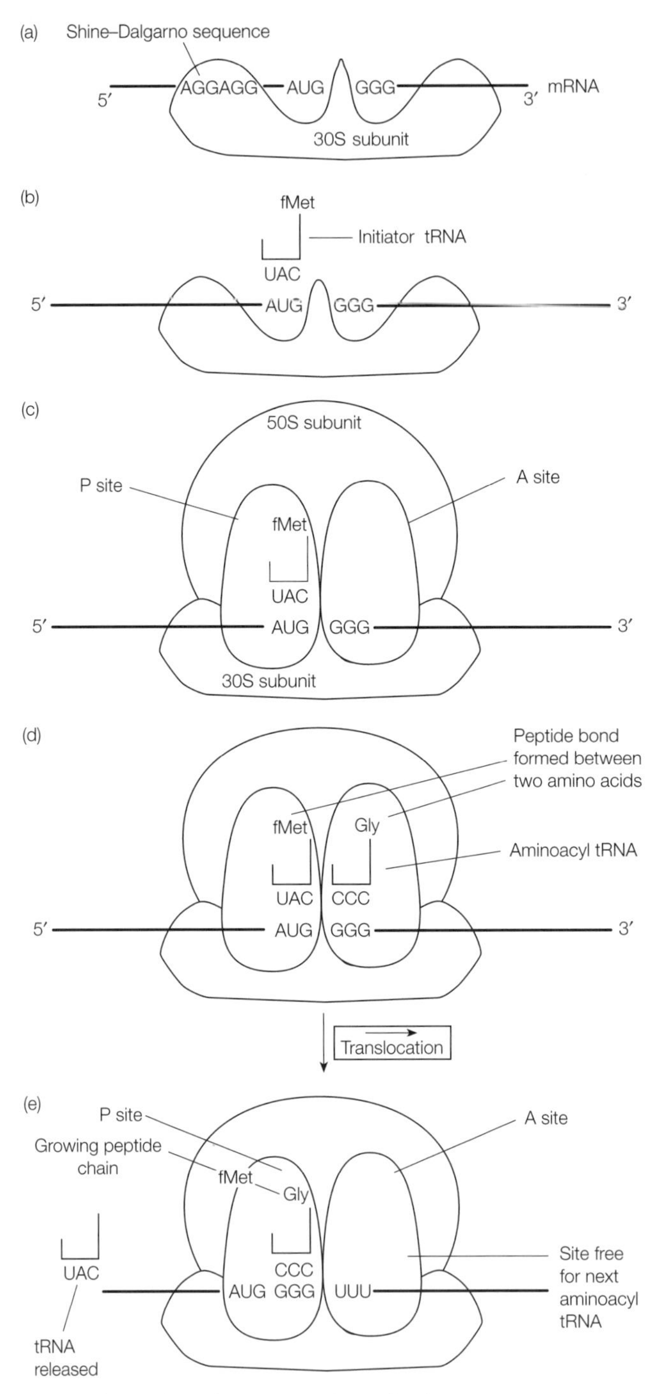

Fig. 3. Stages in prokaryote protein synthesis.

protein synthesis cannot continue. Release factors enter the empty A site and trigger the release of the synthesized polypeptide chain from the ribosome and the dissociation of the ribosome subunits.

Wobble base-pairing

Codons for an amino acid are often in families where only the first two bases in the sequence are important; for example, proline is coded for by CCA, CCU, CCC and CCG. Many tRNA molecules are therefore constructed to allow a mismatch, called a **wobble**, at the third position therefore allowing them to recognize more than one codon:

- Uniquely to the ribosome, U and C can base-pair with G but only at the third position of the codon.
- Some anticodons contain the unusual base inosine (I) in their anticodon which can base-pair with U, C or A at the third position in the codon.

D1 PROKARYOTE TAXONOMY

Key Notes

Classification of prokaryotes

Classification of prokaryotes into groups based on common properties allows prokaryotes to be identified and named. The most important taxonomic rank is the species, and species are grouped further into genera. The name of a bacterium has two parts – the genus name followed by the species epithet, for example *Escherichia coli.*

Characters used in prokaryote identification

Colony morphology, cell morphology, Gram stain reaction, growth characteristics, and biochemical, immunological and DNA tests are all used to classify and identify prokaryotes.

Current classification of prokaryotes

The prokaryotes can be divided into two groups, the Bacteria and the Archaea. The Bacteria can then be further divided into at least 15 phyla based on 16S rRNA sequence analysis.

Bacterial diversity

The 15 phyla of Bacteria exhibit enormous diversity in structure and metabolism.

Related topics

Prokaryote cell structure (D2) Virus taxonomy (J2)
Taxonomy (G1)

Classification of prokaryotes

The main aim of taxonomy is to set up a **classification**: an ordering of prokaryotes into groups based on common properties. This classification can then be used to identify individual bacterial species or strains (**identification**) and name them (**nomenclature**). Classification is also used to study the relationships of different groups of microorganisms based on shared phenotypic properties and probably a common evolutionary history. A wide range of morphological, biochemical and molecular characters are measured and the information derived from this is used to create a hierarchical classification system, into which individual strains of microorganisms can be placed.

The most important rank in the taxonomic hierarchy is called the **species**, which from the time of Linnæus (1707–1778), has been named in a similar way. Related species are grouped into a **genus** (pl. genera), which is usually based around some distinct phenotypic character, and the generic name is used as part of a binomial species name. So, for example, *Escherichia coli* is in the genus *Escherichia* named after Escherich, who first isolated the bacteria in 1885. The specific epithet *'coli'* is in this case an indication as to where the organism was isolated (from i.e. the colon). However, note that the binomial name is a label not a description; it does not have to describe the species in any way. Note too that the genus name has a capital initial letter but the specific epithet (which cannot be used alone) does not, and that the whole binomial name is Latinized and therefore should be italicized or underlined. Normally, once the name of the organism has been written in full, the genus name can be reduced to a letter, or letters, as long as there can be no confusion with other genera. Thus, *Escherichia coli* will subsequently be written as *E. coli*.

The higher taxonomic categories into which the organism can be ranked are, sequentially, family, order, class, division and kingdom (domain). Similarly, species may be divided into subspecies and a culture of a particular organism is usually referred to as a **strain**. Strain is also used to describe a sub-group of a species that shares some feature such as one type of O-antigen (Topic D3) or mutation (Topic E1). There are also informal names, not italicized, which are sometimes useful for describing groups of organisms such as the green-sulfur bacteria or the spirochetes.

The main impetus for prokaryotic classification has been the need to be able to identify disease-causing organisms. However, it should be remembered that many other species have not been well characterized and many have not been isolated or described at all. The authority on prokaryotic classification is *Bergey's Manual of Systematic Bacteriology*, which contains as near as possible complete listings of all prokaryotic species and their distinguishing characteristics.

Characters used in prokaryote identification

A list of the types of characters used to distinguish between prokaryotes is given below. This is not an exhaustive list and equally not all characters are used in either the classification or identification of every prokaryotic species. In many cases of well known microorganisms, a small set of tests may be all that is required to identify a microbe positively, especially if the site from which it has been isolated and the disease it caused are known. A comma-shaped, Gram-negative rod isolated from a patient excreting copious amounts of straw-colored, liquid stools may be enough to identify it putatively as *Vibrio cholera*, the causative agent for cholera. However, it may be necessary to do a wide range of tests to determine the name of an isolate if it is an undescribed microbe.

DNA tests

Comparisons of DNA content and sequence between strains are the most definitive methods for separating organisms into different groups and for measuring the evolutionary relatedness between them (**phylogenetic analysis**). Traditionally, DNA was analyzed by measuring the G-C content or the amount of DNA homology between strains. More recently, the comparison of the actual DNA sequences of conserved cellular molecules including 16S rRNA (see Topic C3), cytochrome c, ATPase and elongation factor has been used. This information has been analyzed by a technique called **cladistics** to trace true phylogenetic lineages that have not been obscured by patterns of convergent evolution.

Colony morphology

The shape, texture and color of colonies of microorganisms growing on solid agar plates can be distinctive. For example, *Staphylococcus aureus* is so named as colonies are of a yellow color; *'aureus'* comes from the Latin for golden.

Cell shape, structures and reaction to stains

Whether Bacteria are Gram-positive or Gram-negative (Topic D3) is one of the most important clinical diagnostic tests. Gram staining and microscopic examination of prokaryotes give information as to the shape, size and arrangement of the prokaryotic cells (Topic D3). Stains may also be used to show other morphological features (see Topic D3), which may be used for classification, such as the presence of:

- spores (malachite green stain);
- unusual cell walls (Ziehl-Neelsen acid-fast stain);
- capsules (Indian ink stain);
- intracellular lipids (Sudan black);
- flagella (flagellar stain);
- metachromatic granules (Albert–Leybourne stain).

Growth characteristics

The temperature, pH and O_2 requirements of an isolate are useful tests for the identification of prokaryotes.

Biochemical tests

There is a wide range of tests available that measure various aspects of prokaryote metabolism:

- what carbon and nitrogen sources the prokaryote can use;
- the end products of their metabolic processes, such as acetoin, tested for in the Vogues–Proskaeur test;
- what enzymes the prokaryote produces, such as decarboxylases, proteases and celluloses;
- the presence of other molecules such as toxins (Topic F6), long-chain fatty acids; or
- resistance to antibiotics.

Immunological tests

Antibodies to cellular components such as the O-side chains of lipopolysaccharide (LPS) (Topic D3) or capsules are frequently used to distinguish between strains of one species.

Current classification of prokaryotes

On the basis of 16S rRNA sequences, it is now clear that within the prokaryotes there are two major groups, the Bacteria and the Archaea. Within the Bacteria there are at least 15 groups, referred to sometimes as phyla, which can be represented on a phylogenetic tree (*Fig. 1*). This is probably a gross underestimate of the total number of groups, as polymerase chain reaction (PCR) methods are currently revealing the vast number of 'unculturable' microbes in the environment that can only be detected as DNA. Within the Archaea there are three groups (see D5).

Bacterial diversity

The diversity in structure and metabolism of the Bacteria and Archaea is enormous even within individual groups and reflects the range of environments from which these microbes are isolated. A very brief overview of each of the groups of bacteria is given below to illustrate this diversity. The groups within the Archaea are described in Topic D5. The different types of metabolism are discussed in Section B.

- **Aquificales** (example *Aquifex aeolicus*) is probably the most ancient group, containing hyperthermophilic bacteria that reduce oxygen using hydrogen, thiosulfate or sulfur as the electron donors.
- **Thermatogales** are extreme thermophiles, capable of growth up to 90°C, isolated from geothermal sediments. One quarter of the *Thermotoga maritima*

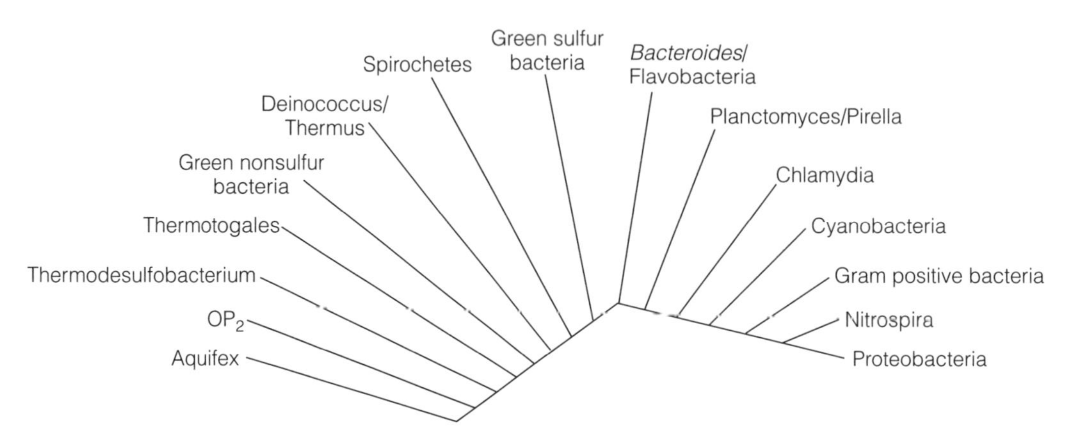

Fig. 1. A phylogenetic tree of Bacteria derived from 16S ribosomal RNA sequences. (Data from the Ribosomal Database project.)

genome is archaeal in origin suggesting that lateral gene transfer may have occurred between Bacteria and Archaea.

- The ***Thermodesulfobacterium*** group (example *Desulfatotherma hydrogeno-philahas)* are sulfate-reducing thermophilic (growth temperature optimum, 70°C) Bacteria with some features in common with the Archaea which may reflect their common ancestry.
- **Green non-sulfur bacteria** (for example *Chloroflexus*) are thermophilic photosynthetic, filamentous bacteria that are anoxygenic, *i.e* they do not release O_2 as a result of photosynthesis (Topic B).
- **OP_2** are a group of microbes whose presence have only been detected using PCR and have never been cultured.
- ***Deinococcus*** cells have extremely complex walls that make them highly resistant to radiation. This genus is in the same group as the thermophilic genus ***Thermus***, which includes *Thermus aquaticus*, the microbe famous for the production of *Taq* DNA polymerase which is used for PCR.
- **Green sulfur Bacteria** are anoxygenic, phototrophic bacteria.
- **Cyanobacteria** are a morphologically heterogeneous group of oxygenic phototrophs. They carry gas vesicles and may contain heterocysts, empty-looking cells with thickened cell walls that act as sites of nitrogen fixation. Green plant chloroplasts probably evolved from this group.
- The **spirochetes** are Gram-negative motile, tightly coiled Bacteria that have one or more flagella folded back from the pole of the cell but held within the periplasm of the cell (endoflagella). A number of serious human pathogens are within this group; for example, *Treponema pallidum* and *Borrelia burgdorferi* the causative agents of syphilis and Lyme disease respectively.
- ***Bacteroides/Flavobacterium/Cytophaga*** are a group of Gram-negative Bacteria associated with animals, water and the soil. *Bacteroides* are obligate anaerobes.
- The ***Planctomyces/Pirella*** group lack peptidoglycan in their cells walls and, instead, have an S-layer type of wall made of protein (Topic D2). *Planctomyces* are stalked cells that can reproduce by 'budding'. A flagellated daughter cell is formed at the opposite end of the cell to the stalk which then buds off.

- **Chlamydia** are obligate, intracellular parasites of animal cells. *Chlamydia trachomatis,* a member of this group causes trachoma, a serious eye disease.
- **Nitrospira** are a group of nitrifying, chemolithotrophic bacteria capable of using reduced inorganic nitrogen as energy for growth. Most of the other nitrifying Bacteria are located in the major Proteobacteria group.
- **Gram-positive Bacteria** are traditionally split into a low and high GC group which reflects the GC content of their DNA. This group contains many bacterial pathogens and industrial microbes such as *Staphylococcus, Streptococcus, Lactobacillus* and the spore-forming *Bacillus* and *Clostridium.*
- The **Proteobacteria** is by far the largest and most diverse group of Bacteria including many of the more commonly known Gram-negative Bacteria. The group is split into five clusters of genera named α, β, γ, δ, and Θ and representatives of practically every type of metabolism can be found. Examples of just some of this group are given below.
  - Bacterial pathogens such as the enterobacteriaceae; the pseudomonads; *Neisseria; Vibrio;* and *Rickettsia,* obligate animal intracellular parasites.
  - Bacteria with interesting life-cycles such as *Bdellovibrio,* predators which replicate within the Bacterial periplasm; sheathed, gliding, stalked and fruiting bacteria, bioluminescent bacteria.
  - Chemolithotrophic bacteria such as the nitrifying bacteria; sulfur and iron-oxidizing bacteria; hydrogen-oxidizing Bacteria.
  - Nitrogen-fixing bacteria such as *Azotobacter.*
  - Photosynthetic bacteria including the purple sulfur and non-sulfur bacteria.
  - Bacteria that can use methane as a source of carbon and energy, methanotrophs and methylotrophs.
  - *Spirillium,* Bacteria that can align in a magnetic field due to the presence of iron-containing magnetosomes.
  - Industrially important species such as the acetic acid Bacteria that produce this acid as an end point of their metabolism.

D2 Prokaryote cell structure

Key Notes

Prokaryotic cells	Prokaryotes are cells whose DNA is not contained within a nuclear membrane. They lack many of the complex organelles and internal structures found in eukaryotic cells.
The cytoplasm and ribosomes	The main structures found in the aqueous cytoplasm of all prokaryote cells are the ribosomes, the sites for protein synthesis in the cell. The Archaeal ribosomes are the same size as those of Bacteria (70S), but differ in sensitivity to diphtheria toxin and some antibiotics.
Prokaryote DNA	DNA is found, generally, as a single circular chromosome (there are exceptions), in the cytoplasm and is sometimes referred to as the nucleoid. Small extra-chromosomal pieces of DNA called plasmids are often found in Bacteria.
Other internal bacterial features	Some prokaryotes contain structures associated with specialist functions. Inclusion bodies and lipid droplets are storage sites in the cell. Membrane systems are associated with the photosynthetic capabilities of some cells, and methane oxidation of others.
Endospores	Endospores are produced by a few genera of bacteria including *Bacillus* and *Clostridium*. They consist of DNA surrounded by a number of layers of protein and peptidoglycan that are highly resistant to drying and heat.
Bacterial cell wall and surface	The prokaryotic cytoplasm is surrounded by a plasma membrane. Surrounding this, Bacterial cells have, with a few exceptions, a rigid cell wall made of peptidoglycan, a substance not found in Archaeal or eukaryotic cells. Gram-negative Bacteria have an additional outer membrane. Other features which may be found associated with the cell surface, in some prokaryotes, are flagella, associated with cell movement, pili (or fimbriae), which have a role in adhesion, and extracellular polysaccharides (sometimes called glycocalyx) or proteins which help to protect the bacteria from their environment.
Archaeal cell wall and surface	The lipids in the plasma membrane of Archaea are branched and long and are ether-linked to glycerol. Cell-wall materials are very diverse and range from a peptidoglycan-like material, pseudomurein, to polysaccharides, protein and glycoproteins.

Related topics

Prokaryote taxonomy (D1)
Bacterial cell envelope and cell wall synthesis (D3)
The Archaea (D5)
Entry and colonization of human hosts (F5)
Eukaryotic cell structure (G2)

Prokaryotic cells

Bacteria and Archaea are small, generally <1 mm–50 mm width/diameter though cells up to 750 mm have been found (*Thiomargarita namibiensis*). They are single-celled microorganisms that belong to a group called prokaryotes so classified because their DNA is not enclosed within a nuclear membrane (Gr. *pro*, before; *karyote*, nucleus). Their internal cell structure is simple (*Fig. 1*), with most of their cellular complexity associated with the cell surface structures. Prokaryotes lack mitochondria and chloroplasts, the organelles associated with energy production in eukaryote cells; and internal membrane structures associated with protein synthesis and processing, such as the endoplasmic reticulum and Golgi apparatus (see Topics A1 and G2). However internal membranes are found in some groups of prokaryotes associated with processes such as photosynthesis or methane oxidation. In spite of this simplicity in structure there is a large amount of variation in the appearance of the cells when observed under the microscope. Two shapes of cell predominate, the coccus (spherical) and the bacillus, or rod, but a wide range of other shapes (morphologies) also can be found as shown in *Table 1*.

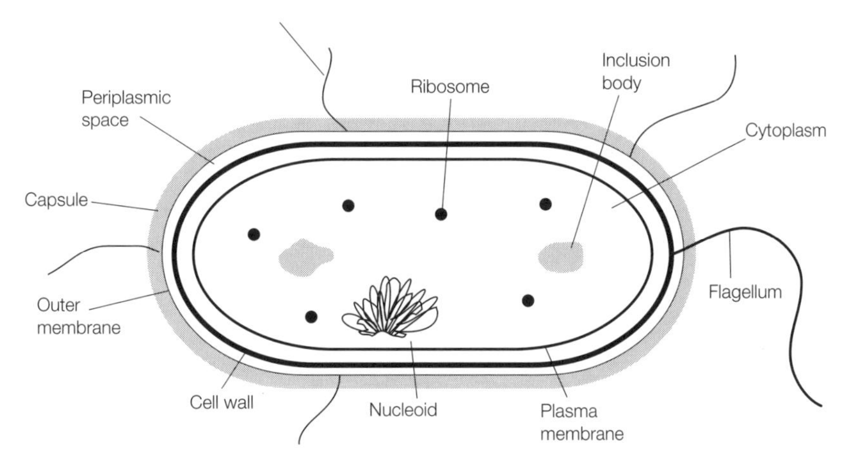

Fig. 1. Diagram of a prokaryote cell (Gram-negative).

Table 1. Shapes of prokaryotes and characteristic associations of cells seen using light microscopy

Shape	Organization of cells	Examples
Cocci – spherical cells	In chains	*Streptococcus*
	In grape-like clusters	*Staphylococcus*
	In pairs (diplococci)	*Neisseria*
	In cubes of cells (packets)	*Sarcina*
Rods/bacilli – rods may be very short or long	Singly	*Pseudomonas*
	In filaments of attached cells	*Bacillus* sp.
Tightly coiled flexible cells – (spirochaetes)		*Treponema*
Comma shaped – called a vibrio		*Vibrio*
Spirillum – long, curved, rigid rods		*Rhodospirillum*
Appendaged – cells have stalk or tube-like extensions		*Rhodomicrobium*
Pleomorphic – bacteria vary in shape		*Corynebacteria*
Branching mycelium-like filaments		*Streptomyces*

Prokaryotes are ubiquitous. They are a highly successful and diverse group of organisms that can obtain energy and carbon from a wide range of sources and therefore can colonize every niche on our planet from deep ocean trenches to volcanic craters. In the 1970s, using DNA sequencing information, it was found that the group we know as the bacteria could be split into two, the **Bacteria** (Gr. *bakterion*, a small rod) and the **Archaea** (Gr. *Archaios*, ancient) and it appears that these two groups evolved away from each other very early in the history of living things (see Topic D1). Members of the Bacteria include some of the more familiar bacteria such as *Escherichia coli* and *Staphylococcus aureus*, the prokaryotes that are best studied and understood. The Archaea are a very diverse group of organisms, which differ from the Bacteria in a number of features having, in particular, very different cell walls and membranes (see *Table 2* and Topic D5). This group includes archaea that are capable of existing in extreme environments, such as hot springs (examples: *Sulfolobus* and *Pyrococcus*) and high salinity (*Halobacterium*), and the methanogens such as *Methobacterium*, which produce methane as a result of metabolism.

Table 2. Some differences between Bacteria and Archaea

Feature	Bacteria	Archaea
Cell wall	Muramic acid present	Muramic acid absent
Lipids	Ester-linked	Ether-linked
Methanogenesis	Absent	Possible
RNA polymerases	One	Several
Initiator tRNA	fMet	Met
Ribosomes	Sensitive to streptomycin and chloramphenicol	Resistant to streptomycin and chloramphenicol
	Resistant to diphtheria toxin	Sensitive to diphtheria toxin

The cytoplasm and ribosomes

The cytoplasm of prokaryotes is aqueous, containing a cocktail of molecules, ribonucleic acid (RNA) and proteins necessary for the cell functions. The main structures found in the cytoplasm, common to all prokaryotes, are the **ribosomes**. The ribosomes consist of a small and a large subunit, made up of a complex of proteins and RNAs, and are the sites of protein synthesis in the cell. The ribosomes in prokaryotic cells, although similar in shape and function to those of eukaryotic cells, are different in the nature of the proteins and RNAs that make up their structure (Topic C7). This has proved to be very useful to the human population as antibiotics that act by inhibiting protein synthesis in prokaryotes are not effective against eukaryotic protein synthesis, thus allowing selective toxicity. This is discussed in more detail in the topics on translation (Topic C7) and antibiotic action (Topic F7). Archaeal ribosomes are the same size as those of the Bacteria but, in some features, they are similar to eukaryotic ribosomes in that they are resistant to the antibiotics streptomycin and chloramphenicol, and sensitive to the action of diphtheria toxin.

Prokaryote DNA

Prokaryote DNA is located within the cytoplasm. It often consists of a single chromosome which varies in size between different species of bacteria (the *E. coli* chromosome is 4.6×10^6 base pairs long). The DNA is circular, tightly supercoiled and associated with proteins (Topic C1). Although the chromosome is not contained within a nuclear membrane, it is often seen as a discrete area within the cell in electron micrographs which may be referred to as the **nucleoid**. Many bacteria also contain small molecules of extra-chromosomal DNA called **plasmids**. These generally carry genes which are not essential to the normal life of the cell but confer an advantage to the cell in certain situations such as antibiotic-resistant plasmids (Topic E5). Chromosomes in Archaea, like that of most Bacteria, are generally single, circular DNA molecules not contained within a nucleus, but the size of the DNA molecule is often smaller than that of *E. coli*.

Other internal prokaryote features

Some prokaryotes contain structures associated with specialist functions. Granular structures, called **inclusion bodies,** can often be seen under the light microscope. These granules are normally used for storage and may be either membrane bound, such as poly-β-hydroxybutyrate (PHB) granules, or found scattered in the cytoplasm, an example being polyphosphate granules (also called **metachromatic granules** as they change the color of basic dyes). **Lipid droplets** can also be seen in some prokaryotes. An interesting inclusion body, the **gas vacuole**, is found in *Cyanobacteria* and other photosynthetic bacteria that live in an aqueous environment. This protein-surrounded vacuole provides buoyancy, allowing the prokaryote to float near the surface of the water.

Although bacteria lack membrane bound organelles, invaginations of the plasma membrane called **mesosomes** are often seen in electron micrographs. There is some debate as to whether these mesosomes really exist and are not just an artefact of the fixation process for electron microscopy. Their function is thought to be in cell division, possibly in laying down new cell-wall material or in the replication of the chromosome and its subsequent distribution to daughter cells. Other, more specialist and complex, intracellular membrane systems are found in photosynthetic Bacteria, such as *Chloroflexus* (green non-sulfur bacteria) and *Rhodopseudomonas* (purple bacterium), associated with the trapping of light energy (Topic B3).

Endospores

A number of Bacteria, mainly studied in the *Bacillus* and *Clostridium* genera, produce a special reproductive structure called the **endospore**. This can be seen in the light microscope using either specialist spore stains such as malachite green or by phase-contrast microscopy. The spore has a number of layers surrounding the genetic material making it incredibly resistant to all kinds of environmental stress such as heat, UV irradiation, chemical disinfectants and drying (*Fig. 2*). As a number of important pathogenic bacteria are spore producers, sterilization measures must therefore be designed to remove these hardy structures some of which can withstand boiling for several hours (Section D).

Bacterial cell wall and surface

The cytoplasm of all Bacteria is enclosed within a **plasma (cytoplasmic) membrane** external to which, in most cases of Bacteria, is a rigid **cell wall** made up of sugars and amino acids called **peptidoglycan**. The role of the cell wall is to protect the cell from lysis resulting from osmotic pressure and it also gives shape to the cell. Some Bacteria, the mycoplasma, do not have cell walls and therefore are unable to survive outside an animal host which provides it with the right osmotic environment. Gram-negative Bacteria have an additional outer membrane containing lipopolysaccharide (LPS). External to this may be other layers of polysaccharide or protein making up a capsule or slime layer. Layers external to the cell wall may be referred to as the cell **envelope.** The structure of the Bacterial cell envelope is discussed in more detail in Topic D3.

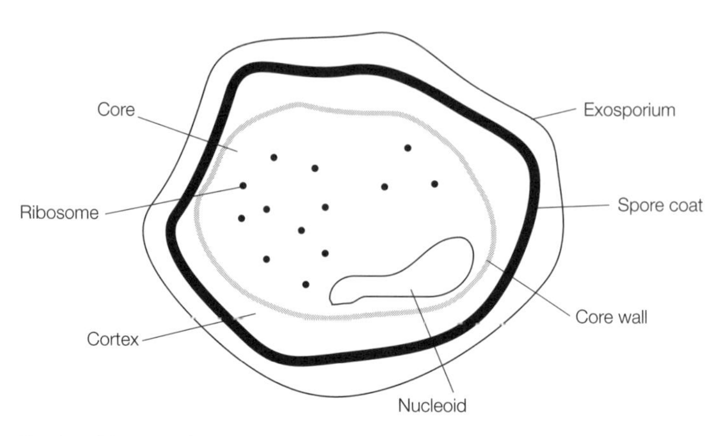

Fig. 2. Diagram of an endospore.

A number of features are found on the surface of the Bacteria which aid the survival of the cell in the environment. Thin hair-like proteinaceous appendages, several micrometers long, called **pili** (s. pilus) or **fimbriae** (s. fimbria) are often seen projecting from the surface of the cell in electron micrographs. These structures are normally associated with the adhesion of Bacteria to surfaces (Topic F5) although there is one particular type of pilus, the sex pilus, which is involved in the transfer of genetic material from one bacterium to another (Topic E6). Motile prokaryotes have one or more **flagella** (s. flagellum) which are slender, rigid structures up to 20 μm long. Rotation of the flagellum in relation to the cell allows the movement of the bacterium. Normally, prokaryotes move in response to external stimuli; an example of this is **chemotaxis** when the microbe move towards nutrients or away from harmful substances as a result of chemical gradients (see Topic D5).

Archaeal cell wall and surface

One of the distinctive features of Archaea is the nature of the lipids in the plasma membrane which, unlike the ester-linked lipids of Bacteria and eukaryotes are ether-linked to glycerol. They are also long chained and branched. Archaeal cell walls and envelopes show great diversity and complexity in structure. They do not contain peptidoglycan although some do have a similar compound called **pseudomurein** which contains *N*-acetylalosaminuronic acid in place of muramic acid (see Topic D3). Another common type of cell wall is the S-layer, a two-dimensional, paracrystalline, array of protein or glycoprotein on the cell surface. Others have thick polysaccharide walls outside their plasma membrane.

D3 BACTERIAL CELL ENVELOPE AND CELL WALL SYNTHESIS

Key Notes

The Gram stain

The Gram stain is one of the most commonly used stains for Bacteria as it distinguishes between two groups of Bacteria on the basis of their cell-wall structure. The envelope of Gram-positive cells consists of a plasma membrane surrounded by a thick peptidoglycan layer. Gram-negative Bacteria have only a thin layer of exterior peptidoglycan but on the outside of this is a second, outer, membrane. Gram-positive Bacteria retain a crystal violet and iodine complex on washing with alcohol and therefore appear purple under light microscopy. Gram-negative Bacteria stain pink as they lose this complex and take up the paler carbol fuschin counter-stain.

Lipid bilayer

The membranes of the cell are made of phospholipids which are amphipathic molecules containing hydrophilic (water-loving) and hydrophobic (water-hating) regions. In the membrane, the phospholipids are arranged in a bilayer so that their polar headgroups, the hydrophilic regions, are on the surface of the membrane and their hydrophobic, fatty acid tails, are pointed into the interior of the membrane. Consequently, the membrane acts as a selective permeability barrier to bulky or highly charged molecules, which cannot easily pass through the hydrophobic interior of the lipid bilayer.

The plasma (cytoplasmic) membrane

The plasma membrane is a phospholipid bilayer in which are embedded integral proteins associated with transport, energy metabolism and signal reception. Other, peripheral proteins, are loosely associated with the membrane by charge interactions. The lipids and proteins in the membrane are mobile in relation to each other.

Transport across membranes

Three types of active transport systems are found in bacteria. Proton motive force driven transport, ATP-binding cassette transport systems which are associated with high affinity periplasmic binding proteins and group translocation which involves the chemical modification of a substrate as it moves through the membrane.

The bacterial cell wall

The cell wall is made up of a peptidoglycan, a polymer of *N*-acetyl glucosamine (NAG) and *N*-acetyl muramic acid (NAM) which has side chains of alternating D- and L-amino acids. This is a highly cross-linked molecule which gives the cell rigidity, strength and protection against osmotic lysis. Peptidoglycan contains many unique features such as D-amino acids, which makes it a useful target for antibiotics. Gram-positive cell walls also contain teichoic acids.

The periplasmic space

The periplasmic space is an aqueous region between the two membranes of Gram-negative Bacteria. It contains proteins associated with nutrient transport and nutrient acquisition.

The outer membrane

Peculiar to Gram-negative Bacteria, the outer membrane contains a number of unique structures. Braun's lipoprotein attaches the outer membrane to the peptidoglycan; porins allow the passive movement of nutrients through the membrane, and LPS gives protection to the cell. LPS, also known as endotoxin, is highly toxic to mammals.

Extracellular layers

The surface of cells is often covered by layers of polysaccharide or proteinaceous material known as capsules or slime layers, depending on their density. The polysaccharide layers are sometimes called the glycocalyx. These layers play a role in protecting the cell from desiccation and toxic compounds and also in the adhesion of bacteria to surfaces.

Synthesis of peptidoglycan

Precursor subunits of peptidoglycan are synthesized in the cytoplasm of the cell, translocated across the cytoplasmic membrane on a hydrophobic carrier and, finally, polymerized together on the outside of the membrane. Cross-linking of the peptide side chains occurs by transpeptidation.

Insertion of peptidoglycan into existing cell walls

Autolysins create breaks in the existing peptidoglycan to provide sites where new precursors may be inserted. Sites for the insertion of new cell wall material tend to fall into one of two patterns: either a few areas of insertion near the septal region, as seen in Gram-positive cocci; or multiple sites throughout the cell, as seen in many rod-shaped bacteria.

Related topics

Bacterial cell envelope and cell wall synthesis (D3)
Bacterial movement and chemotaxis (D4)
Prokaryote growth and cell cycle (D7)
Bacterial toxins and human disease (F6)
Control of bacterial infection (F7)

The Gram stain

The Bacteria are frequently divided into two groups on the basis of their reaction to a stain devised by Christian Gram in 1884. The differential reaction to the staining procedure is because of the structure of the cell envelope in these two groups of bacteria. **Gram-positive** Bacteria have a single membrane called the **cytoplasmic** (or **plasma)** membrane, surrounded by a thick layer of **peptidoglycan** (20–80 nm). The **Gram-negative** Bacteria have only a thin layer of peptidoglycan (1–3 nm) but on the outside of this there is a further **outer membrane** which acts as an additional barrier (*Fig. 1*).

The procedure for the Gram stain is as follows. Fixed cells are stained with a dark stain such as crystal violet, followed by iodine which complexes with the stain in the cell wall of the bacteria. Alcohol is added, which washes the dark stain of crystal-violet–iodine complex out of the thin-walled cells but not from

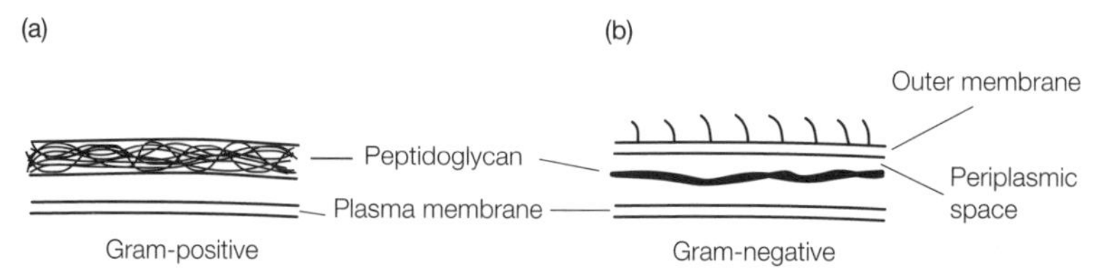

Fig. 1. Structure of the cell surface of a Gram-positive (a) and a Gram-negative (b) bacterium.

those that have thick cell walls. Finally, a paler stain such as carbol fuschin, called a counter-stain, is added which stains the decolorized cells pink. This light stain is not seen on the cells that retained the first darker stain. The cells that retain the crystal-violet–iodine complex (with thick cell walls) are called Gram-positive and appear dark purple under light microscopy. The ones that lose the stain (with thin cell walls and an outer membrane) are called Gram-negative and stain pink or pale purple.

Lipid bilayer

The plasma membrane is a semi-permeable lipid bilayer which acts as a barrier between the cytoplasm and the surrounding environment. The membrane is made of phospholipids (*Fig.* 2), such as phosphatidyl choline, which consists of a polar headgroup attached via glycerol to two long chains of non-polar fatty acids. Such molecules which have both polar and non-polar groups are called **amphipathic**. The polar headgroups are **hydrophilic** (Gr. *hydro*, water; *philic*, loving) and the fatty acids are **hydrophobic** (Gr. *phobis*, hating). This means that in an aqueous environment phopholipid molecules orientate themselves to ensure that the polar headgroups are associated with the water molecules and the hydrophobic chains are tucked away in the interior of the bilayer away from the water. Hence, a lipid bilayer can be formed, called a membrane (*Fig.* 2), about 5–10 nm thick. The interior of the membrane is highly hydrophobic, therefore acting as a barrier to anything that is bulky, such as glucose, or highly polar, for example, ions. Small molecules like water can diffuse through and hydrophobic compounds such as benzene can also move through by dissolving in the hydrophobic interior. Bacterial membranes do not contain sterols such as cholesterol which provide stiffness in higher organisms. However, recently it has been shown that sterol-like molecules called **hopanoids** are found in Bacterial membranes and methanotrophs contain a large number of sterols in their internal membrane systems.

The plasma (cytoplasmic) membrane

Embedded in the lipid bilayer are proteins of various functions (*Fig.* 2) including transport proteins (see below), proteins involved in energy metabolism (Topics B1 and B2), and receptor proteins that can detect and respond to chemical stimuli (Topic D4). **Integral** proteins are those that are fully associated with the membrane and may penetrate all the way through. These proteins therefore contain hydrophobic amino acids in the regions which are buried in the lipid

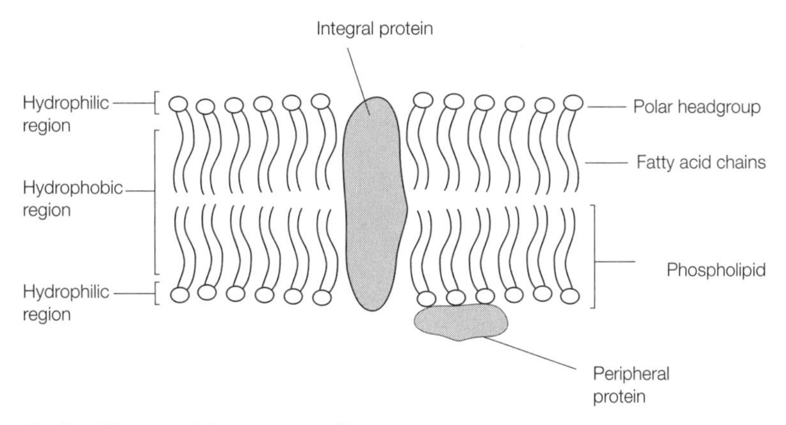

Fig. 2. Diagram of the structure of the plasma membrane.

(Topic C6). **Peripheral** proteins are ones that are only loosely associated by charge interactions with the positively charged polar headgroups of the phospholipids or on integral proteins and can be removed from purified membranes by washing with salt solutions. The lipids and proteins are fully mobile and move in relation to each other. This widely accepted model of membrane structure is called the fluid mosaic model.

Transport across membranes

Due to the hydrophobic nature of the lipid bilayer, only hydrophobic and small uncharged molecules can easily move across the cytoplasmic membrane. Most other solutes need a carrier (transport) protein to assist their movement into the cell and, if these molecules are to be accumulated at higher concentrations inside the cell compared to the outside (against a concentration gradient), then some form of energy must be provided.

Three different types of energy-driven transport systems have been found in bacteria.

- Proton-motive force driven transport which involves a single membrane spanning protein. Energy is provided by the proton gradient across the cytoplasmic membrane. An example is the lactose permease protein in *E. coli*
- ABC (ATP-binding cassette) transport systems which consist of three components.
  - A high-affinity substrate-binding component which may be a periplasmic protein in Gram negatives or a protein bound to the cytoplasmic membrane in Gram positives.
  - A membrane-spanning transport protein which accepts the substrate from the binding protein and moves it across the membrane, a result of a conformational change.
  - An ATPase which provides the energy that drives the movement of the substrate.

  An example of this type of mechanism is the dipeptide transport system in *E. coli.*
- Group translocation is unique to bacteria and involves the chemical modification of the substrate during transport. The most studied system is the glucose phosphotransferase system in *E. coli* in which glucose is phosphorylated as it moves across the membrane. Group translocation of a number of sugars have been found.

The bacterial cell wall

Outside the plasma membrane is the **peptidoglycan** cell wall which is made of sugars and amino acids. This is sometimes called the murein layer. The role of the cell wall is to provide rigidity and strength, preventing the cell from osmotic lysis when placed in dilute environments. The structure of peptidoglycan from *E. coli* is shown in *Fig. 3*. It consists of long polymers of two sugar derivitives, NAG and NAM with side chains of four alternating D- and L-amino acids attached to the NAM. Rigidity is achieved by crosslinks between the amino acid chains; normally from the third amino acid in one chain to the fourth amino acid in another chain. The nature of the amino acid side chains, and the links that join them, vary between bacterial species; however, the third amino acid is always a diamino acid (i.e. it has two amino groups, so it can form an extra peptide bond) and the fourth amino acid is normally D-alanine (D-Ala). In *E. coli* and most other Gram-negative Bacteria there is a direct link between the diamino acid *meso*-diaminopimelic acid and D-Ala. In Gram-positive Bacteria the linkage

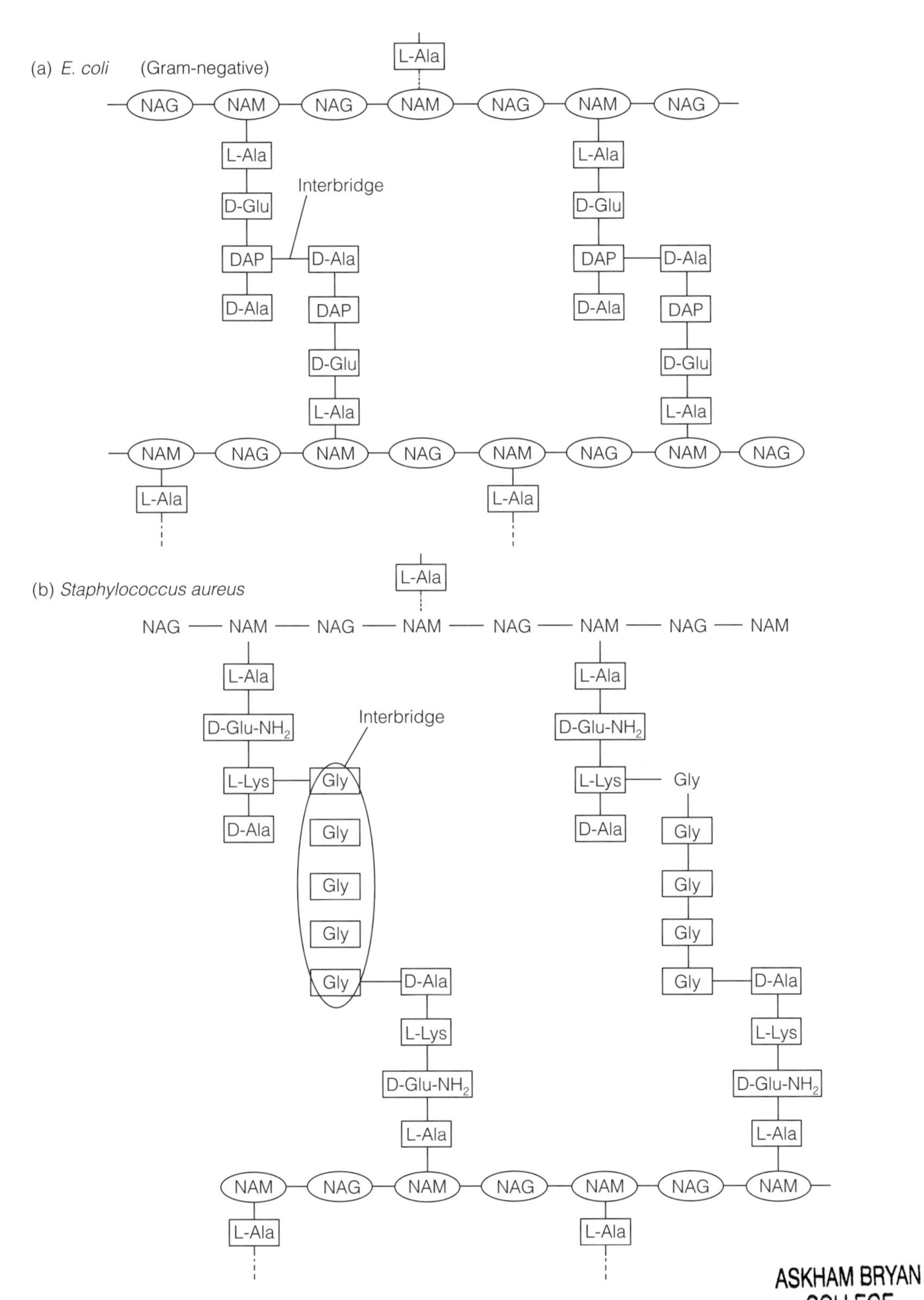

Fig. 3. Structure of the E. coli *peptidoglycan showing crosslinks between two strands. NAG:* N-acetylglucosamine; *NAM:* N-*acetylmuramic acid;* D-*Ala:* D-*alanine; DAP: meso-diaminopimelic acid;* D-*Glu;* D-*glutamic acid.*

often contains a bridge of amino acids: for example, in *Staphylococcus aureus* there is a bridge of five glycines between D-Ala and L-lysine. Multiple crosslinks within and between chains makes peptidoglycan a very strong and rigid structure.

There are several unique features of peptidoglycan including:

- the presence of NAM; this sugar is not found in eukaryotic cells;
- the presence of D-amino acids; L-amino acids are normally found in proteins.

These features make the peptidoglycan a target for antimicrobial agents that destroy prokaryotic cells specifically, but do not harm eukaryotic cells; an example of this is the antibiotic **penicillin** (Topic F7). **Lysozyme**, a natural antibacterial agent found in tears and natural secretions (Topic F4), breaks down the β(1→4) linkage between NAM and NAG. Removal of the cell wall under conditions where the osmolarity of the medium is the same as the inside of the cell (isotonic solution) results in the formation of round **protoplasts** (Gram-positives) or **spheroplasts** (Gram-negatives) which survive as long as the isotonicity is maintained. These structures lyse if placed in a more dilute medium, illustrating the importance of peptidoglycan to the cells' survival.

Gram-positive cell walls also contain large amounts of another polymer, called **teichoic acid**, made up of glycerol or ribitol joined by phosphate groups. D-Ala, glucose or sugars may be attached to the glycerol or ribitol and the polymers are attached either directly to the NAM in the peptidoglycan or to lipids in the membrane (in this case they are called **lipoteichoic** acids). The function of these molecules is unclear but they may have a role in maintaining the structure of the cell wall and in the control of **autolysis**.

Cell walls of a number of genera including *Mycobacterium*, *Corynebacterium* and *Nocardia* contain waxy esters of mycolic acids, which are complex fatty acids.

The periplasmic space

The outer membrane of Gram-negative Bacteria acts as an additional barrier protecting the peptidoglycan from toxic compounds such as lysozyme which act on the cell wall. It creates an aqueous space between the two membranes called the **periplasmic space** which is thought to have a gel-like structure with a loose network of peptidoglycan running through it. Estimates as to the width of the peptidoglycan vary from 1–71 nm but it has proved to be difficult to obtain a real definitive value. The periplasmic space contains a range of proteins associated with:

- *c*-type cytochromes and other electron-transporting proteins;
- transport of nutrients into the cell;
- enzymes that are involved in nutrient acquisition such as proteases;
- enzymes that defend the cell against toxic chemicals such as β-lactamases that destroy penicillin (Topic F7).

In Gram-positive cells some of these enzymes (called exoenzymes) are normally secreted into the surrounding medium. The presence of the outer membrane in Gram-negative Bacteria allows the cell to keep the enzymes close to itself rather than losing them into the medium.

The outer membrane

The outer membrane of Gram-negative Bacteria is made up of phospholipids but it also contains some unique features (*Fig. 4*).

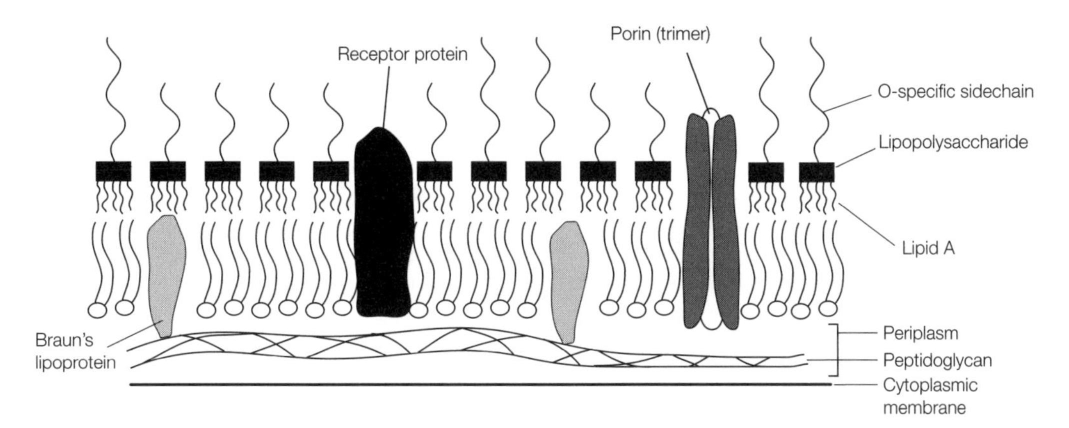

Fig. 4. Diagram showing the structure of the outer membrane of a Gram-negative bacterium.

- Pores formed by proteins called **porins** such as OmpF and OmpC allow the passive diffusion of small molecules into the periplasmic space.
- An abundant small lipoprotein called Braun's lipoprotein that is covalently bound to the peptidoglycan and is embedded in the outer membrane by its hydrophobic lipid, therefore holding the peptidoglycan and outer membrane close together.
- LPS molecules are found in the outer membrane projecting into the surrounding medium.

LPS consists of three parts:

- Lipid A which is embedded in the membrane.
- A core polysaccharide which contains a number of unusual sugars including 2-keto-3-deoxyoctonate (KDO).
- A side chain of repeating sugars called the O-antigen or O-side chain. The composition of the O-side chain varies between strains and antibodies to the O-antigen are often used for typing bacteria in the laboratory.

LPS molecules are responsible for a number of features of the Gram-negative Bacteria:

- they cause a net negative charge on the surface of the cell;
- they may hinder the access of toxic molecules to the surface of the cell therefore play a protective role (Topic F5);
- the long side chains are capable of variation in structure, and so may play a role in allowing Gram-negative Bacteria to evade the immune response (Topic F4).

Most importantly, the lipid A portion of the LPS molecule is frequently highly toxic to mammals. Called endotoxin, its presence in the blood stream, even at very low concentrations, leads to toxic shock and death (Topic F6). This is true of pathogenic as well as non-pathogenic Bacteria.

Extracellular layers

Different words are used to describe the layers of material often seen on the surface of cells. A well organized dense structure is called a **capsule** whereas if the material is diffuse and easily lost it is called **a slime layer**. **Glycocalyx** is often used to describe both capsules and slime layers that are made up of polysaccharides; however, some bacteria have proteinaceous capsules. These additional surfaces layers have a number of functions:

- they act as permeability barriers to the cell surface;
- they protect the Bacteria from phagocytosis;
- they protect the cell from desiccation;
- they aid in the attachment (adhesion) of prokaryotes to surfaces.

These roles will be discussed in more detail in Topic F5 in the context of microbial colonization of a host.

Synthesis of peptidoglycan

Research into peptidoglycan biosynthesis has been of major importance as this molecule is unique to Bacteria and is therefore a major target for antibiotics, the β-lactams and vancomycin being just two examples (see Topic F7). The process has been best studied in *Staphylococcus aureus* but the principles are probably true for most Bacteria. The synthesis of peptidoglycan begins with the formation of precursor subunits inside the cell. These are transported through the cytoplasmic membrane either into the periplasm of Gram-negative Bacteria or the outside of the Gram-positive cell. There, the subunits are joined together to form polymers by **transglycosylation** and the peptide side chains are crosslinked by **transpeptidation** to give the macromolecule rigidity. In the growing cell, the new peptidoglycan must be inserted into the existing wall without forming any weak areas, which might cause the cell to lyse from internal turgor pressure.

Peptidoglycan synthesis involves two carriers, uridine diphosphate (UDP) which carries the growing precursor in the cytoplasm of the cell and a lipid carrier called **bactoprenol,** which, being hydrophobic, is able to transfer the basic subunit across the cytoplasmic membrane. The stages in the synthesis of peptidoglycan are as follows.

1. UDP-*N*-acetylmuramic acid (NAM) is synthesized in the cytoplasm.
2. L-Ala is isomerized to D-Ala.
3. Two molecules of D-Ala are joined to form a dipeptide.
4. Amino acids are added sequentially to the UDP-NAM ending with the addition of D-Ala-D-Ala to form a UDP-NAM-pentapeptide (*Fig. 5*).
5. The NAM-pentapeptide is transferred to the bactoprenol-phosphate carrier located in the cytoplasmic membrane.
6. *N*-acetylglucosamine (NAG) is transferred from UDP-NAG to the NAM-pentapeptide to give the final peptidoglycan repeating subunit. Other peptide side chains may also be added at this point such as the pentaglycine interbridge found in *Staphylococcus aureus* peptidoglycan.
7. The repeating subunit is translocated to the external surface of the membrane.
8. The NAG-NAM-pentapeptide is transferred from bactoprenol onto the growing end of peptidoglycan chain by **transglycosylation** and the bactoprenol carrier is recycled back to the inside of the membrane. This stage is sensitive to the antibiotic vancomycin.
9. Finally, cross-links are formed between the peptide side chains by **transpeptidation**. These can be either directly, as in *Escherichia coli* where diaminopimelic acid (DAP) is linked to D-Ala on another side chain with the loss of the terminal D-Ala (*Fig. 5a*) or, via a pentaglycine bridge as occurs in *Staphylococcus aureus* (*Fig. 5b*). It is this stage that is sensitive to the β-lactam antibiotics.

(a)

−NAG−NAM−NAG−
L-Ala
D-Glu
DAP
D-Ala → NH_2 − DAP
D-Ala
D-Ala
D-Ala
D-Glu
L-Ala
−NAG−NAM−NAG−
Bond is transferred from D-ala to DAP

−NAG−NAM−NAG−
L-Ala
D-Glu
DAP D-Ala
D-Ala− DAP
D-Glu
L-Ala
−NAG−NAM−NAG−

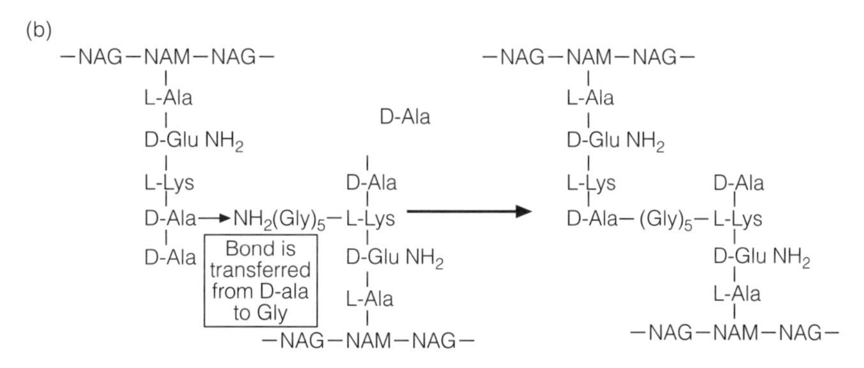

Fig. 5. Transpeptidation. Crosslinking between peptide side chains in (a) Escherichia coli *and (b)* Staphylococcus aureus.

Insertion of peptidoglycan into existing cell walls

In order to protect the cell from lysis by turgor pressure within the cell, cross-linked peptidoglycan must be continuous across the cell surface. Gaps in the peptidoglycan therefore need to be kept to the minimum and carefully controlled by the cell. A carefully controlled process allows the creation of breaks in the existing peptidoglycan by a group of enzymes called **autolysins**. These can either break the polysaccharide chains or the peptide cross-links and are controlled by **autolysin inhibitors**. New peptidoglycan is then added onto the existing peptidoglycan. In some Bacteria, such as the Gram-positive cocci, only a few sites for the insertion of new material exist in the cell surface and the principal site is where the cell will eventually divide (the area of septum formation). In many rod-shaped Bacteria, both Gram-positive and -negative, although there is major insertion of peptidoglycan in the area of septum formation, there are also multiple sites around the cell.

D4 Bacterial movement and chemotaxis

Key Notes

Motility

Many prokaryotes use a special structure, the flagellum, for motility. The prokaryote flagellum is a long structure that is free at one end and attached to the cell at the other end.

Flagella structure

Flagella are composed of helically arranged protein subunits; the protein is called flagellin. The portion of the flagellum embedded in the membrane is surrounded by two pairs of rings (basal body). In Gram-negative Bacteria, the outer pair of rings associated are attached to the LPS and peptidoglycan layers of the cell wall, and the inner pair of rings is located within or just above the plasma membrane. In Gram-positive Bacteria, only one pair of rings is present.

Flagellar movement

Flagella are rigid structures which do not flex but rotate. The rotary motion of the flagellum is driven by the basal body, which acts like a motor. The direction of flagellar rotation determines the type of movement. Prokaryotes with a single flagellum move forward during counterclockwise rotation and tumble when the flagellum rotates clockwise. Where there are more than one flagellum, they behave as a single bundle during counterclockwise rotation and thus move forward; however, during clockwise rotation the flagella act independently and the organism tumbles.

Chemotaxis

Chemotaxis is the movement of an organism towards or away from a chemical. Positive chemotaxis is movement towards a chemical (attractant); negative chemotaxis is movement away from a chemical (repellent). Prokaryote movement is controlled by the presence of these compounds such that where there is no gradient of attractant or repellent in the environment, the organism moves in a random way. However, in the presence of a concentration gradient, the net movement of the bacterium is in one direction. Prokaryotes detect the presence of a gradient through the action of membrane-bound chemoreceptors.

Related topics

Electron transport, oxidative phosphorylation and β-oxidation of fatty acids (B2)
Prokaryote cell structure (D2)
Bacterial cell envelope and cell wall synthesis (D3)

Motility

The bacterial flagellum is a long, thin (14 nm) structure that is free at one end and attached to the cell at the other end. The position and number of flagella are often used as characteristics for classification. **Polar flagella** are positioned at one or both ends of the cell. Where flagella are found at various sites around the cell these are termed **peritrichous**, and where several are located at one end of the cell they are termed **lophotrichous**.

Flagella structure

Flagella are composed of subunits of a protein called **flagellin**. The structure of a typical flagellum is shown in *Fig. 1*. The portion of the flagellum embedded in the membrane is surrounded by two pairs of rings called the basal body. In Gram-negative Bacteria, an outer pair of rings is associated with the LPS and peptidoglycan layers of the cell wall, and an inner pair of rings is located within or just above the plasma membrane. In Gram-positive Bacteria, which lack the outer LPS layer, only the inner pair of rings is present.

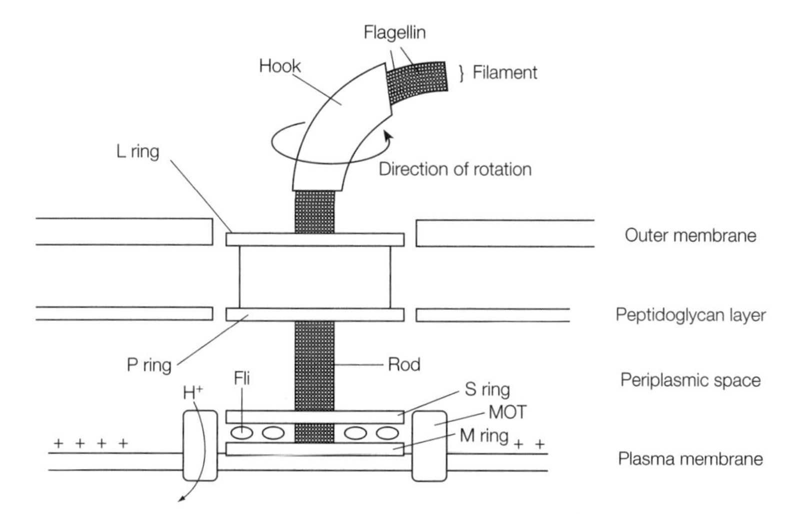

Fig. 1. Structure of the flagellum basal body and hook in Gram-negative Bacteria. Fli = motor switch proteins; MOT = flagellar driver proteins; Fli + MOT + S ring + M ring = basal body.

Flagellar movement

Flagella are rigid structures which do not flex but rotate. The rotary motion of the flagellum is driven by the basal body, which acts like a motor. The rings within the structure are thought to act together in generating rotational movement and the driving force for this comes from PMF (see Topic B2). Dissipation of the proton gradient releases energy which causes rotation of the flagellum. The direction of flagellar rotation determines the type of movement. Prokaryotes with a single flagellum move forward during counterclockwise rotation and tumble when the flagellum rotates clockwise. Flagella of peritrichous organisms behave as a single bundle of flagella during counterclockwise rotation and thus move forwards; however, during clockwise rotation the flagella act independently and the organism tumbles (*Fig. 2*).

Chemotaxis

Chemotaxis is the movement of an organism towards or away from a chemical. **Positive chemotaxis** is movement towards a chemical (attractant, i.e. a substrate). **Negative chemotaxis** is movement away from a chemical (repellent). Chemotaxis is thus a response to chemicals in the environment and requires that the cell has some form of sensory system. Prokaryote movement can be divided into runs where the organism moves in one direction (caused by counterclockwise flagellar rotation) and 'twiddles' where the organism randomly tumbles (caused by clockwise rotation of the flagella). Where there is no gradient of attractant or repellent the organism moves in a random way

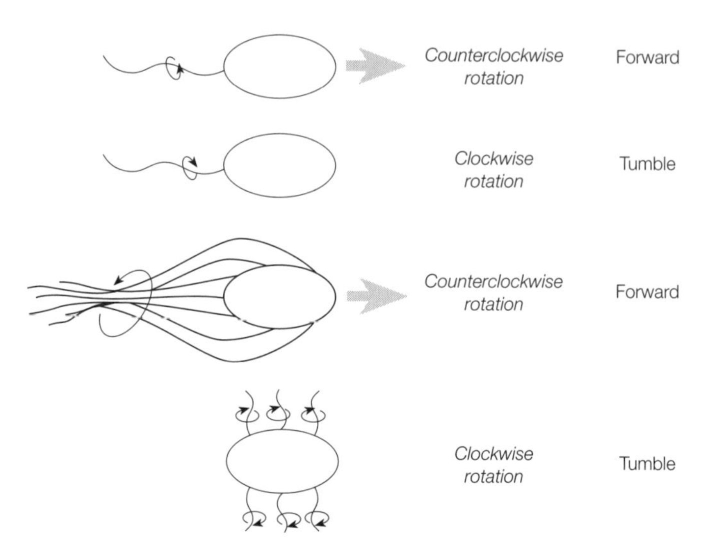

Fig. 2. Effects of flagellar rotation on prokaryote movement.

with a large number of twiddles; however, in the presence of a gradient, the runs become longer and twiddles less frequent as the organism experiences higher concentrations of attractant or repellent. The link between flagella movement and concentration of attractant or repellent involves the action of chemoreceptors which are proteins located in the periplasm. Although a chemoreceptor is fairly specific for the compound which it combines with, this specificity is not absolute. For example, the galactose chemoreceptor also recognizes glucose and fructose, and the mannose chemoreceptor also recognizes glucose. **Methyl-accepting chemotaxis proteins** (MCPs, also called **transducers**) are involved in translating the chemotactic signals from chemoreceptors to the flagellar motor.

In *E. coli* four types of MCP proteins have been identified – MCPs I, II, III and IV. Each MCP responds to different attractants and repellents. MCPs are so named because they become methylated or demethylated during chemotactic events. MCPs are transmembrane proteins that interact with attractants or repellents either directly or by way of chemoreceptors. A maximum of four methyl groups can be added to each MCP. Following sensory recognition of an attractant, an MCP becomes increasingly methylated and this causes the transmission of an excitatory signal to the flagellar motor, causing it to rotate in a specific direction. If rotation of the flagellum is counterclockwise, the cell continues to move. The longer the counterclockwise movement, the longer the run. When the chemoreceptor can no longer sense an increasing concentration gradient, a process called **adaptation** occurs. This is a result of the complete methylation of the MCPs and leads to an approximately 100-fold decrease in sensitivity to the attractant/repellent. This action results in clockwise flagellar movement and the cell twiddles. Demethylation of the MCP returns it to a state in which it can excite the flagellar motor. In the presence of a repellent, increasing concentration causes increasing demethylation. The link between the MCPs and flagella motion shown in *Fig. 3* is not completely understood.

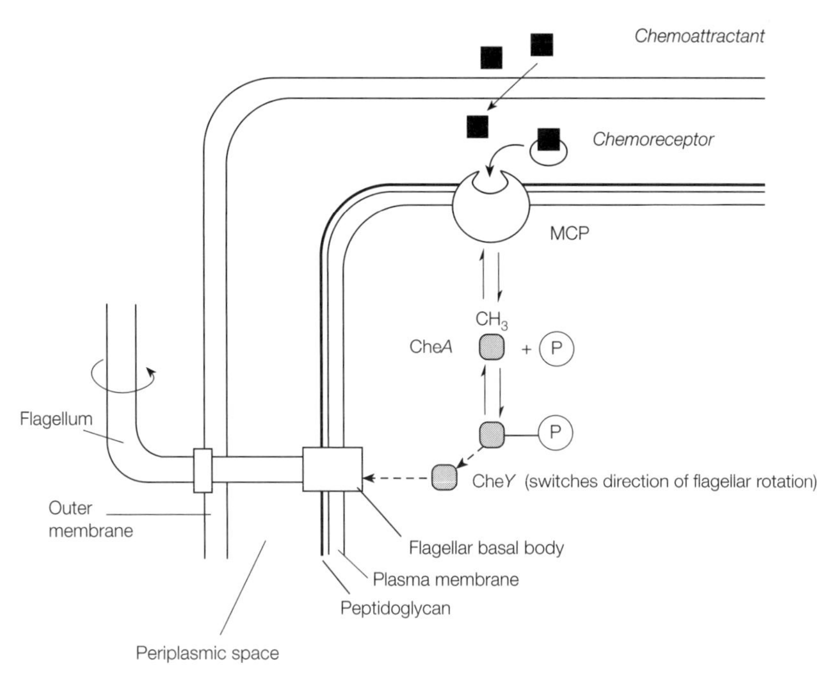

Fig. 3. Interaction between a chemoattractant and flagellar rotation. CheA and CheY are small proteins involved in flagellar rotation; methylation of MCP affects phosphorylation of CheA, which in turn affects the activity of CheY.

D5 The Archaea

Key Notes

Current taxonomic status of the Archaea — The Archaea are subdivided into the **Euryarchaeota**, **Crenarchaeota** and the Korarchaeota. The Korarchaeota are viable but non-culturable microorganisms.

Comparison to the Bacteria and the Eukarya — Archaea have a unique physiology and molecular biology with some elements similar to the Bacteria and some similar to the Eukarya.

Extreme habitats — Archaea have been found growing at 113°C, at pH 2 or at pH 10.5 but can also be found in many other habitats.

Methanogenesis — Most biochemical pathways found in the Archaea have similarities to those in the Eukarya and/or Bacteria, but methanogenesis is only performed by organisms of this Kingdom.

Biotechnology — The ability of the Archaea to grow on extreme habitats make them ideal sources of unusual thermostable, acid stable or salt-stable enzymes.

Current taxonomic status of the Archaea

Initially, microorganisms such as *Sulfolobus* were classified as members of the Bacteria on the basis of their morphology. Carl Woese and his co-workers first proposed an entirely new Kingdom of life in 1977, using **16S ribosomal RNA** sequence as a phylogenetic '**molecular clock**'. This Kingdom was first known as the **Archaeabacteria** and latterly as the **Archaea** and its existence has been supported by a wealth of biochemical and genetic studies (see *Table 1*). The kingdom is subdivided into three divisions: the Euryarchaeota (including genera such as *Methanococcus* and *Pyrococcus*), the Crenarchaeota (including genera such as *Sulfolobus* and *Pyrolobus*) and the Korarchaeota (*Fig. 1*). The Korarchaeota are unusual in that no pure culture of any member of this division exists. Their existence has been proposed solely on the basis of 16S rRNA

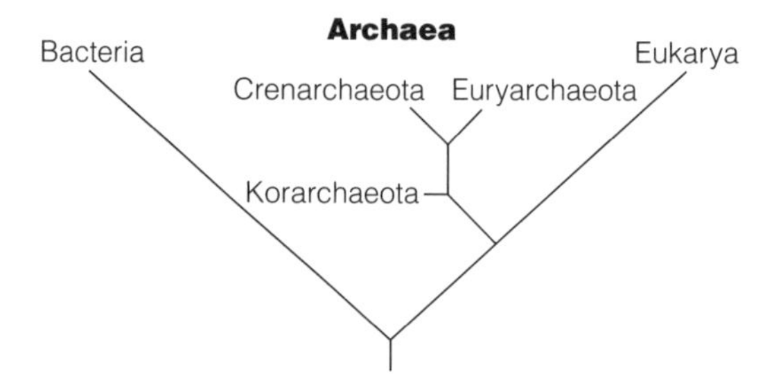

Fig. 1. Simplified phylogenetic tree of the Archaea, showing the three main subdivisions of the Kingdom and their relationship to Bacteria and the Eukarya.

Table 1. Selected features of the Archaea, compared to the Eukarya and Bacteria

	Bacteria	Archaea	Eukarya
Cell wall lipids	Isoprenoids ester bonded to glycerol	Isoprenoids ether bonded to glycerol	Isoprenoids ester bonded to glycerol
Light mediated ATP synthesis	Photosynthetic	Bacterioruberins catalyzed system (e.g. Bacteriorhodopsin)	Photosynthetic
Chromosome(s)	At least one circular, rarely linear	Circular	Linear
Nuclear membrane	Absent	Absent	Present
Histones	Absent	Present	Present
DNA replication	Unique	Similar to Eukarya	Similar to Archaea
MRNA	May be polycistronic	May be polycistronic	Monocistronic
RNA polymerase holoenzyme	$\alpha_2\beta\beta'\sigma$	Up to 13 subunits	More than 33 subunits
rRNA	5S, 16S, 23S	5S, 16S, 23S	5S, 18S, 23S
Transcription	Unique. Generally initiated by derepression	Simplified Eukarya-like mechanism	Generally initiated by activation
Translation initiation	Via Shine Delgarno sequence	Sometimes via Shine Delgarno sequence	Scanning mechanism

sequences obtained from mixed communities of microorganisms. As such they represent the most extreme example of **viable but non-culturable** (VBNC) microorganisms, whose presence can be detected in natural habitats, but cannot be grown in the laboratory.

Comparison to the Bacteria and the Eukarya

The genomes of the Archaea share features in common with both the Bacteria and the Eukarya (*Table 1*). The similarities to Bacteria include circular chromosomes and production of polycistronic mRNA from clusters of genes with similar function (operons). The genes within an operon may share orthology to either those of the Bacteria or the Eukarya, but possess inteins. **Inteins** are in frame coding regions that which removes themselves from the maturing polypeptide by self-splicing, in a manner reminiscent of the eukaryotic feature of having introns and exons. The replication of the archaeal genome is poorly understood, but involves DNA polymerases, helicases and gyrases more like those of the Eukarya than the Bacteria.

Extreme habitats

The best known examples of the Archaea are **extremophiles**, surviving in places where physical or chemical factors restrict the diversity of life. Archaea have been found growing in extreme conditions of temperature. *Pyrolobus fumarii* (a thermophile) will not grow below 90°C, but divides at temperatures up to 113°C. The **cryophilic** Archaea are also thought to make up about 30% of marine Antarctic microorganisms, growing at temperatures below 4°C. Conditions of extreme salinity (**halophiles** such as *Halobacterium salinarum*) or pH (**acidophiles** such as *Acidianus breyerlii* grow at pH 2, **alkaliphiles** such as *Thermococcus* *alcaliphilus* grow at pH 10.5) are also tolerated. However, archaeabacteria are not restricted to growth in these extreme conditions, and have been found in temperate soils.

Methanogenesis

Energy metabolism in the Archaea has many similarities to either the Bacteria or the Eukarya, subject to slight modifications (e.g. the Entner-Doudoroff pathway for glucose catabolism). Pathways cannot be generalized because of the metabolic diversity of the Archaea, but the metabolism of carbon sources resulting in the release of methane (methanogenesis) is unique. Substrates such as carbon monoxide, formate and carbon dioxide are metabolised anaerobically in the presence of hydrogen:

$$CO_2 + 4H_2 \rightarrow CH_4 + 2H_2O$$

C1 compounds such as methanol, methylamine and dimethylsulfide are used when hydrogen is an electron donor in the following way:

$$CH_3OH + H_2 \rightarrow CH_4 + H_2O$$

However, some methanogens can use methanol in the absence of hydrogen:

$$4CH_3OH \rightarrow CH_4 + CO_2 + 2H_2O$$

Compounds such as acetate are cleaved in an acetotrophic process:

$$CH_3COO^- + H_2O \rightarrow CH4 + HCO_3^-$$

All these reaction are chemiosmotically linked to ATP synthesis. The acetotrophic reaction yields the least amount of energy per mole substrate, whereas the CO_2-type substrates yield the most.

The environmental impact of methanogenesis is manifested in the evolution of methane from rubbish pits and in coal mines ('firedamp'). Methane from marshland Archaea may ignite to give the phenomenon 'Will o'the Wisp'.

Biotechnology

The extremeophilic lifestyles of the Archaea make them good sources for enzymes which work under harsh conditions of temperature or pH. The enzymes from the mesophiles are poorly active or completely inactive under such conditions. The best commercial success has been in the sale of ***Pfu*** **DNA polymerase for PCR**. Originally from *Pyrococcus furiosus*, an organism isolated from a geothermal vent on the deep ocean floor, the polymerase has better proofreading properties than *Taq* from *Thermus aquaticus* (a thermophile of the kingdom Bacteria) and can work at higher temperatures.

D6 Growth in the laboratory

Key Notes

Growth media

Microbes are grown (cultured) in the laboratory in either liquid or solid medium. Prokaryotic growth media should contain all the essential requirements for growth: an energy, carbon and nitrogen source as well as the essential ions, phosphate, sulfate, sodium, calcium, magnesium, potassium and iron, and various trace elements. Prototrophs require no additional ingredients in the medium, while auxotrophs require additional growth factors such as amino acids, vitamins and nitrogenous bases. The media used may be defined (synthetic) media where all the ingredients are known or complex media containing a mixture of undefined nutrients. Agar is used to solidify a liquid medium. Selective or differential media are used to detect the presence of particular groups of microbes.

Environmental conditions for growth

Each bacterial species has an optimum temperature, oxygen concentration, pH and water activity for growth and will survive in a range of conditions around these optima.

Related topics

Heterotrophic pathways (B1)
Autotrophic metabolism (B3)
Techniques used to study microorganisms (D8)

Growth media

In the laboratory, prokaryotes may be grown (**cultured**) in either **liquid** or **solid medium.** The liquid medium is contained in flasks, bottles or large culture vessels, called fermenters. Solid medium is usually found in **Petri dishes** which are normally plastic, sterile dishes with lids. Introduction of microbes into or on to these media is called **inoculation**. The nature of the medium depends on the microorganism's natural environment and on the reason why it is being grown. Normally, to grow large numbers of microbes, or their products, liquid culture is used but solid medium is used for the isolation of individual species (Topic D8) and for storage. As prokaryotes are a very disparate group of microbes that can exist in a wide range of environments, they can be split into a number of different nutritional groups with very different prerequisites for growth (see Topics B1 and B3). The ingredients in the medium will therefore depend on the requirements of an individual species or nutritional group. The common necessities for the growth of all prokaryotes are: water; a source of energy; carbon; nitrogen; essential inorganic ions such as phosphate, sulfate, sodium, calcium, magnesium, potassium and iron; and a number of trace elements (Zn, Mn, Mo, Se, Co, Cu, Ni, W). For heterotrophs, energy and carbon can be derived from the same molecule. Chemolithotrophs and phototrophs will have different requirements.

Liquid medium (or **broth**) can be converted into solid medium by the addition of agar (1–2%) which is isolated from seaweed. It is a particularly useful polysaccharide polymer because, once it is melted by boiling, it does not begin to harden until it is cooled to 40–42°C which allows the addition of temperature-sensitive ingredients, such as proteins, before it sets. Having set, it will not melt again until it has been heated to 80–90°C. Agar is also not normally degraded by bacteria. Petri dishes containing a nutrient agar are often referred to as agar plates.

Prokaryotes that can synthesize all they require from these basic ingredients are called **prototrophs** and most microorganisms that survive in the outside environment can do this. Microbes that have become adapted to life in a situation rich with nutrients such as the human body may require other growth factors to be provided, such as vitamins, amino acids or nitrogenous bases. These organisms are called **auxotrophs**. *Leuconostoc mesenteroides*, for example, requires over 40 additional growth factors. Prokaryote growth media are therefore designed to suit the organism of interest and may be either **defined** or **complex media**.

- A **defined,** or synthetic **medium,** contains known ingredients which are usually those necessary for the growth of a particular microbe. These are normally simple (**minimal**) media providing only the minimum requirements for growth. Examples of defined media for photosynthetic *Cyanobacteria* (note the lack of a carbon source) and heterotrophic *E. coli* are shown in *Table 1*.
- A **complex medium** contains undefined ingredients, such as proteolytic digests of meat (peptones), and meat and yeast extracts, which provide enough ingredients (amino acids, vitamins, sugars and bases) to sustain the growth of a wide range of microbes. Complex media such as nutrient broth (NB) and tryptic soya broth (TSB; *Table 1*) are routinely used for cultivation of prokaryotes in the laboratory and are particularly useful for growing prokaryotes whose growth requirements have not been defined. Blood is frequently added to media used for the isolation of human pathogens, as it provides many of the essential nutrients for the growth of fastidious human pathogens such as *Streptococcus.*
- Specialist **selective** and **differential** media are frequently used to isolate particular groups of microbes. These media are designed to select for a particular group of microbes or to differentiate between two species. They therefore

Table 1. Examples of growth media for prokaryotes

Defined media		Complex medium Tryptone soya broth (TSB)
For *Escherichia coli*[a]	For *Cyanobacteria*	For many different types of bacteria
Glucose	NH_4Cl	Enzymic digest of casein (tryptone)
Na_2HPO_4	$NaNO_3$	Enzymic digest of soybean meal (peptone)
KH_2PO_4	Na_2HPO_4	Glucose
NH_4Cl	NaH_2PO_4	NaCl
NaCl	$MgSO_4$	K_2HPO_4
$MgSO_4$	$FeSO_4$	
$FeSO_4$	Trace elements	
$CaCl_2$ (optional)	$ZnSO_4$	
	H_3BO_3	
	$MnSO_4$	
	MoO_3	
	$CoSO_4$	
	$CuSO_4$	

[a]Normally enough trace elements are present in the water.

contain ingredients which suppress the growth of unwanted organisms, encourage the growth of those of interest and, in the case of differential media, some means of distinguishing between two different types of microbe.

- A medium containing only acetate as a carbon source would be selective for organisms that grow on acetate.
- Blood in agar will allow the detection of those Bacteria capable of lysing red blood cells for example β-hemolytic streptococci. Lysis is seen as zones of clearing (hemolysis) around the colony.
- MacConkey agar is one of many selective media used to isolate or identify *E. coli* from environmental or medical samples. The medium contains bile salts and dyes that suppress the growth of Gram-positive Bacteria but allows Gram-negative organisms to grow. It is also a differential medium in that it allows the distinction between lactose-fermenting organisms (e.g. *E. coli*) and other non-fermenters (e.g. *Shigella* spp.). Lactose in the medium is fermented by *E. coli* producing acid which causes an indicator dye to change color to red. Colonies that ferment lactose are therefore red whereas non-fermenters are white.

Enrichment media are often selective media in that they contain (or lack) some ingredient which promotes the growth of a specific group of prokaryotes. For example, blood is used as an enrichment for human pathogens and a medium containing cellulose would enrich for Bacteria capable of using this polysaccharide as a carbon source.

Many specialist media are available for use in diagnostic microbiology. Sometimes referred to as **characteristic** media, they contain ingredients which allow the detection of a particular metabolic activity of the inoculated prokaryotes (Topic D1). For example, carbohydrate fermentation media are used to investigate whether a microbe has the ability to ferment various carbohydrates.

Environmental conditions for growth

Temperature

Most prokaryotes have characteristic temperature ranges of growth with a maximum, minimum and optimum growth temperature. **Psychrophiles** such as *Bacillus psychrophilus* are those that have become adapted to living at temperatures as low as –10°C and have an optimum growth temperature around 20°C. **Mesophiles** grow at temperatures between 15°C and 45°C with an optimum around 37°C. Bacterial **thermophiles** typified by *B. stereothermophilus* grow between 30°C and 75°C with an optimum of 55°C, but there are also groups of Bacteria and Archaea, with optimum growth temperatures greater than 80°C, termed **hyperthermophiles**. Many of these hyperthermophiles have growth minima above 60°C.

Oxygen concentration

Prokaryotes vary in their requirements for oxygen, depending on the nature of their metabolism (Topic B2). **Aerobes** are capable of growing in the presence of oxygen; **anaerobes** are prokaryotes that do not require oxygen for growth. Within this range are **obligate anaerobes** that die in the presence of O_2; **facultative anaerobes** such as *E. coli* which grow much better in the presence of O_2 but can grow anaerobically; **aerotolerant** anaerobes that ignore the presence of O_2 and grow equally well in its presence or absence; and **microaerophilic** bacteria that are damaged by normal atmospheric concentrations of 20% and only survive at much lower concentrations of O_2.

pH

Prokaryotes that can grow at high pH (8.5–11.5) are called **alkaliphiles. Acidophiles** are those that grow at low pH (0–5.5).

Water activity

Like all organisms microbes are sensitive to the osmolarity of the surrounding medium: at low osmolarity water will be accumulated in the cell and at high osmolarity water will be lost. Lysis of the cell at low osmolarity is prevented by the presence of the cell wall. Growth at high osmolarity is dependent on the ability of the microbe to maintain a high osmolarity within the cell without damaging cellular metabolism. Compounds such as betaine, choline and potassium ions (called **compatible solutes**) are used by prokaryotes to maintain this osmotic balance. Bacteria, such as *Staphylococcus aureus*, that are capable of growing in 3 M NaCl, are called **osmotolerant** whereas prokaryotes that have become adapted to growth at very high concentrations of salt (2.8–6.2 M NaCl), typified by the archaeon *Halobacterium*, are called **halophiles**.

In the laboratory, water activity and pH are controlled by the composition of the medium. Incubators are used to provide the correct temperature. Aeration to provide adequate O_2 levels is normally achieved by shaking, or stirring, liquid cultures and by providing large surface areas of solid medium. Anaerobic conditions can be obtained by simply filling a bottle to the top with medium so there is not much room for air, but for more stringent anaerobes a reducing agent such as thioglycolate is added to the medium. Specialist anaerobic jars or growth cabinets where O_2 is eliminated or replaced are also used.

D7 PROKARYOTE GROWTH AND CELL CYCLE

Key Notes

How bacteria grow

Prokaryote growth is exponential as cells divide by binary fission (1→2→4). The time it takes for the cells in the population to double, the generation (g) or doubling time, depends on the growth rate of the organism, media composition and environmental conditions.

Bacterial cell cycle

The cell cycle is the sequence of events between the formation of one cell and the next. The process can be divided into distinct phases: the C (chromosome replication), G (variable) and D (division) phases. The start of DNA replication is controlled by the mass of the cell. DNA segregation and division are controlled by the length of the cell.

Rapid growth

The complete cell cycle in *E. coli* must take a minimum of 60 min; however, in favorable conditions, *E. coli* can grow with a doubling time of approximately 20 min. A new round of DNA replication must therefore be initiated every 20 min although previous cell cycles have not been completed. Several cycles may therefore happen in the cell at the same time.

The bacterial growth curve

Prokaryote growth in liquid medium follows a typical pattern often called the prokaryote growth curve. Immediately after inoculation there is a lag phase where the prokaryote adapt to the medium. Exponential (or log) growth follows which is seen as a straight line on a plot of $\log_{10}$ viable count of cells against time. Once nutrients become exhausted or toxic metabolites accumulate, the net increase in cell numbers stops. This is called the stationary phase, which may be followed after a period of time by the death phase in which the number of cells decreases due to lysis.

Continuous culture

Continuous prokaryote growth can be maintained in systems, such as the chemostat, where fresh medium is constantly fed into a culture vessel. Removal of excess liquid by an overflow mechanism ensures that the culture volume is kept constant.

Related topic

DNA replication (C2)

How bacteria grow

When a prokaryote cell is inoculated into (placed on or in) medium, containing all the essential ingredients for growth, the cell will: accumulate nutrients; synthesize new cell constituents; grow in size; replicate its genetic material; lay down new cell wall; and, eventually, divide in two. Consequently one cell becomes two and then, after another period of time, these divide to become four. This type of cell division is called **binary fission** and this type of doubling growth is called **exponential growth**. A population of prokaryotes growing in this way will double in number during a particular length of time called the generation or doubling time.

$$\text{Generation time } (g) = \text{time } (t)/\text{number of generations } (n)$$

The number of generations can be calculated, if the original (N_o) and final number (N) of cells is known, using the formula

$$n = 3.3 (\log N - \log N_o)$$

The rate at which a population grows (the number of generations per unit time) is expressed as mean, or specific growth rate constant, and this is measured using the following equation:

$$\text{Mean growth rate } (\mu) = 0.69/g$$

From this formula it can be seen that as the specific growth rate increases, the generation time will decrease.

The rate at which bacteria grow and divide depends on the nature of the microbe, the ingredients of the medium in which it is grown, and the environmental conditions. For example, *E. coli*, when grown in a rich medium, with plenty of aeration at 37°C is capable of dividing every 20 min. This rate of cell division decreases if the prokaryotes are placed in a minimal medium where they are required to synthesize essential macromolecular precursors such as amino acids and bases (see Topic B4). In contrast, *Mycobacterium tuberculosis* has a maximum doubling time of about 18 h and will take much longer than *E. coli*, for example, to form colonies on an agar plate.

Prokaryote cell cycle

The sequence of events extending from the formation of a new cell to the next division is called the **cell cycle**. In this cycle, an *E. coli* cell will grow in length, with little change in diameter, until it reaches a critical size, twice a unit cell length. Cell division is initiated: a contractile ring is formed in the middle of the cell, septation proteins synthesize new cell wall and two new cells are formed, each one containing at least one copy of the bacterial DNA. Consequently, during this time, a copy of the chromosome must be synthesized and the two chromosomes segregated into the two progeny cells. DNA replication occurs during the **C (chromosome replication) phase** and chromosome segregation occurs in the **G (gap) phase**, which may be of variable length. The mechanism by which chromosomes segregate is still unclear. Finally, a cross wall (septum) is laid down between the two chromosomes and the cell divides into two (**D phase**). Cell division and DNA replication have to be coordinated. Initiation of DNA replication (see Topic C2) at the origin (*oriC*), a short adenine and thymine rich sequence, is dependent on the cell reaching a critical mass (initiation mass) and requires a number of protein initiation factors. DNA segregation and division, however, are controlled by the length of the cell which must reach a particular threshold length before the chromosomes are partitioned and cell division initiated. A multitude of cellular and environmental factors control the process.

Rapid growth

When conditions for growth are favorable, *E. coli* can grow with a generation time (see Topic C3) of approximately 20 min. However, the time it takes to synthesize a complete copy of the *E. coli* chromosome is 40 min, under optimum conditions, and segregation of the DNA and division takes another 20 min.

Thus, the shortest cell cycle and, therefore, generation time for *E. coli* should be 60 min. This is obviously not the case. For cells to divide faster than every 60 min, DNA replication must begin in one cycle and finish in another. When cells are growing quickly (generation time <60 min), initiation of replication occurs, as normal, producing two replication forks which move bidirectionally round the chromosome to the termination point (Topic C2). However, the origins on these new strands then initiate further rounds of replication before the previous round of DNA replication has finished (*Fig. 1*). Thus, when cell division occurs the DNA in the daughter cells is already replicating. The faster the cell growth rate, the more replication forks are formed such that the DNA in new cells may have multiple replication forks.

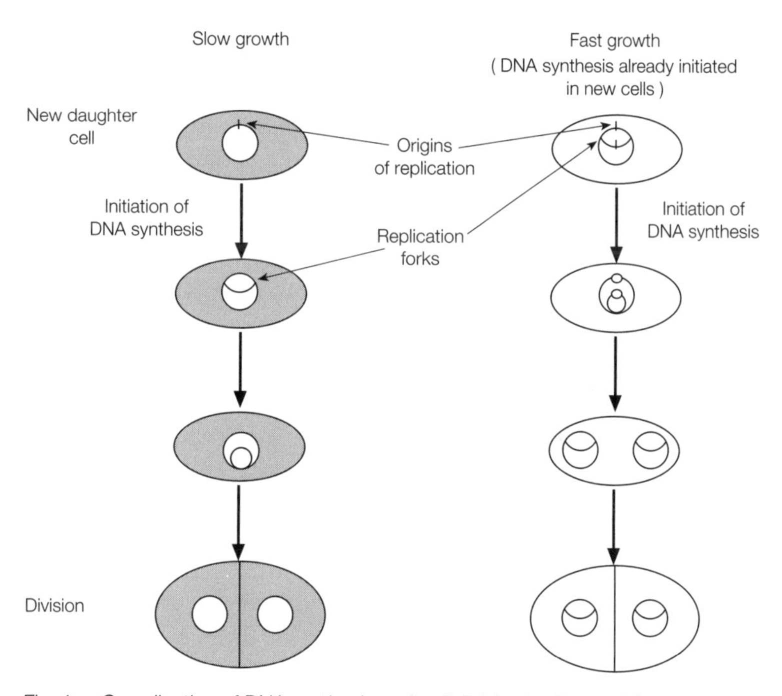

Fig. 1. Coordination of DNA synthesis and cell division in slow and fast growing bacteria.

The bacterial growth curve

The best way of producing large numbers of microbes or their natural products is to grow them in a liquid medium. Normally, the technique we use is called **batch culture** in which the cells are inoculated into flasks of a suitable medium and grown at an appropriate temperature and degree of aeration. Prokaryotes grown in this way show a particular pattern of growth which is referred to as the bacterial growth curve (*Fig. 2*). The number of viable bacterial cells is measured over time and is plotted as a graph of the $\log_{10}$ viable cell numbers against time. This is called a **semi-logarithmic** plot. A logarithmic scale is used to plot prokaryote growth owing to the large numbers of cells produced and to reveal the exponential nature of microbial growth. If an arithmetic scale is used to plot the increase in the number of cells, a curve of increasing gradient would

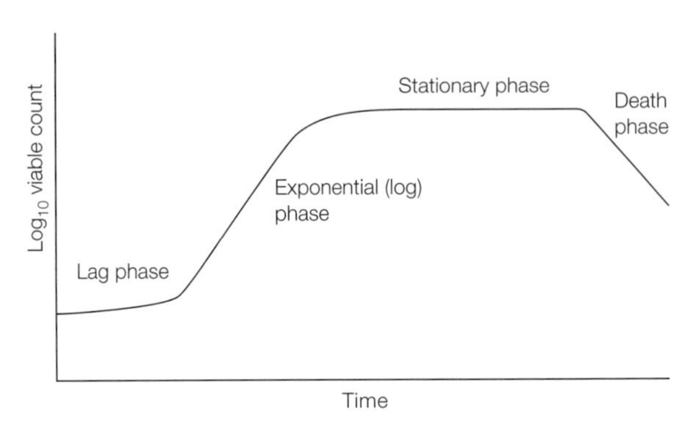

Fig. 2. A typical prokaryote growth curve.

be seen. This is converted into a straight line when a logarithmic scale is used. The generation time for the prokaryote can be read directly from the graph. The prokaryote growth curve reveals four phases of growth.

- **Lag phase.** When bacteria are first inoculated into a medium there is a period in which no growth occurs. During this phase the cells are adapting to the new environment, synthesizing new enzymes as required and increasing cell size ready for cell division. The length of this time depends on the nature of the inoculum. If this comes from a fresh culture in the same medium, the lag phase will be short, but if the inoculum is old or the medium has been changed (especially moving bacteria from a rich medium to a poor one) the lag phase will be longer.
- **Exponential (logarithmic) phase.** Once the prokaryotes start to divide, the numbers increase at a constant rate which reflects the generation (doubling) time of the prokaryote. This is seen as a straight line in this part of the graph (*Fig. 2*).
- **Stationary phase.** As the prokaryotes increase in numbers they use up all the available nutrients and accumulate growth inhibitors. Eventually a point is reached where there is no net increase in cell numbers, seen as a flattening-off of the growth curve. During this state of equilibrium, cells are still functioning. There is some cell death which is balanced by some small amounts of controlled cell division.
- **Death phase.** After a while the rate of cell death becomes greater than cell division and the number of viable cells drops. Cells lyse and the culture becomes less turbid.

Continuous culture

The problem with batch culture is that eventually cell growth stops due to the exhaustion of nutrients or the accumulation of toxic products. If these problems can be prevented by replacing the spent medium with fresh, then growth-inhibiting microbes can be maintained indefinitely. Continuous culture systems such as the **chemostat** do just that. They are flow systems in which fresh medium is constantly fed into a culture vessel. The vessel is kept at a constant volume by an overflow system, that removes the excess of liquid (*Fig. 3*) so the microbes are kept in the exponential phase of growth. Consequently, once the system has reached equilibrium, cell numbers are kept constant and a **steady-state** growth rate is reached. This growth rate can be manipulated by altering conditions such as medium composition and flow rate. A major advantage of these systems is in the provision of constant conditions for physiological studies. This is impossible in batch culture as the growth of the bacteria changes the environmental conditions such as pH and nutrient concentration of the medium.

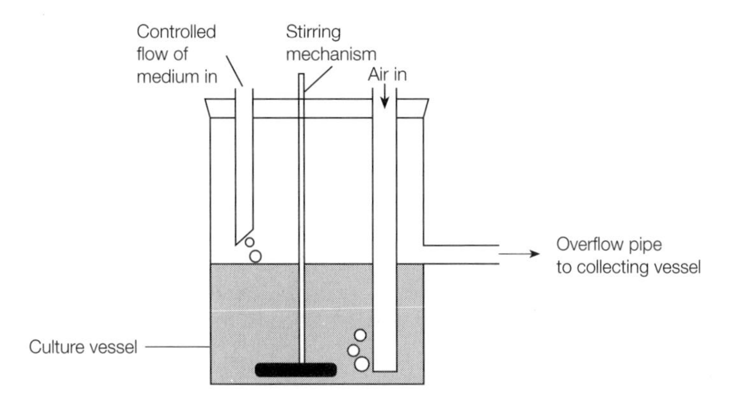

Fig. 3. A simplified diagram of a continuous culture system.

D8 TECHNIQUES USED TO STUDY MICROORGANISMS

Key Notes

Aseptic (sterile) technique

The methods used to protect the worker from the microbes with which they work, and microbial cultures from contamination with other microbes from the environment, are called aseptic or sterile techniques. Sterilization is the process that eradicates all microbes whereas disinfection implies only the removal of potential disease-causing organisms. Heat (wet or dry), chemicals, radiation and filtration are all used to sterilize materials used in microbiology.

Obtaining bacterial colonies

Pure cultures of microorganisms are normally required in the laboratory where all the cells have arisen from the same original bacterium. Streaking, spread plates and pour plates are all methods used to inoculate microorganisms on to agar plates so that the individual cells are separated from each other. After incubation at an appropriate temperature, each cell may form a colony, visible to the naked eye, which may contain up to 10^9 cells.

Counting bacteria

The number of cells in a culture can be estimated by: counting the individual cells using some form of counting chamber (total counts); diluting and inoculating the culture on to solid medium so that each original cell may develop into a colony which can be counted (viable count); measuring the light scattered by a culture using a nephelometer or spectrophotometer; weighing wet or dried cells; using biochemical methods to measure cell components; and using specialist electronic methods.

Related topics

Growth in the laboratory (D6) Prokaryote growth and cell cycle (D7)

Aseptic (sterile) technique

When handling any microorganisms in the laboratory it is very important to behave as if that microbe is potentially harmful. It is therefore essential that the methods used ensure that there is no or minimal contact between the worker (and others in the laboratory) and the microbe, either directly or as aerosols. Equally, it is important that the microbes being worked with are kept pure and are not contaminated with those from the environment. The skills and methods used to prevent contamination are called **aseptic** or **sterile techniques**. This essentially means that all the tools and media used are free from the presence of any living material (**sterile**) and experiments are carried out in such a way as to prevent the accidental introduction of unknown microbes (**contamination**). **Sterilization** is a process which ensures the complete eradication of all living cells and viruses. **Disinfection** is a vaguer term normally used to indicate the removal of potentially harmful microorganisms but not necessarily all other microbes. Physical (heat and radiation) and chemical methods are used to sterilize or disinfect materials and the method used is dependent on the nature of the materials to be sterilized (see *Table 1*).

Heat, either moist or dry, is one of the most commonly used techniques for sterilization of media and equipment but obviously, this cannot be used for any heat-labile materials such as proteins and some plastics. In addition, bacterial spores are very heat stable (Topic D2) and cannot be removed by boiling. Therefore routine sterilization using moist heat is normally done at higher temperature and pressure in an **autoclave**. These machines can operate at 121°C and 15 pounds per square inch (psi) which is sufficient to kill most microbes with the exception of spongiform encephalopathic agents and some Archaea. The length of time used for sterilization depends on the volume of the materials. Dry heat is routinely used in the laboratory to sterilize wire loops and remove microbes from the rims of culture vessels by **flaming** (heating) in a Bunsen flame.

Heat-labile materials can be sterilized by filtration (liquids) or by radiation (solids). Chemicals such as sodium hypochloride (chlorine bleach), 70% alcohols and other commercial disinfectants are normally used to decontaminate surfaces.

Obtaining prokaryote colonies

Prokaryotes generally grow as complex mixtures of microorganisms in their natural environments. However, in the laboratory, we normally wish to work with **pure cultures** of one particular strain of prokaryotes. The routine way to

Table 1. Methods of sterilization

Method	Effectiveness	Use
Moist heat		
Boiling/steam treatment	Kills most vegetative bacteria, fungi and viruses – not effective against endospores	Dishes, basin, heat resistant equipment
Autoclaving – 15 pounds per square inch (121°C)	Kills all vegetative cells and endospores. The length of time depends on the number of microbes present	Media, solutions, utensils, any item that can withstand heat and pressure
Pasteurization		
71.7°C for 15 sec	Heat treatment for liquids that kill pathogens but not necessarily all prokaryotes. Minimizes changes to flavor	Milk, juices, etc.
Tyndalization		
Three treatments of 90–100°C for 10 min with 24 h gaps in between	Vegetative cells are killed on day 1. Any spores germinate during the gap periods and are killed by the next heat treatment	Media that cannot be autoclaved but are relatively heat resistant
Dry heat		
Flaming	Red-hot direct heat	Inoculating loops
Hot air	200°C for 2 h will destroy cells and endospores	Glassware
Filtration		
Filters – pores size of 0.22–0.45 μm	Does not remove viruses	Liquids that cannot be heat treated
Radiation		
UV light – non-ionizing	Does not penetrate materials well	Work surfaces and air
γ-rays – ionizing	Penetrates well	Pharmaceuticals Plastic ware

isolate a single strain is to inoculate the culture on to plates using a method called **streaking**. A wire (or sterile plastic) loop is used to streak microorganisms on to plates of solid nutrient medium, as shown in *Fig. 1*, in order to dilute the cells and disperse them over the surface of the medium. The plates are then incubated at an appropriate temperature and each cell replicates to form a **colony** that is visible to the naked eye. A Bacterial colony contains up to 10^9 copies of the original cell. The appearance of the colony is often distinctive and can be used to distinguish one species from another by its morphology (see Topic D1). An alternative method of isolating single colonies is to use a **spread plate** method. In this <200 cells in a liquid medium are spread over the surface of a plate using a glass spreader sterilized by flaming in alcohol. After incubation, individual colonies may be seen. This method is frequently used for counting the number of viable cells in a culture as described below. An alternative method is to use spread plates or pour plates where the microbes are mixed with molten agar which has been cooled to 45°C and then poured into Petri dishes.

Counting bacteria

Total counts

The total number of cells in a culture can be determined directly by counting the number of cells under the microscope using specialist counting chambers. These chambers consist of a grid of known area etched on to a depression in a glass slide. A rigid coverslip is placed over the depression creating a precise volume between 0.02 and 0.1 μl. The number of microbes in that volume can be counted under the microscope. This method is difficult to use for small cells and does not distinguish between live and dead cells.

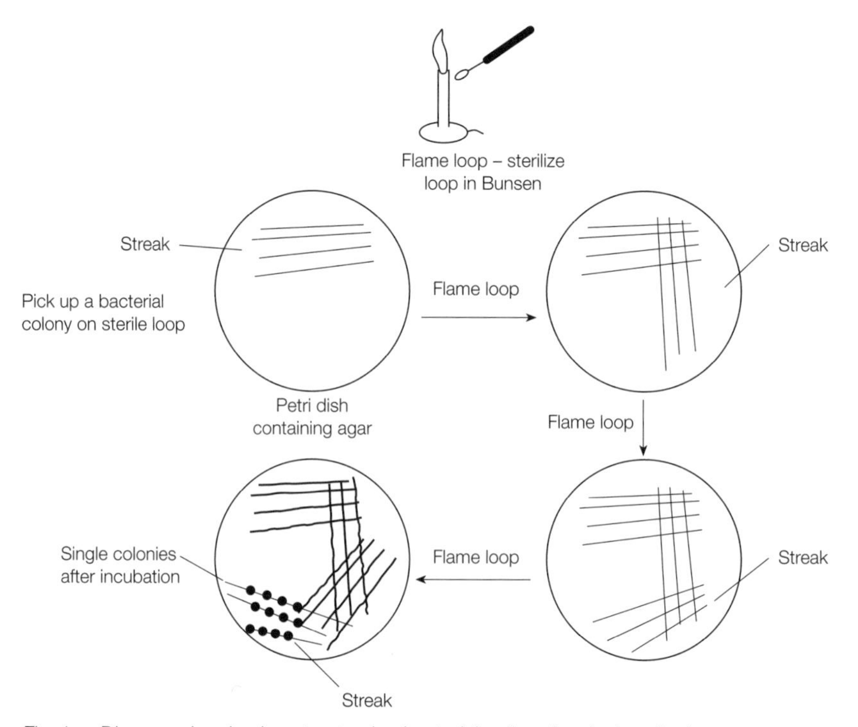

Fig. 1. Diagram showing how to streak a bacterial culture for single colonies.

Viable counts

In this method the culture is inoculated on to solid medium using the spread-plate technique or pour-plate method. The number of live cells in the original culture can be estimated from the number of colonies growing on the plate. As the number of microbes in the original culture may be as high as 10^{10} ml^{-1}, and the ideal number of countable colonies on a plate is between 30 and 200 ml^{-1}, it is standard procedure to dilute the culture in a series of 10-fold dilutions (1 ml of culture plus 9 ml of diluent) in order to achieve a countable number of cells in 0.1 ml. As the precise number of colonies in the original culture is normally unknown, 0.1 ml of a number of different dilutions are spread on to plates, to ensure that one of the dilutions will provide a suitable number of colonies to count (*Fig. 2*).

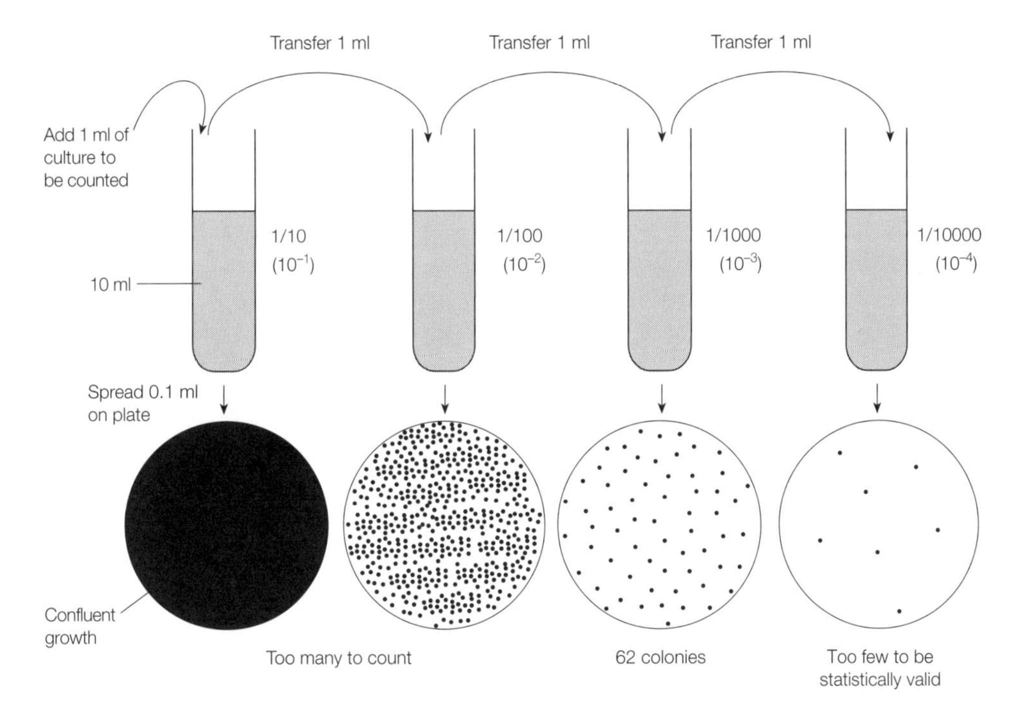

Fig. 2. Diagram showing serial dilutions for the estimation of a viable count of a microbial culture.

Turbimetric measurements

A very useful method for estimating cell numbers in liquid culture is to measure the turbidity of a suspension of the microbes. The microbial culture appears cloudy due to the scattering of light by the cells. The greater the number of cells, the greater the turbidity (i.e. light scattering is a **colligative** property dependent on number of particles in solution and not on mass of those particles). Routinely used machines are the nephelometer, which measures scattered light directly, and the spectrophotometer which measures light lost from the culture due to scattering; this is called absorbance and is measured as absorbance (A) or optical density (OD) units. The amount of light scattered by the suspension is proportional to the number of cells present (at low absorbance levels), but is also influenced by the size, shape and nature of the cells. A standard curve relating turbidity to cell numbers is therefore required for each individual species under specific growth conditions if an accurate estimation of cell numbers is needed. However, for monitoring microbial growth it is often sufficient to measure the changes in turbidity or absorbance of the culture with time of incubation.

Other methods for monitoring growth

Wet and dry weight of cells; biochemical analysis of cellular components such as protein, nucleic acids or ATP; change of conductivity (used in the Coulter counter) and changes in electrical impedance are just some of the alternative methods for measuring microorganisms.

D9 THE MICROSCOPE

Key Notes

Parts of a microscope

A microscope consists of a source of illumination, light or electrons; a mechanism for focusing the light onto a specimen, the condenser; somewhere to put the specimen; a system for magnifying the image in order to see the maximum amount of detail possible, the objective; and a way of visualizing the image, the eyepiece or screen.

Types of light microscopy

The type of image seen under the light microscope can be varied using different types of lenses and filters. The most common types used in microbiology are bright-field; phase-contrast; dark-field microscopy; and fluorescent microscopy. Confocal scanning laser microscopy, atomic force microscopy and differential interference contrast microscopy are used to provide three-dimensional images of cells.

Light versus the electron microscope

Resolution – the ability to see two adjacent points distinctly – is the property which limits the effective magnification of a microscope. Resolution is a function of the numerical aperture of the lens which depends on the wavelength of the radiation used and the nature of the lens. The maximum resolution of the light microscope is about 0.2 μm whilst that of the electron microscope is 0.2 nm.

Manipulation of the specimen for microscopy

Untreated microbial samples may be observed directly by light microscopy if they are thin enough and clear. However in order to see internal detail the sample may need to be sectioned, fixed and stained to increase contrast. Fixation and staining processes frequently introduce artifacts into the sample.

Related topics

Prokaryote taxonomy (D1)

Bacterial cell envelope and cell wall synthesis (D3)

Parts of a microscope

As, by definition, microbes are generally very small and cannot be observed in any detail by the naked eye, microbiologists generally need to use microscopy to magnify their subjects. Both light and electron microscopes are used frequently in microbiology: light microscopes being used for magnifications up to 1000-fold whereas electron microscopes can provide magnifications in excess of 100 000-fold. However the principles behind both types of microscopes are the same, the differences being in the nature of the electromagnetic radiation used and the nature of the lens.

The common features of light and electron microscopes are shown in *Fig. 1.* Each consists of a source of radiation, visible light in the case of the light microscope and electrons emitted from a heated tungsten filament for the electron microscope. The radiation is focused onto a specimen, mounted on a **stage,** using lenses which are made of glass in the light microscope and electromagnets in the electron microscope. The lenses used to focus the radiation on the

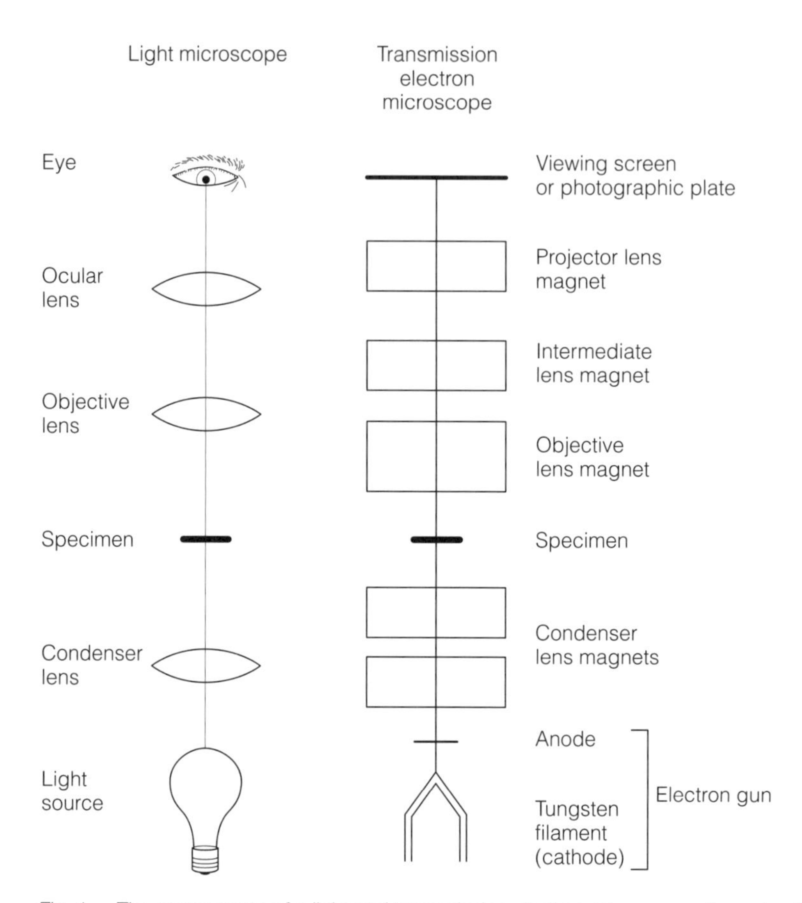

Fig. 1. The components of a light and transmission electron microscope drawn to show the similarities between the two.

specimen are called the **condenser** lenses and those which provide the magnification of the specimen are called **objective** lenses. Normally in microbiology laboratories a light microscope will have a number of different objective lens giving a range of different magnifications between 4-fold and 100-fold. In addition, the light microscope has an eyepiece or **ocular** lens that also provides magnification (normally 10-fold). The overall magnification is calculated by multiplying the two magnifications together. In the electron microscope the image is focused onto a fluorescent screen or photographic film for viewing. In addition, the whole of the electron microscope is held in a strong vacuum, as electrons cannot travel very far in air.

Types of light microscopy

There are many different types of light microscopy and this section will only cover a few that are normally used in microbiology. Basic **bright-field microscopy**, consisting of a simple set of lenses, is used routinely to view cells and sections of samples. An image is seen if there is a contrast between the sample and the surrounding medium. This contrast is due to the differing ability of various materials to scatter light. Colored materials therefore provide more contrast then transparent tissues or cells. Consequently bright-field microscopy often requires the staining of microbiological materials in order to provide a clear image and detail within the sample. Fixing and staining of material normally leads to cell death which is not always desirable when examining microbiological specimens.

Phase-contrast microscopy is a technique which improves the contrast in non-stained samples therefore allowing the visualization of wet-mounted live cells. Cells differ in refractive index from their surroundings and hence bend some of the light that passes through them. Although the velocity and wavelengths of light remain constant, the image appears out of phase to the observer, and this enhances contrast between the cell and its background. Light waves that are in phase have matching peaks and troughs and are seen as bright light. Light that is out of phase is darker. The optics of a phase-contrast microscope allow these differences in phase to be detected and a highly contrasting image can be seen under the microscope. In **dark-field microscopy** the specimen is illuminated only from the side so the only light that can reach the objective lens is light scattered by the specimen. The cells therefore appear bright on a dark background. Ultraviolet light is used in **fluorescence microscopy** to excite fluorescence in components either intrinsic to the sample (such as chlorophyll which naturally fluoresces), or as a result of the addition of a specific fluorescent dye or immunolabel. This method is used mainly to detect structures and molecules in cells.

Modern day computing technology has increased the power of the light microscopy allowing the production of three-dimensional images of an object. In **confocal scanning laser microscopy**, a laser beam is used to illuminate the specimen and the image is processed by computer software to provide detailed images of thick samples such as microbial biofilms. **Differential interference microscopy** using polarized light and **atomic force microscopy** are two other examples of three-dimensional imaging microscopes.

Light versus the electron microscope

The limit of the light microscope is not the magnification that can be achieved but a property called **resolution**. This is the ability to see two adjacent points distinctly, i.e. the ability to see fine detail. The resolving power of any system is dependent on the optical instrument itself and the recording system, for example the human eye or a television screen. The actual resolution is a function of a property of the lens called the **numerical aperture** (its ability to gather light) and the wavelength (λ) of the electromagnetic radiation used. A formula, **0.5λ/numerical aperture**, may be used to calculate the diameter of the smallest resolvable detail. In practice, this means that at the highest magnification in the light microscope, two objects will not be seen as distinct if they are separated by a distance of less than 0.2 μm. The use of oil between the lens and the specimen increases the light-gathering capacity of the lens, and therefore its numerical aperture, which is why **oil-immersion lenses** are frequently used in microbiology.

The resolving power of the electron microscope is much greater than that of the light microscope as the wavelength of the electron beam is much shorter than that of visible light. Maximum resolution in the electron microscope is approximately 0.2 nm which means that even molecules such as nucleic acids and proteins may be visualized. Two types of electron microscopy are commonly used. In transmission electron microscopy (TEM), electron beams are focused on very thin sections to see internal detail within cells. Special fixing, slicing and staining procedures are required to provide sufficient electron scattering to give contrast in the sample.

In scanning electron microscopy dried samples are normally coated with a thin film of a heavy metal such as gold and are scanned by a beam of electrons.

These induce the production of secondary electrons from the surface of the specimen, which are focused into a scintillation detector. The detector converts the electrons into light flashes that are then sent as a signal to the screen for viewing. Contrast is provided by the differing amounts of electrons, and therefore light, to be produced from the surface of the specimen. Raised areas appear bright and depressions are dark resulting in a good three-dimensional image. This method provides very useful information on the external features of organisms and their structures.

Manipulation of the specimen for microscopy

In light microscopy samples are mounted on glass slides, for electron microscopy they are placed on copper grids. The amount of preparation required depends on the nature of the sample and the detail required in the image. For light microscopy, samples may be placed in a drop of water, covered with a coverslip and viewed directly under the microscope. This is called a **wet-mount**. Simple stains such as methylene blue and crystal violet may be added to give additional contrast. These positively charged dyes bind strongly with the negative charges on cellular constituents. For more complex staining procedures microbes are usually dried and fixed onto the glass slide using direct heat. These can then be treated with specialist and specific strains such as the Gram stain (Topic D3) or malachite green stain for spores (Topic D1).

In electron microscopy, live specimens cannot be used as samples are placed in high vacuum. Specimens are chemically fixed and stabilized using aldehydes such as glutaraldehyde, then stained with an electron dense stain such as osmium tetroxide. Once the sample has been dehydrated in alcohol, it is embedded in a resin before being sliced into ultra-thin sections (50–100 nm thick) using specialist glass knives in an ultramicrotome. The sections are floated onto grids and may be stained again with heavy metals such as lead and uranium. **Negative staining** may also be used to observe small structures, such as viruses, without sectioning. In this case the specimen is overlaid with a plastic film, a drop of stain such as uranyl acetate is applied and pulled over the specimen onto the background. Under the microscope the particles can be seen in contrast to the dark background. For scanning electron microscopy, samples are fixed and dried and then coated with gold or other heavy metals using a form of vacuum evaporation called **sputter coating.**

As can be seen from the above descriptions, in many cases, the microbes have been extensively treated before they are actually observed in the microscope. This can lead to the introduction of structures or features called **artifacts** that are not actually part of the cell. It is therefore important that care is always taken when interpreting features seen on micrographs and electron micrographs in particular.

E1 Mutations

Key Notes

Genotype/phenotype

The genotype of an organism is the information stored in its genetic material, that is, the sequence of the bases in its DNA. The phenotype of an organism is the expression of that genotype as it can be observed or measured; for example, whether or not it produces a capsule or is resistant to an antibiotic. The most common form of an organism is often referred to as the wild-type.

Mutation

Any change to the sequence of bases in the DNA of an organism is a mutation. The change may be a single-base change called a point mutation or involve a much larger sequence of DNA in which case it is called a multi-site mutation. DNA may be deleted, inserted, inverted, duplicated or substituted.

Point mutations

Replacement of a purine with another purine (or a pyrimidine with a pyrimidine) is called a transition. The change of a purine to a pyrimidine or vice versa is a transversion. The consequence of a point mutation on the phenotype of the cell depends on how it alters the protein sequence. The possibilities are same-sense, mis-sense, nonsense and frameshift mutations.

Isolation of mutants

Mutant cells resistant to toxic compounds can be selected for by growth in the presence of that agent. Mutations leading to altered growth requirements may be screened for by replica plating.

Replica plating

Colonies of microbes are transferred, using a sterile velvet pad, to two sets of media; one on which the mutant will not grow and the other on which it will. The mutant colony is identified by its lack of growth on the former medium but purified from the latter.

Conditional mutants

Mutations which are expressed only under certain conditions are called conditional mutants. The most common type are temperature-sensitive mutants which only express their phenotype at high temperature.

Reversion of mutations

The effect of a mutation may be reversed by either a true back mutation, to give the original phenotype, or by a suppressor mutation at a different site, which overcomes the effect of the first mutation.

Related topics

Structure and organization of DNA (C1)
DNA replication (C2)
Structure of proteins (C6)
Translation (C7)
Recombination and transposition (E3)
DNA repair mechanisms (E4)

Genotype/phenotype

The genetic information of an organism is the exact sequence of nucleotides in its DNA, often referred to as its **genotype**. This sequence dictates which proteins or

RNA molecules can be produced by the organism, how the genes may be regulated and the functioning of the protein once it is synthesized. Not all proteins are made at the same time and the full genotype of an organism cannot be discovered unless the genome is sequenced. What is observable and measurable is the **phenotype** of the organism; the manifestation of the genotype in the form of characteristics such as the ability to produce a capsule or resistance to an antibiotic. Alterations to the genotype can therefore be seen as changes in the phenotype. Different nomenclatures/abbreviations are used for describing genes and mutations in different organisms, but in prokaryotes the following convention is used:

- Phenotype is described using a three-letter abbreviation of roman letters with the first letter capitalized. A superscript + or – is used to designate the presence or absence of this phenotype; for example, Phe^- indicates the inability to synthesize phenylalanine or Lac^+ indicates the ability to utilize lactose as a carbon source. The superscripts 's' and 'r' may also be used to designate sensitive or resistant.
- Genotype is indicated by three lower case letters (usually reflecting the phenotype) with a fourth upper case letter to indicate specific genes involved, all of which are italicized, for example, *pheA* or *lacZ*.

Although, naturally, there may be a large number of individuals of a species with slightly different phenotypes, microbiologists and in particular microbial geneticists try to work with organisms that are genetically identical. Consequently, in the laboratory, we work with **pure cultures** where microbes are descendent from the same original parent to minimize the genetic variability. That parent may often be referred to as the **wild-type.**

Mutation

A **mutation** is any inheritable change in the nucleotide sequence of the DNA in a cell. This change may be seen as an alteration in the phenotype of the cell but it may be silent if the change occurs in a non-essential part of a protein sequence, or in a non-coding part of the genome. Mutations in essential genes such as those necessary for DNA replication may be lethal and therefore never isolated. The changes in the DNA sequence range from single base changes, normally called **point mutations**, to large rearrangements of the genome, sometimes called multi-site mutations, which involve short or long stretches of DNA and may affect a number of genes. These multi-site mutations may be deletions, insertions, inversions, substitutions and duplications of sequence as shown in *Fig. 1* and occur as a result of recombination or transposition in the genome as described in Topic E3.

Point mutations

Point mutations may be the change of a purine for another purine (A ↔ G) or a pyrimidine for another pyrimidine (C ↔ T), called **transitions,** whereas the changes of a purine to a pyrimidine and vice versa are called **transversions**. The effect of a base change in the genotype on the phenotype depends on the nature of the mutation and where it occurs on the genome. There are four types of point mutation illustrated in *Fig. 2* and described below.

1. **Same-sense mutations** are due to the redundancy in the genetic code; the same amino acid is inserted into the protein so no change in the phenotype is seen (see the genetic code in Topic C7).
2. **Mis-sense mutations** occur when a different amino acid is inserted into the chain. The consequence of this change is dependent on whether the alter-

Fig. 1. Possible changes to DNA sequence that may result in mutation.

Original sequence

5′-AUG CCU UCA AGA UGU GGG CAA-3′
Met Pro Ser Arg Cys Gly Gln

Same sense mutation – same amino acid inserted
5′-AUG CCU UCA AGA UGU GGA CAA-3′
Met Pro Ser Arg Cys Gly Gln

Mis-sense mutation – different amino acid inserted
5′-AUG CCU UCA GGA UGU GGG CAA-3′
Met Pro Ser Gly Cys Gly Gln

Nonsense, the creation of a termination codon – protein synthesis stops
5′-AUG CCU UCA AGA UGA GGG CAA-3′
Met Pro Ser Arg STOP

Frame shift mutation – from the point that a base is inserted or deleted the amino acids are altered

Deletion or insertion of one or two bases

↓

5′-AUG CCU UCA AG U GUG GGC AA-3′
Met Pro Ser Ser Val Gly etc.

Fig. 2. The nature of point mutations as seen in the mRNA sequence and their consequence on protein sequence.

ation is in an essential or non-essential part of the protein. In the first case, protein function may be altered or lost but in the latter no change to the phenotype may be observed.

3. **Nonsense mutations** result when the change in the base sequence alters an amino acid-encoding codon to a STOP codon (UAA, UAG, UGA). Protein synthesis terminates prematurely, leading to the production of a truncated protein.
4. **Frameshift mutations** occur when one or two nucleotides are deleted or inserted into the DNA sequence, therefore altering the translational reading frame and producing a completely altered amino acid sequence from the point of change.

Isolation of mutants

Although mutations occur spontaneously in bacteria, this is normally at a very low frequency, so it is often advantageous to use a mutagen, such as UV light, to increase the chance of isolating the mutation of interest (Topic E2). Identification of the presence of that mutation depends on the nature of the gene of interest, but generally these fall into three categories:

1. Mutants that are resistant to some toxic compound, such as an antibiotic, or to infection by a bacteriophage (Topic E7). These mutations can be **selected** by growing in the presence of the toxic agent. Only the resistant mutant strains will grow.
2. Auxotrophic mutants that are unable to synthesize a compound essential for growth such as an amino acid or vitamin (Topic D7). These mutants cannot be isolated directly but can be **screened** for by replica plating as described below.
3. Mutants that are unable to use particular substrates such as lactose or maltose for growth. These may also be identified using replica plating.

Replica plating

This is a method used to screen a large number of colonies for the presence of a particular mutation. Prokaryotes are spread on to plates, containing medium on which both the parent and the mutants will grow, at a dilution such that individual colonies can be seen on the plate (Topic D8). After incubation, the colonies are transferred, using a sterile velvet pad, to plates containing media that will allow detection of the mutation (*Fig. 3*). The nature of that media depends on the desired mutation. In the case of auxotrophic mutations, the colonies may be replica plated on to minimal medium with, and without, a particular growth factor. Mutant strains, that are unable to synthesize that factor, will not grow in its absence but will grow on the medium in which it is provided. Having identified the mutant colony by its lack of growth, it may then be picked off and purified from the plate on which it has grown.

Conditional mutants

The products of many genes in the cell are essential for cell growth and division; examples being the enzymes necessary for DNA replication. It is therefore impossible to isolate mutations, which totally inactivate the gene function, in these genes, as the cell would die. **Conditional** mutants can be used in these circumstances. These are mutations which are expressed only under certain circumstances, the most common being high growth temperature. These **temperature-sensitive** mutants have a wild-type phenotype at low temperatures but, when switched to a higher temperature, the mutation is manifested, which allows the effect of the mutation to be studied.

Reversion of mutations

The effect of a mutation can be reversed by a number of means. The simplest is by reversion where a **back** mutation occurs to give back the original base

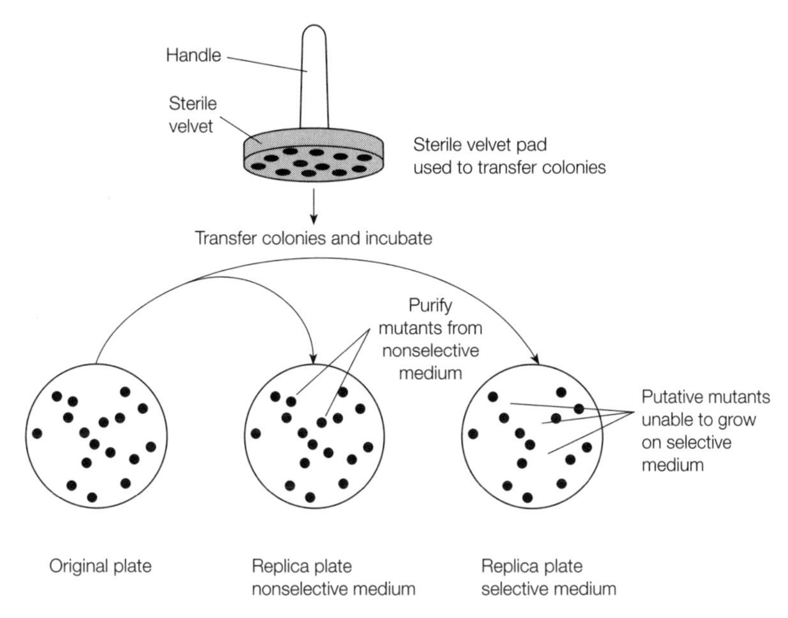

Fig. 3. Replica plating for the isolation of mutant colonies.

sequence (the wild-type). Selection for reversions is a good test to indicate if the original mutation was a point mutation as deletions do not revert easily.

Another way of achieving a wild-type phenotype is by a mechanism called suppression. Here, a second mutation occurring at a different point in the genome, can compensate for the effect of the first mutation. The second mutation may be at a different site in the same gene in which case it is called an **intragenic suppressor** or in a completely different gene when it is termed an **intergenic suppressor.**

E2 Mutagenesis

Key Notes

Spontaneous mutations

Mutations constantly occur in the genome virtually at random, although 'hot spots' exist where mutations can occur at much higher frequencies. Such spontaneous mutations are a result of tautomerization of bases, oxidative deamination of cytosine or strand slippage during DNA synthesis at sites where there are short repeated nucleotide sequences. The movement of, and recombination between, transposable elements may also be responsible for mutations in the genome.

Mutagens

Mutagens are agents that increase the frequency of mutation by interacting with the DNA molecule. Chemical agents include base analogs which are incorporated into the DNA molecule during replication, intercalating agents which insert between the base pairs of the helix, or DNA-modifying agents which alter the structure of DNA. The most important physical DNA-damaging agent is short-wave radiation especially UV light. Transposable elements are mutagenic agents and are responsible for a range of genomic rearrangements in the cell. Finally, in the laboratory, specific mutations can be introduced into DNA by *in vitro* mutagenesis when the DNA sequence is known.

Related topics

Structure and organization of DNA (C1)
DNA replication (C2)
Mutations (E1)
DNA repair mechanisms (E4)

Spontaneous mutations

Mutations occur throughout the genome all the time, but some areas of the genome are more prone to change than others. These are called **hot spots.** The rates for mutations in a gene can vary from one mutation per gene every 10^4 rounds of replication to one every 10^{11}, with an average of about one mutation per gene every 10^6 rounds of replication. Mutations can arise for a number of reasons including: errors made by DNA polymerase during DNA replication (Topic C2); physical damage to the DNA; chemical damage; recombination; and transposition (Topic E3). However, it is important to be aware that DNA is protected by a large number of DNA-repair systems geared to minimize the mutation rate (Topic E4).

One main cause of spontaneous mutations results from the fact that bases can exist as different forms called **tautomers**, which are capable of forming different base pairs. So during replication, adenine, which normally base-pairs with thymine in its normal amino form, switches to its rare imino form (**tautomerism**) and base pairs with cytosine meaning that a cytosine is inserted into the DNA instead of a thymine. If this is not repaired before the next round of replication the A–T (adenine–thymine) base pair will be changed to a G–C base pair as shown in *Fig. 1*.

Spontaneous mutations can also arise as a result of slippage of DNA strands at short repeated nucleotide sequences during DNA replication. This leads to

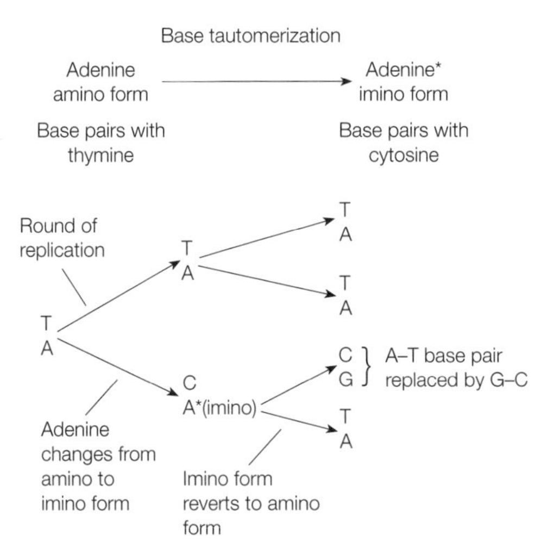

Fig. 1. Spontaneous mutation of an A–T base-pair to G–C as a result of adenine base tautomerization.

the insertion or deletion of a short length of DNA. Very rarely, a base may be removed from a nucleotide leaving a gap called an **apurinic** or **apyrimidinic** site which will not be able to base-pair normally at the next round of replication. A common reason for this is the natural deamination of cytosine to form uracil. This is recognized by DNA repair systems as an error and the uracil is removed to leave an apyrimidinic site which can be filled with the wrong base. Finally, major causes of spontaneous mutations are transposable elements which can insert at random in the genome, resulting in mutation. In addition, if two or more copies of the element are present, they can act as sites for homologous recombination leading to deletions, duplication or inversions of sections of the genome (Topic E3).

Mutagens

Chemical and physical agents that bind to DNA can increase the mutagenesis rate. Chemical mutagens can act in a number of ways:

1. **Base analogs** such as 5-bromouracil and 2-aminopurine are incorporated into the DNA during normal DNA replication. These undergo tautomeric shifts similar to the naturally occurring bases, but at much higher frequencies, leading to mutation.
2. **Intercalating agents** are flat, three-ringed compounds similar in shape to a base pair. They can slot into the DNA, leading to a distortion of the helix which leads to slippage during DNA replication and the insertion and deletion of bases. Ethidium bromide and acridine orange are representatives of these agents.
3. **DNA-modifying chemicals** are compounds that alter a base in the existing DNA, leading to mis-pairing at the next round of replication. Examples include hypoxanthine which converts cytosine into hydroxylaminocytosine which pairs with adenine, leading to the transition of G–C to A–T.
4. Ionizing radiation can be a potent cause of DNA damage. The principal damage caused by UV light is the formation of pyrimidine dimers, the most common being **thymine dimers**, between adjacent bases which causes a

distortion of the DNA helix. This in itself is not mutagenic but mutations occur when the cell tries to repair the damage using an error-prone repair system called the SOS repair system discussed in Topic E4.

5. *In vitro* mutagenesis is frequently used now to change sequences precisely in the DNA when the DNA sequence is known. A copy of a mutated sequence is chemically synthesized and used to replace the wild-type sequence in the genome.

E3 Recombination and transposition

Key Notes

General recombination

General (or homologous) recombination is mediated by RecA protein which causes the exchange of genetic information between DNA strands with very similar sequences. Recombination proceeds via the production of a cross-shaped heteroduplex molecule, sometimes called a chi form which is resolved by endonuclease action to produce recombinant DNA molecules.

Site-specific recombination

Non-RecA protein-dependent recombination between specific sequences, generally resulting in phage genome integration into bacterial chromosomes, is mediated by phage-encoded proteins and host-encoded integration host factor (IHF).

Transposition

Transposable elements move between or within DNA molecules, mediated by a transposase enzyme encoded by the element. There are a number of different types of transposable element in bacteria: the simplest are insertion sequences; more complex are transposons which carry additional genes, often antibiotic resistance genes; transposable phage, such as Mu, use transposition as a mechanism for replication and the most recently discovered are conjugative transposons in Gram-positive bacteria.

Related topics

Mutations (E1)
Mutagenesis (E2)
DNA repair mechanisms (E4)
F plasmids and conjugation (E6)
Replication of bacteriophage (E8)

General recombination

General (or homologous) recombination is the exchange of genetic information between two double-stranded DNA molecules with identical or very similar sequences. In diploid cells this type of recombination is responsible for the exchange of genetic information between homologous chromosomes during meiosis (Topic G3). In haploid organisms, such as bacteria, there are also occasions when homologous regions of DNA may be present in the cell, for example during chromosome replication or when plasmid or phage DNA is present. This allows the opportunity for homologous recombination. In bacteria, genetic exchange is mediated by a protein called **RecA** which has a number of pivotal roles in the cell (see also DNA repair, Topic E4). During recombination RecA has two important abilities. It binds to single-stranded DNA and aligns the DNA to which it is bound with a homologous target, that is a piece of double-stranded DNA with the same sequence. It then has the ability to insert that single-stranded DNA into the target DNA displacing a strand of the DNA (*Fig. 1a*) allowing the incoming strand to base-pair with its complementary sequence in the intact strand. This is called strand invasion and leads to the formation of a **D-loop**. The exchange of two single strands between two homologous sequences of double-stranded DNA leads to the formation of a cross-over between the DNA duplexes. DNA ligase seals the single-stranded gaps. The resulting **heteroduplex** is called a **Holliday structure** after the person who first identified it (*Fig. 1b*).

Recombination may also be initiated by the RecBCD protein which has helicase and nuclease activity. It is thought to bind to linear double-stranded DNA, migrate along the DNA to specific sequences called **chi sites**, where it makes a single-stranded nick about 56 nucleotides to the 3′ side of that site. With the association of RecA protein, a single strand capable of invading a homologous duplex is created which leads to strand invasion and the formation of a Holliday structure as before.

In both cases, once formed, the Holliday structure can move down the molecule by a process called branch migration, creating either short or long lengths of hybrid DNA duplexes.

The resolution of the two DNA duplexes to produce recombinants is achieved by endonucleases cutting the DNA molecule. The nature of the products depends on where the cutting of the strands occur, as the hybrid molecule can also be drawn as a cross-like structure called a **chi form** (*Fig. 2*). If cuts are made at points a and b, two DNA molecules will be resolved in which only a single strand of DNA has been exchanged (*Fig. 2*). Any change in the DNA molecule will therefore be small as the two original molecules were very similar. If the cuts occur at points c and d (*Fig. 2*), the consequences are much greater in that effectively the two DNA molecules are joined together.

The effect of recombination on DNA molecules depends on whether they are linear or circular and whether recombination occurs at one or two sites (*Fig. 3*). One recombination event (cross-over) would lead to the joining of two linear molecules (*Fig. 3a*) or two circular molecules (*Fig. 3b*). Two recombination events at two different sites would lead to the integration of a new stretch of DNA into a DNA molecule (*Fig. 3c*).

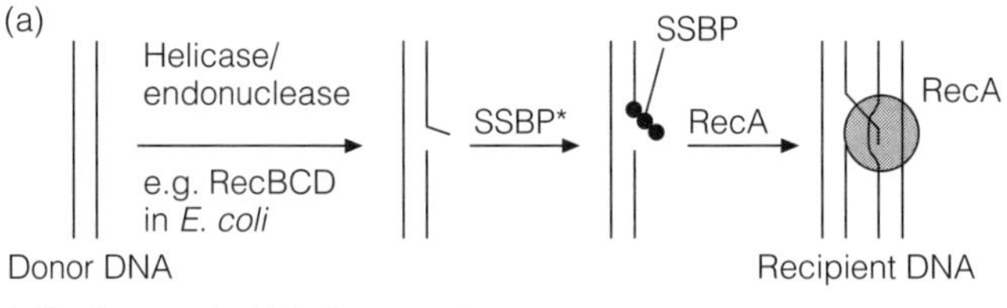

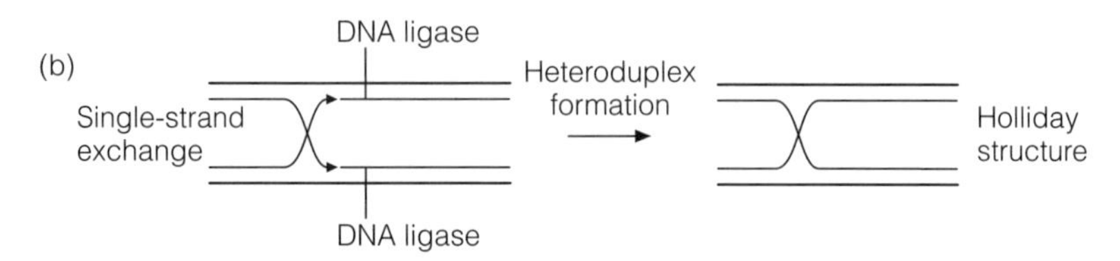

Fig. 1. The Holliday model for general (homologous) recombination. (a) Single-stranded invasion mediated by RecA protein. (b) Formation of a Holliday structure.

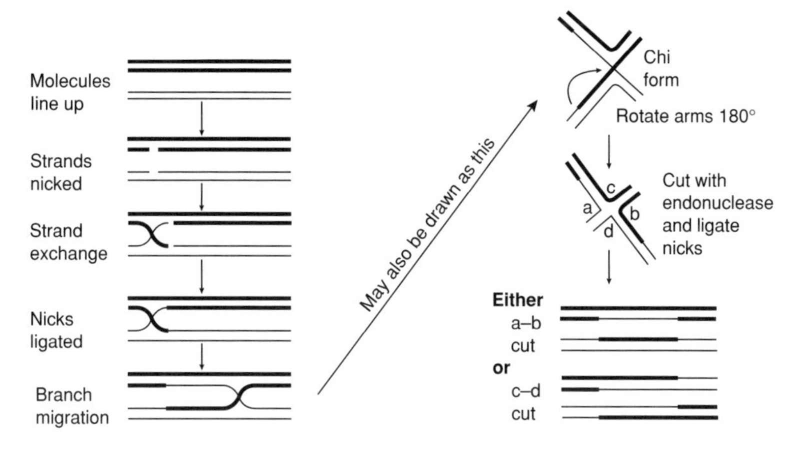

Fig. 2. The Holliday model for general (homologous) recombination.

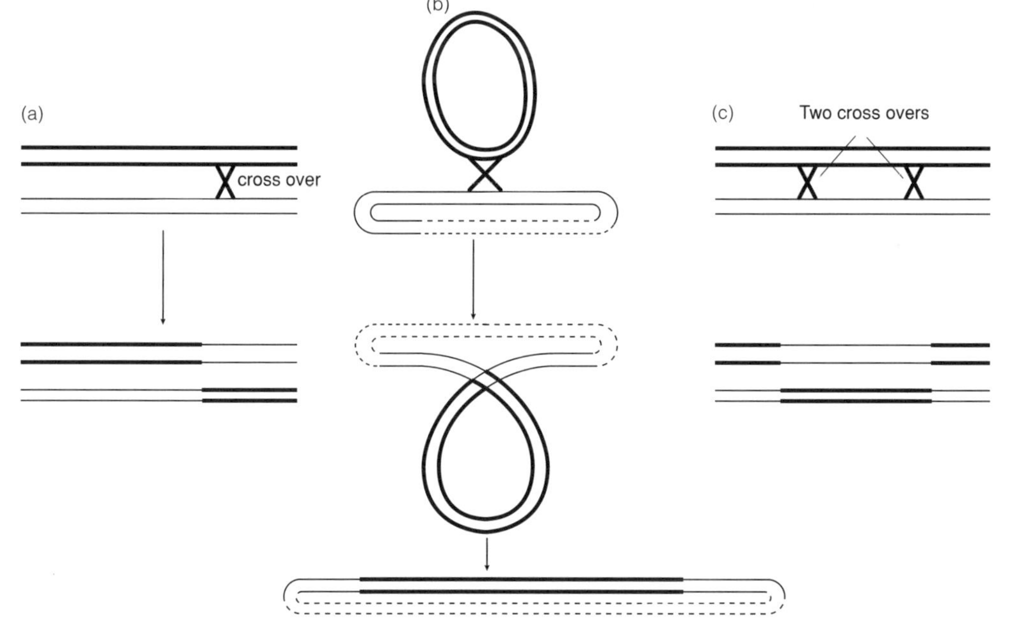

Fig. 3. The results of recombination between two molecules of DNA: (a) the joining of two linear molecules of DNA; (b) the joining of two circular molecules; (c) the integration of a DNA sequence.

Site-specific recombination

Site-specific recombination occurs in a number of specialist situations such as the integration of phage, such as λ, into the genome of bacteria (Topic E7). RecA protein is not required; the integration is mediated by a phage-encoded enzyme and a bacterial enzyme called integration host factor (IHF). These proteins interact with short homologous sequences, called *att* sites, found on the bacterial chromosome and on the phage molecule, forming a Holliday structure. Genetic exchange at these sites leads to integration of the phage into the genome (see *Fig. 3* in Topic E7). In a similar fashion, excision of phage occurs as a reverse of this process.

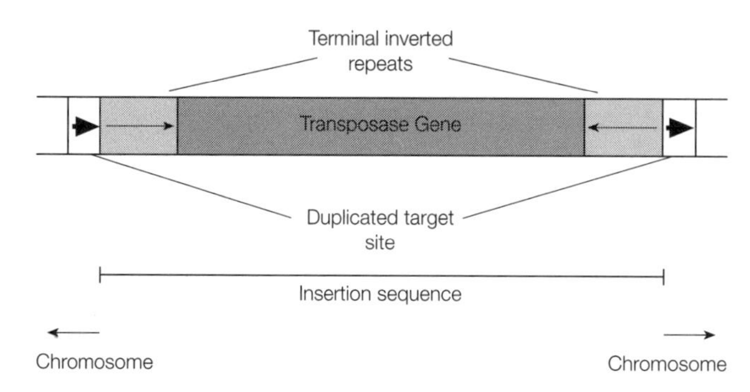

Fig. 4. The structure of an E. coli *insertion sequence showing the duplication of a chromosomal target sequence which occurs during transposition.*

Transposition

Transposable elements are mobile pieces of DNA that can move within a DNA molecule or between two DNA molecules at a low frequency (between 10^{-7} and 10^{-2}/generation). This movement does not require the presence of RecA protein and is pseudo-random. The ability to move is mediated by an enzyme called **transposase** which is carried on the transposable element. These elements are found in all kinds of organisms and have probably had a major role in the evolution of genomes by carrying genes between DNA molecules.

The simplest kind of transposable elements are **insertion sequences (IS elements)** found in bacteria. These are around 1 to 2 kb in size and consist of a stretch of DNA which encodes the transposase enzyme with a pair of inverted repeats, 9–41 bases in size, at either end (*Fig. 4*). These sequences can insert at random in the genome by a mechanism which leads to the duplication of a short sequence within the target site. There are a number of copies of the different IS elements in the genome and plasmids of bacteria, for example IS2 is present as five copies on the *E. coli* K12 chromosome and as one copy on its F plasmid (Topic E6).

A second group of transposable agents found in bacteria are called **transposons** which carry other genes as well as the transposase enzyme. Frequently, these genes are antibiotic resistance genes (Topic F7), which may explain why they spread rapidly through populations of bacteria leading to the appearance of strains that are resistant to a wide range of antibiotics. Transposons can be split into two groups:

- **Composite transposons** that are made up of a central region containing the additional genes between two copies of an IS element. Examples of these are Tn*10* which carries tetracycline resistance genes and Tn*1681* which codes for a heat stable enterotoxin.
- **Tn*3* type transposons** do not have IS elements but are capable of transposition by a mechanism which involves duplication of the element so that one copy remains on the original site and a new copy is inserted at another site. An example of this is Tn*3* itself which carries penicillin resistance genes.

Another interesting group of transposable elements, found in bacteria, are a number of bacteriophage typified by Mu and D108 which replicate by transposition (see Topic E8).

A new group of conjugative transposons have been found, mainly in Gram-positive bacteria. These transposons have the ability to transfer between individual bacterial cells (and across genera boundaries) via a plasmid-like intermediate (see Topic E6). These transposons frequently carry antibiotic resistance genes and may contribute to the rise in antibiotic-resistant bacteria in the population (Topic F7).

Finally, transposable elements are mutagenic. If they insert into the middle of a gene they will normally inactivate its function and therefore cause a mutation (Topics E1 and E2). They are often present on a genome as more than one copy and therefore can act as sites for homologous recombination. This can lead to various types of chromosome rearrangements such as duplications, inversions and deletions, which may cause large changes to the cell's genotype.

E4 DNA REPAIR MECHANISMS

Key Notes

Overview

As DNA damage is potentially lethal to the cell, a large number of DNA-repair systems have evolved to repair damage by mutagens. The best understood mechanisms are those that repair UV damage in *E. coli*; there are at least four different systems.

Photoreactivation

Photolyase converts the UV-induced thymine dimers back into monomers in the presence of light.

Excision (dark) repair

The uvrABC endonuclease recognizes and removes damaged nucleotides from the DNA. DNA polymerase I fills in the missing nucleotides and DNA ligase joins the gap. This mechanism of repair is used for other forms of DNA damage as well as UV damage.

Recombinational (post-replication) repair

If replication occurs before the DNA has been repaired, there will be gaps in the DNA due to the inability of DNA polymerase to insert bases opposite a pyrimidine dimer. These gaps can be repaired by recombination with the sister DNA molecule that has an intact DNA sequence.

SOS-repair systems

A large number of genes are induced as a result of DNA damage. Among these emergency repair systems is an error-prone repair system which causes DNA polymerase III to insert any base opposite a pyrimidine dimer, allowing replication to continue. The SOS system is induced by RecA protein, activated by interaction with damaged DNA, which, in turn, inactivates a repressor of transcription, LexA protein.

Related topics

Structure and organization of DNA (C1)
DNA replication (C2)
Control of gene expression (C5)
Mutagenesis (E2)

Overview

Errors during DNA replication and DNA damage, due to exposure to mutagens, happen all the time. Microorganisms have evolved very complex systems to recognize and repair the damage. In the best situation the repair directly reverses the original damage, causing no change to the base sequence, but, sometimes, the amount of damage is so great that DNA synthesis cannot continue. In these situations an **SOS-repair system** is induced that allows quick and extensive repairs but at the expense of accuracy, leading to the accumulation of mutations. The repair of UV damage in *E. coli* is the best understood and will be the main subject covered in this topic. UV light induces the formation of a cyclobutane pyrimidine dimer and a consequent distortion of the DNA helix (Topic E2). These dimers are unable to form normal base pairs so DNA synthesis stalls when DNA polymerase III reaches these adducts (Topic C2). A normal *E. coli* cell can cope with approximately 30 hits per molecule, and at least four different repair systems have evolved to cope with this damage and overcome the problem.

Photoreactivation

In a light-dependent repair mechanism, the enzyme photolyase, with the addition of flavin adenine dinucleotide (FAD), splits the cylobutane ring between the adjacent pyrimidines to restore the original monomers. This is the front-line defense against UV damage, and is specific for pyrimidine dimers.

Excision (dark) repair

Excision repair is often called dark repair to distinguish it from photoreactivation. It is a general system used to repair a number of different types of DNA damage including pyrimidine dimers and mis-matched base pairs. The products of the *uvr ABC* genes combine to form an endonuclease which recognizes distortions of the DNA helix caused by changes in the bases. The UvrABC endonuclease nicks the DNA on either side of the lesion (*Fig. 1*) and DNA polymerase I removes and replaces the bases. DNA ligase fills in the gap (Topic C2).

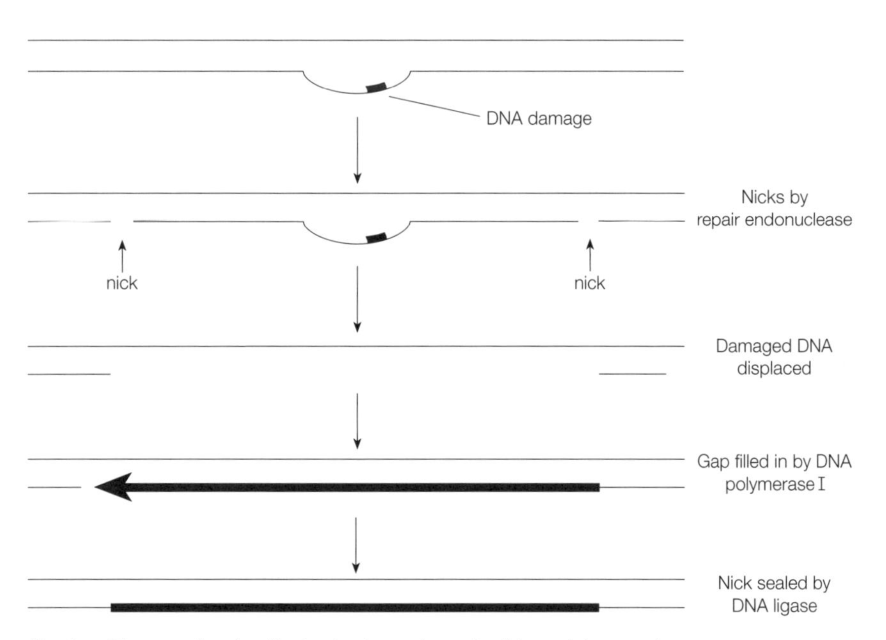

Fig. 1. Diagram showing the basic stages in nucleotide excision repair.

Recombinational (post-replication) repair

Pyrimidine dimers, formed by exposure to UV light, cannot be replicated. However, instead of stalling at these sites, DNA polymerase can jump over the lesion and resume DNA synthesis further down the DNA template. This leaves a gap in the newly synthesized DNA strand opposite the dimer. The gap cannot be repaired by excision repair as this requires an intact strand to be used as the template. Instead it can be filled by RecA-mediated recombination with the sister DNA helix, which contains the intact sequence that should be opposite the dimer (*Fig. 2a*). Although recombination does not repair the damage it does create the situation where both new DNA molecules can be repaired by excision repair.

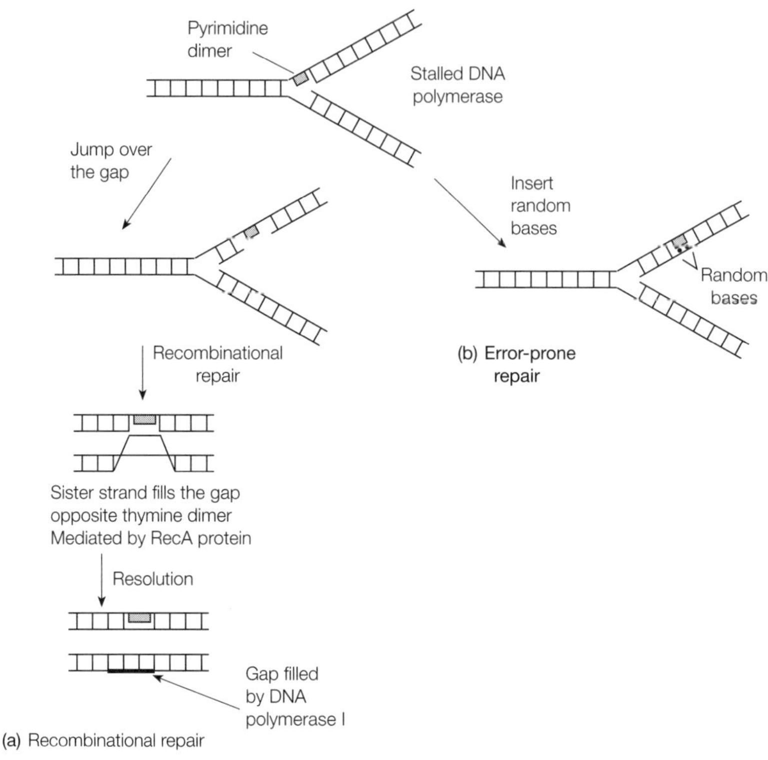

Fig. 2. Alternatives for coping with pyrimidine dimers during replication. (a) Recombinational repair and (b) error-prone repair of UV-damaged DNA during replication.

SOS-repair systems

There are a number of genes and operons in the cell whose products deal with DNA damage. Called the SOS-repair systems, they are coordinately regulated by RecA protein and a repressor of transcription, LexA (see Topic C5). Expression of SOS-repair genes is induced by the presence of DNA damage in the cell. These repair mechanisms include an error-prone repair mechanism which interacts with DNA polymerase allowing it to continue replication of DNA past a pyrimidine dimer (*Fig. 2b*). The 3′ → 5′ proof-reading ability of the enzyme is inhibited so that random bases can be inserted opposite the dimer without accurate base-pairing (see Topic C2).

The SOS-repair systems are induced by RecA protein which is activated by the presence of damaged DNA. The activated RecA protein causes LexA protein to be proteolytically cleaved so that it no longer acts as a repressor of transcription. The genes in the system may then be expressed as long as the DNA damage is present in the cell. Once the damage is removed, RecA protein is no longer activated as an inducer of LexA proteolysis, so LexA protein accumulates in the cell and inhibits transcription of the SOS-repair genes.

E5 Plasmids

Key Notes

Structure of plasmids

Plasmids are extra-chromosomal molecules of DNA that replicate independently of the bacterial chromosome. They are normally covalently closed, circular, supercoiled molecules but there are also instances of linear plasmids in both yeast and a few bacterial species. They range in size from about 2.2 to 210 kb.

Function of plasmids

Plasmids generally carry genes encoding functions that may be useful to the cell in certain circumstances but are not essential for normal cellular activities. Examples include antibiotic resistance genes, virulence genes and genes that allow bacteria to utilize unusual carbon sources such as catechol. Some plasmids are called cryptic as they do not carry any function detectable in the laboratory.

Replication of plasmids

Control of plasmid replication may be stringent, resulting in one or two copies of the plasmid per cell or relaxed, leading to multiple copies, in some cases as high as 40 copies per cell. Large plasmids tend to have low copy numbers.

Plasmid incompatibility

Plasmids that are similar to each other cannot survive in the same cell. This allows a useful method for classification of plasmids based on this incompatibility, called Inc groups. Plasmids in the same Inc group cannot be maintained in the same cell over a number of generations.

Conjugative plasmids

A number of plasmids have the ability to transfer themselves to other bacteria. These are called fertility (F) plasmids or resistance (R) plasmids, if they also carry genes for antibiotic resistance.

Related topics

Structure and organization of DNA (C1)
DNA replication (C2)
F plasmids and conjugation (E6)
Control of bacterial infection (F7)

Structure of plasmids

Prokaryotic cells often contain one or more types of extra-chromosomal DNA called plasmids. These are generally covalently closed, circular, supercoiled molecules of DNA that are capable of replication independently from the chromosome. To date a very large number of plasmids has been isolated from virtually every species of bacteria and a number of yeasts. Linear plasmids have also recently been found in *Borrelia*, *Streptomyces* and yeasts. The sizes of plasmids range from very small, an example being pUC8, an artificially constructed cloning vector, which is 2.1 kb (1.8 MDa in weight), to very large. The Ti plasmid of *Agrobacterium tumefaciens* is 213 kb (142 MDa) in size.

Function of plasmids

Plasmids frequently carry genes that are useful to the cell but are not essential for the normal growth and division. Examples of the different types of bacterial genes carried by plasmids are shown in *Table 1*. A number of plasmids have been isolated from bacteria which do not appear to have any function. These plasmids are frequently called **cryptic** plasmids.

Table 1. The types of genes that may be carried on plasmids and some examples of natural plasmids

Types of genes carried by plasmids	Plasmid	Host
Fertility – ability to self-transmit	F	*E. coli*
	RP4	*Pseudomonas* sp.
Resistance to compounds toxic to bacteria		
Resistance to antibiotics	RP4	*Pseudomonas aeruginosa*
	RP1	*E. coli*
Resistance to heavy metals	FP2	*P. aeruginosa*
Production of compounds toxic to other organisms		
Colicins (antibacterial compounds)	ColE1	*E. coli*
Antibiotic production	SCP1	*Streptomyces coelicolor*
Degradative enzymes that allow the breakdown of unusual carbon sources		
Toluene degradation	pWWO	*P. putida*
Camphor degradation	CAM	*Pseudomonas* sp.
Virulence factors – compounds that enhance a microbe's ability to infect and colonize a host		
Tumor production in plants	Ti	*Agrobacterium tumefaciens*
Adhesion to mucous membranes	K88, K99, CFA	*E. coli*
Enterotoxin production	Ent (P307)	*E. coli*
	pZA10	*Staphylococcus aureus*
Symbiosis	Sym	*Rhizobium* sp.
Self defense		
Restriction and modification enzymes	R1	*E. coli*
UV light resistance	Col1b	*E. coli*

Replication of plasmids

Very small plasmids use the host's replication machinery to make copies of themselves (Topic C2). All they need is an **origin of replication** to which DNA polymerase and the DNA synthetic machinery can bind in order for the plasmid to be replicated. Larger plasmids may carry additional genes whose gene products are necessary for that plasmid's replication. Some plasmids, called **episomes,** can integrate into the chromosome and are replicated when the chromosome is copied.

The number of copies of a plasmid in the cell varies depending on the mechanism that controls plasmid replication. Some plasmids are under a **relaxed** form of control which allows multiple rounds of replication of the plasmid before cell division, leading to the presence of a **high copy number** of the plasmid being present in the cell at one time (up to 40 copies). It is generally small plasmids, especially some of the genetically manipulated cloning vectors now available, that exist at these high numbers. Larger plasmids tend to be **stringently** controlled, ensuring that only one to three copies of the plasmid are present in the cell at one time. At cell division large plasmids are distributed to the daughter cells by attachment to membrane sites by a mechanism similar to chromosome segregation (Topic D7). Multi-copy plasmids tend to be distributed to progeny cells by random segregation.

Plasmid incompatibility

Similar plasmids cannot survive in the same cell unless a very strong selection pressure is maintained to ensure that they do. They are described as **incompatible**. The reason for this is still not fully understood but it is thought that incompatible plasmids might share a common mechanism to control replication, which means that after a few rounds of replication one of the plasmid types is lost from the cell. This incompatibility between plasmids has led to a useful classification system in which plasmids that cannot coexist with each other are placed in the same incompatibility group. There are at least 30 different incompatibility groups found in *E. coli*. On the other hand, plasmids from different incompatibility groups can exist in the same cell and up to seven different types of plasmid have been found in one *E. coli* cell.

Conjugative plasmids

An important group of plasmids is the conjugative plasmids, which have been found in a number of the Bacteria. These are plasmids which are capable of transferring themselves between prokaryotes by a mechanism called **conjugation**. At their simplest they are represented by the **F-factor** from *E. coli*, isolated originally because it caused the transfer of genes between cells (Topic E6). Other conjugative plasmids, such as the R plasmids, carry additional genes such as antibiotic resistance genes. The conjugative plasmids are generally large plasmids whose replication is stringently controlled with the exception of a plasmid, RK6, which is maintained as a relaxed plasmid. Many conjugative plasmids only transfer themselves between cells of the same species, but one important group is capable of transferring itself among a wide range of Gram-negative Bacteria. These are called **promiscuous** plasmids and are typified by the antibiotic resistance plasmid, RP4, from *Pseudomonas putida*. All R plasmids, but particularly the promiscuous ones, cause concern in a medical context because of their potential to spread antibiotic resistance genes through a wide range of clinically important Bacteria (Topic F7).

E6 F PLASMIDS AND CONJUGATION

Key Notes

Conjugation

Conjugation is the mechanism by which genetic material is transferred between two bacterial cells by plasmids. A cell containing a conjugative plasmid (F^+, fertility$^+$) forms a mating pair with a cell that does not contain a conjugative-plasmid (F^-) by means of an F-pilus on the surface of the cell. The pilus contracts, pulling the two cells into contact, and the DNA is transferred from the plasmid-containing cell, the donor, to the recipient. The *tra* genes, carried on the F plasmid, contain all the information for the conjugative process.

DNA transfer

The DNA that is transferred is single stranded and is synthesized by rolling-circle replication. The plasmid is nicked in one strand at a site called *oriT*, origin of transfer. The intact strand acts as a template for DNA synthesis while the nicked strand is displaced and transferred into the recipient cell. Once inside the recipient cell, a complementary strand of DNA is synthesized to form a new F factor.

Integration and Hfr strains

F factors can integrate into specific sites in the *E. coli* chromosome by homologous recombination between insertion sequences found on the chromosome and the plasmid. Even though integrated, the plasmid is still capable of initiating the rolling-circle mechanism of replication at *oriT,* followed by the transfer of single-stranded DNA into a recipient. The whole chromosome therefore acts like a large F plasmid, allowing the transfer of chromosomal genes to another bacterium. Genes closest to the plasmid are transferred first. In the recipient, the donor chromosomal DNA is integrated into the recipient chromosome by recombination. Strains which have an F plasmid integrated into the chromosome are called Hfr (high-frequency recombination) strains.

F′ factors

F plasmids integrated in the chromosome normally excise themselves as precisely as they entered to give an intact F plasmid. However, in some cases, the excision is not precise and a piece of the chromosome is integrated into the F plasmid. These are called F′ factors.

Related topics

DNA replication (C2)
Recombination and transposition (E3)
Plasmids (E5)

Conjugation

An important mechanism by which DNA can be transferred between cells is **conjugation**, mediated by some plasmids, which requires direct contact between cells. Conjugative plasmids, typified by the F plasmid of *E. coli,* have the ability to transfer themselves between cells and, in some cases, also to transfer pieces of chromosomal DNA. F plasmids and their relatives carry a group of genes called the *tra* genes which encode all the proteins required to form a mating pair with another bacterium that does not contain an F plasmid. DNA is then transferred from the plasmid-containing cell, called the donor or F^+ cell, to the recipient, F^- cell. A map of the F plasmid of *E. coli* is shown in *Fig. 1*. The plasmid is large, 95 kb in size, and normally exists as one or two copies in the cell. Vegetative replication of the plasmid at cell division is by normal semi-conservative, bi-directional DNA synthesis from *oriV* followed by segregation of the plasmid into the daughter cells (see Topics C2 and E5). The F plasmid also contains a number of insertion sequences (Topic E3) distributed throughout the molecule (*Fig. 1*).

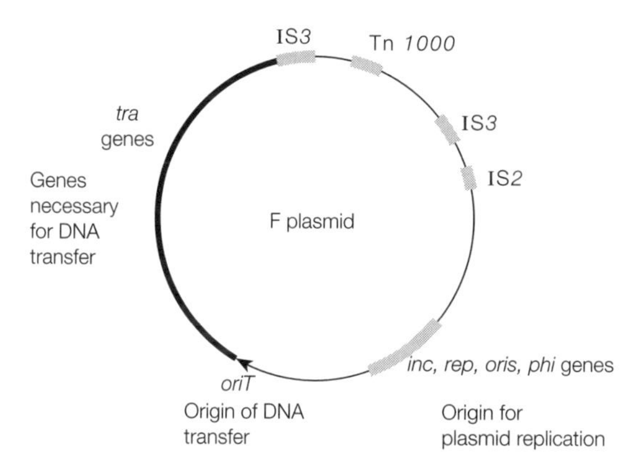

Fig. 1. A simplified map of the F plasmid of E. coli.

Cells containing F plasmids (F^+) produce specialist F-pili on their surface which are encoded by the *tra* genes. These are proteinaceous, filamentous structures that attach to the surface of F^- cells to form a mating pair (*Fig. 2a*). Mating pairs between two F^+ cells cannot be formed due to a phenomenon called **surface exclusion**, mediated by two *tra*-gene encoded proteins located in the outer membrane of F^+ cells. Once a mating pair forms, the F-pilus retracts, pulling the two cells into close contact.

DNA transfer

A single-stranded copy of the F plasmid is transferred into the recipient cell. It is produced by rolling-circle replication (see also Topic C2). A single-strand nick is made in the DNA at a site called *oriT* (origin of transfer) which is located beside the *tra* genes (*Fig. 1*). The 3′ end of the nicked DNA is used as a primer to start synthesis of a new strand of DNA, using the intact strand as a template by rolling-circle replication (see *Fig. 6*, Topic C2). The nicked strand is displaced and transferred through a pore into the recipient cell (*Fig. 2b*). Synthesis of a complementary strand of DNA occurs in the recipient cell. The order of transfer of the F plasmid is such that the last genes to be transferred are *tra* genes. The plasmid circularizes in the recipient to give an F^+ cell and the mating-pair separates (*Fig. 2c*).

Integration and Hfr strains

As mentioned previously, the F plasmid has a number of insertion sequences scattered about the molecule, as does the *E. coli* chromosome. Occasionally, recombination between these homologous sequences leads to the integration of the F plasmid into the genome at specific sites (*Fig. 3*). Strains that have an F plasmid integrated into the chromosome are called **Hfr** strains (standing for

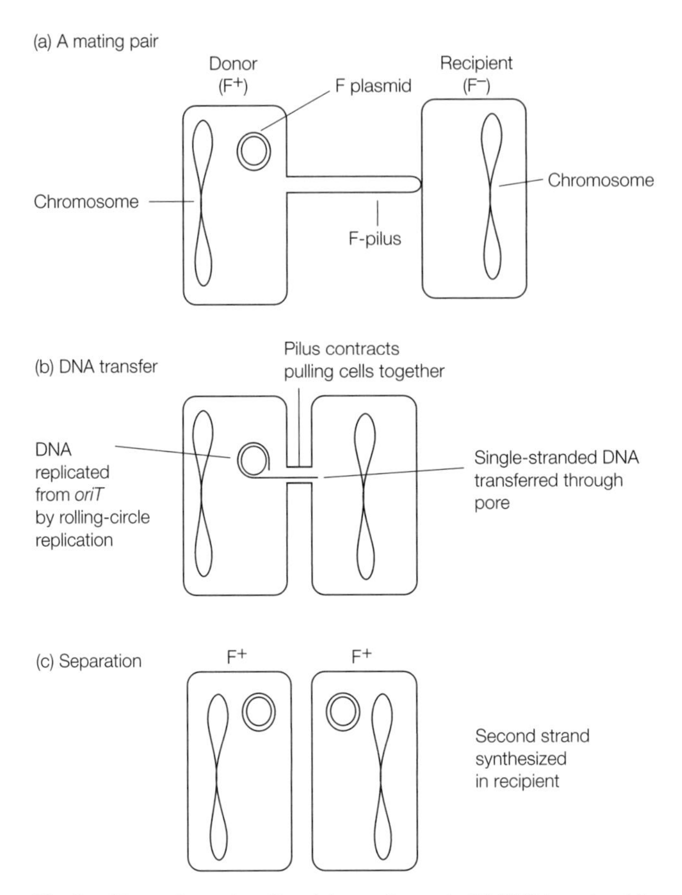

Fig. 2. Stages in conjugation: (a) a mating pair; (b) DNA transfer; (c) separation.

high-frequency recombination for the reasons described below). The F plasmid, once integrated, is still capable of forming mating pairs and initiating transfer replication at *oriT*. However, in this case, the chromosome plus plasmid acts like a large F factor; so F-plasmid DNA followed by the *E. coli* chromosome is transferred into the recipient (*Fig. 4*). Chromosomal genes closest to the *oriT* are transferred first. The order of transfer can be clockwise or anticlockwise, depending on the orientation in which the F plasmid is inserted. The number of genes transferred depends on how long the mating pair is held together but, in theory, it takes 100 min to transfer the whole *E. coli* chromosome. The last genes to be transferred are the *tra* genes. In practice, the mating pairs break at random and, normally, only part of the DNA is transferred. The incoming single-stranded, chromosomal DNA cannot recircularize to form a plasmid or a second chromosome. However, it can integrate into the recipient chromosome by homologous recombination (Topic E3), thus allowing the transfer of genes from donor to recipient. This is why the strains are named high-frequency recombination strains as they allow the transfer of chromosomal genes from one strain to another at high frequency. Hfr strains are used in genetic mapping for determining the position of genes on the chromosome of *E. coli* as the order of genes can be determined by the time at which they enter a recipient during mating.

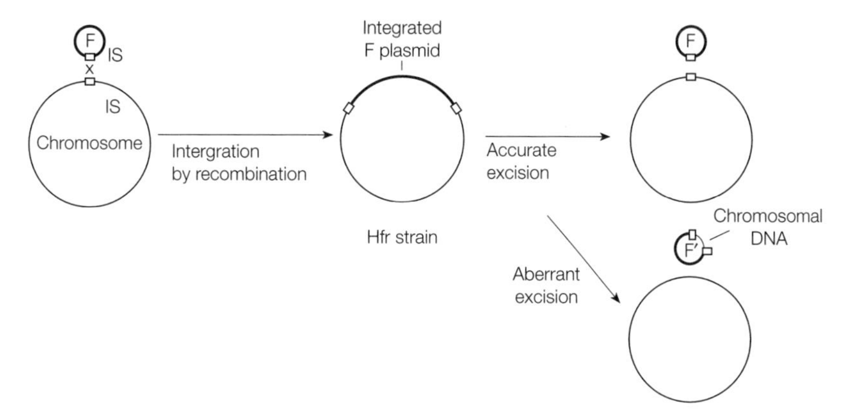

Fig. 3. Integration and excision of F plasmids from the chromosomes showing the production of F′ factors by aberrant excision.

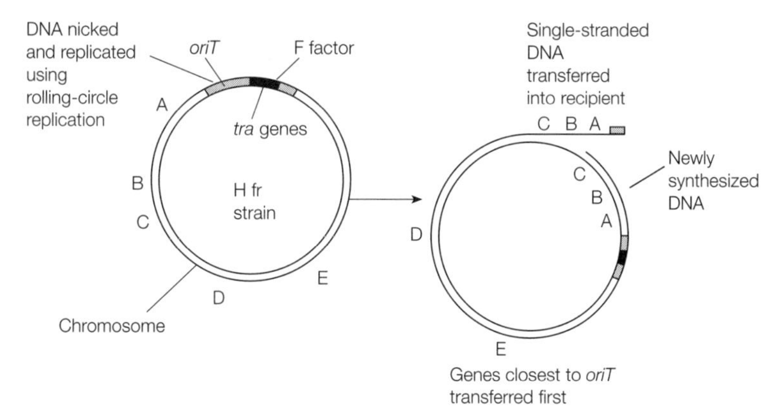

Fig. 4. Transfer of chromosomal genes to a recipient by Hfr strains.

F′ factors

F factors excise themselves from the chromosome by a process which is the reverse of integration. Normally this is an accurate process but occasionally an error is made and the plasmid picks up a piece of the adjacent chromosome to form an F′ factor (*Fig. 3*). These F′ factors can be used to transfer chromosomal genes into a recipient cell and are used by Bacterial geneticists to create partial diploids of genes in a cell in order to investigate dominance/recessive relationships between alleles of the same gene.

E7 Bacteriophage

Key Notes

Introduction

Bacteriophage, normally called phage, are viruses that infect prokaryotes. They are obligate intracellular parasites that are capable of existence as phage particles outside the host cell but can only reproduce inside the cell. They consist of a nucleic acid genome surrounded by a protein coat called a capsid. Phage possess the ability to infect a prokaryotes and redirect the cell to synthesize phage components.

Structure of bacteriophage

Phage, like viruses, have a number of shapes. They may be icosahedrons, almost spherical shapes consisting of 20 triangular faces, or filamentous or complex structures consisting of icosahedral heads with helical tails. The genome can be DNA or RNA, single-stranded or double-stranded, circular or linear. Some phage are very small such as the *E. coli*, single-stranded RNA phage, MS2, whose genome contains information for only four proteins, while others are large like T4 whose genome encodes around 135 different proteins.

A typical phage life cycle

A typical life cycle of a phage starts by the adsorption of the phage on to a specific receptor on the bacterial cell surface followed by penetration of the genetic material into the host. Prokaryotes contain restriction-modification systems that are designed to degrade foreign DNA, so many infections are unsuccessful. Enzymes encoded by the phage genome are synthesized, which is followed by nucleic acid replication. Finally, phage capsid proteins are made and assembled into a new phage coat, at the same time packaging a copy of the genome. The phage are then released, normally by lysis, into the surrounding medium.

Lysogenic life cycle

Some phage, a typical example being phage λ, instead of starting the lytic cycle on entry into the cell, produce repressor proteins that stop phage replication. The phage then enters a state called lysogeny in which its genome is replicated at the same time as the host chromosome and is passed from one generation to the next. The phage may be induced from this lysogenic state spontaneously and enter into the lytic cycle. In order to replicate with the host, most lysogenic (sometimes called temperate) phage integrate into the chromosome to form a prophage but some, like P1, exist in the cytoplasm in a form similar to plasmids.

Growth and assays for phage

The number of infectious phage particles can be measured using a plaque assay to count the number of plaque-forming units (pfus) in a sample. A plaque is a clear area in a turbid lawn of bacteria, where phage released by a bacterium have gone on to lyse bacteria in the surrounding area. A plaque is therefore equivalent to one phage in the original sample.

Related topics

Control of gene expression (C5)
Recombination and transposition (E3)
Replication of bacteriophage (E8)
Transduction (E9)
Virus structure (J1)

Introduction

Bacteriophage, normally shortened to **phage**, is the name given to viruses that infect prokaryotes. They are genetic entities consisting of a nucleic acid genome and a protein coat, called a **capsid**, but they are not cellular. Although capable of existence as particles outside a cell, the phage genetic material must enter a cell in order for it to replicate. Bacteriophage are therefore obligate, intracellular parasites of prokaryotic cells. There are many types of bacteriophage, which specifically infect different types of prokaryotes. This specificity of infection is based on the presence of receptors for that phage on the surface of the prokaryote (the **host**). Once inside the host, the phage genome directs the synthesis of components for new phage particles using the host's biosynthetic machinery. These new phage particles are then assembled and released. This is called the **lytic** cycle.

Bacteriophage are easy to handle and have made good model systems for studying viruses in general. One does not need expensive tissue culture, or to work with whole animals and plants, to study phage organization, replication and assembly. Phage have always been important tools for microbial geneticists for moving genes around and for gene mapping by transduction (Topic E9). It is important to note that most of our knowledge is about *Escherichia coli* phage, but phage do infect most other types of prokaryote.

Structure of bacteriophage

Bacteriophage are made up of a nucleic acid genome surrounded by a protein coat called a capsid whose role is to protect the genetic material and to aid in the infection of a new host. Some phage may also carry additional enzymes with the nucleic acid inside the capsid. The coat is made up of protein subunits arranged in highly ordered structures which give the phage a distinctive shape. The number of different types of protein that go to make up the capsid may range from one in simple phage like MS2 to >25 in a complex phage like T4.

There are three morphological shapes associated with bacteriophage structure (*Fig. 1*):

Icosahedron

This is an almost spherical shape consisting of 20 triangular faces. Icosahedrons are very common shapes in nature as they are very efficient ways to create an enclosed shell from subunits.

Filamentous

These are long protein tubes formed by capsid proteins assembled into a helical structure.

Complex

Some phage consist of icosahedral heads attached to helical tails. In the case of a group of phages called the T-even phages the head is actually an elongated icosahedron. The tails can be contractile or non-contractile, sheathed or non-sheathed and they may also have base plates and tail fibers associated with them. The role of the tail is to assist the injection of the genetic material into the cell.

Bacteriophage range in size from about 25 nm (icosahedron diameter) for MS2 to 110 nm × 85 nm (head size) for T4 to which is attached a 25 nm x 110 nm tail. The filamentous phage such as M13 have measurements in the region of

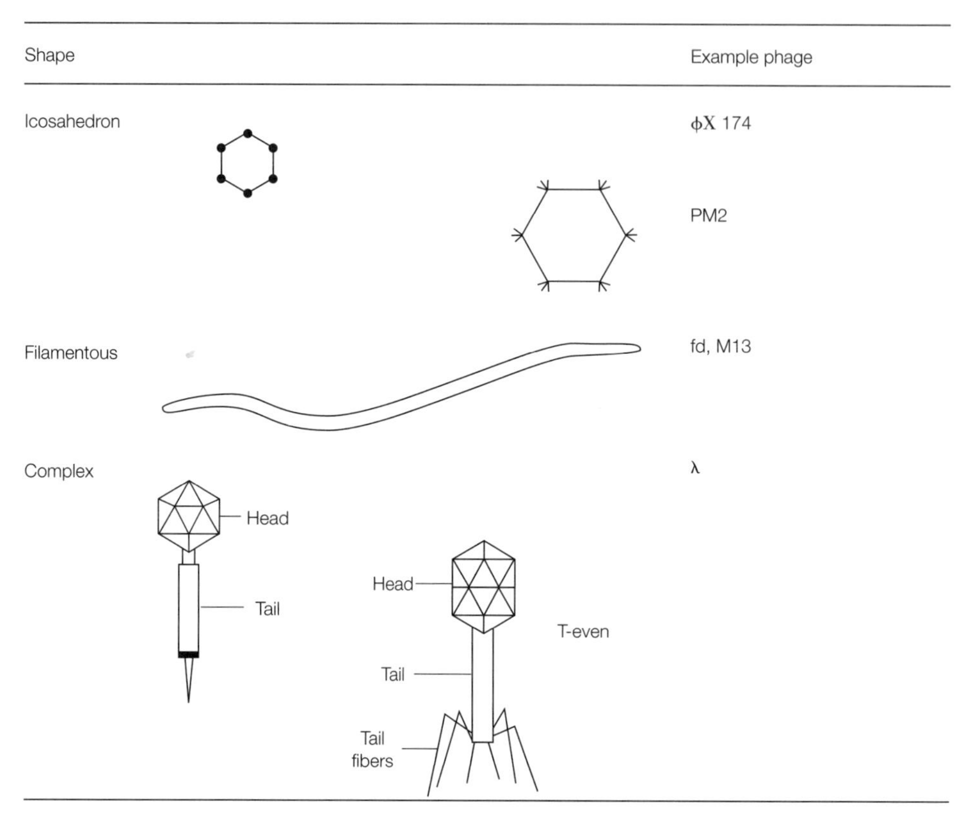

Fig. 1. Typical bacteriophage shapes (not drawn to scale).

6 nm wide by 860 nm long. A few bacteriophage, such as φ6, have membrane envelopes but, in general, they are not as common as in animal viruses (Topic J1).

The genetic material inside the phage capsid may be RNA or DNA, which can be single- or double-stranded, linear or circular (*Table 1*). The mechanism by which these different types of nucleic acid molecules are replicated is described in Topic E8.

Table 1. The nature of phage genomes

Nucleic acid type		Structure	Example
DNA	Single-stranded	Circular	φX174 M13
	Double-stranded	Linear Circular	T phage PM2
RNA	Single-stranded	Linear	MS2
	Double-stranded	Linear	φ6

A typical phage life cycle

Although there are differences between the life cycles of bacteriophage they all tend to follow a similar pattern, which is outlined below and in *Fig. 2*.

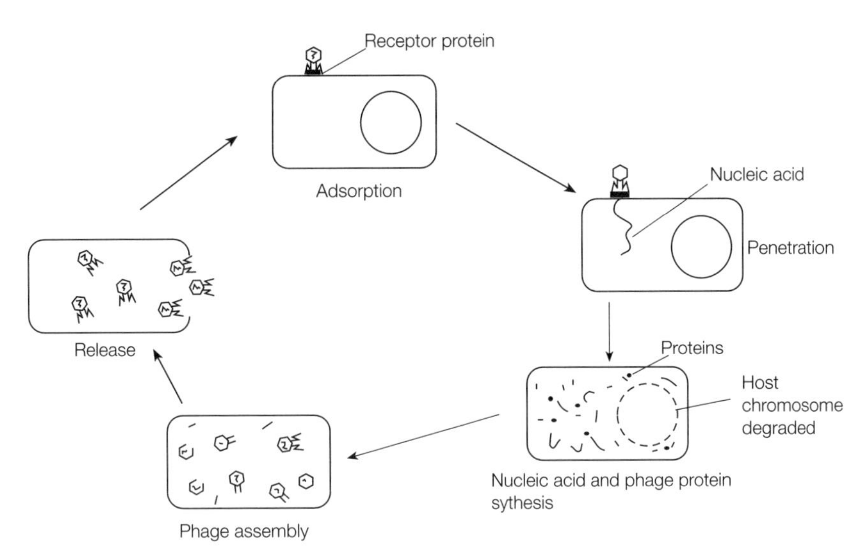

Fig. 2. The life cycle of a typical lytic phage.

- **Adsorption (attachment).** Phage infection of a host starts by attachment to specific receptors on the surface of the cell. These receptors vary in nature, being proteins or polysaccharides, which may be present on the surface all the time or only produced under certain conditions. For example, phage T4 binds to lipoproteins in the outer membrane of *E. coli* whereas phage λ binds to a maltose-transport protein only present in the outer membrane when bacteria are grown in maltose-containing medium.
- **Penetration.** Generally, only the genetic material is injected into the host, leaving the empty protein coat on the surface of the cell. Lysozyme associated with the phage head is used to dissolve the peptidoglycan (Topic D3). The mechanism of penetration varies from phage to phage depending on its structure. In the case of T4, which has a contractile tail, a highly complex mechanism has evolved. After attachment by its long tail fibers and contact of the base plate with the cell wall, the tail sheath contracts and the DNA is injected into the cell. In many cases the incoming DNA may be degraded by endonucleases, which are part of the host's **restriction-modification** defense systems, but occasionally DNA molecules will survive to set up an infection. These defense endonucleases, called **restriction enzymes,** bind to specific sequences in foreign DNA and cut the DNA at these sites. The microbe's own DNA is modified, normally by methylation, to prevent it being degraded by these restriction enzymes. Some phage such as the T-even phages have evolved mechanisms to modify their own DNA by glycosylation or methylation to prevent them being degraded by restriction enzymes on injection.
- **Nucleic acid replication.** Once inside the host the nucleic acid is transcribed and translated (or just translated in the case of RNA viruses) to produce enzymes that direct the synthesis of new phage nucleic acids. These are sometimes called **early proteins.** Many phage switch off the synthesis of host proteins and degrade the host genome, thus ensuring that all the cellular biosynthetic machinery is directed to the production of phage components. After a short time there is usually a switch to the production of large

amounts of phage structural proteins, scaffold proteins required for phage assembly and proteins required for lysis and phage release. These are referred to as **late proteins.**

- **Phage assembly**. Once sufficient phage capsid components and nucleic acid has been synthesized the new phage particles are assembled spontaneously while, simultaneously, packaging the nucleic acid into the capsid.
- **Release**. Many phage are released by lysing the prokaryote cell wall which is why this life cycle is often called the lytic cycle and the phage are referred to as virulent phage. Enzymes soften the cell wall and the phage are released as bursts of between 50–1000 phage per cell. The time from infection to release is about 22 min at 37°C for T4. Other phage, such as the filamentous phage M13, release phage through the cell wall without damaging the cell so phage particles can be released over a long period of time. The host cells continue to grow but at a reduced rate.

Lysogenic life cycle

A number of phage, typified by phage λ, have an alternative to a lytic life cycle inside the host. On entering the host these phage do not enter the lytic cycle. Instead they enter into a state called **lysogeny** in which they either integrate into the host chromosome to form a **prophage** (e.g. phage λ) or exist in the cytoplasm like a plasmid (e.g. phage P1). Consequently, these lysogenic (sometimes called **temperate**) phage replicate at the same time as the host chromosome and are passed on to daughter cells (*Fig. 3*). The protein in λ that is responsible for the

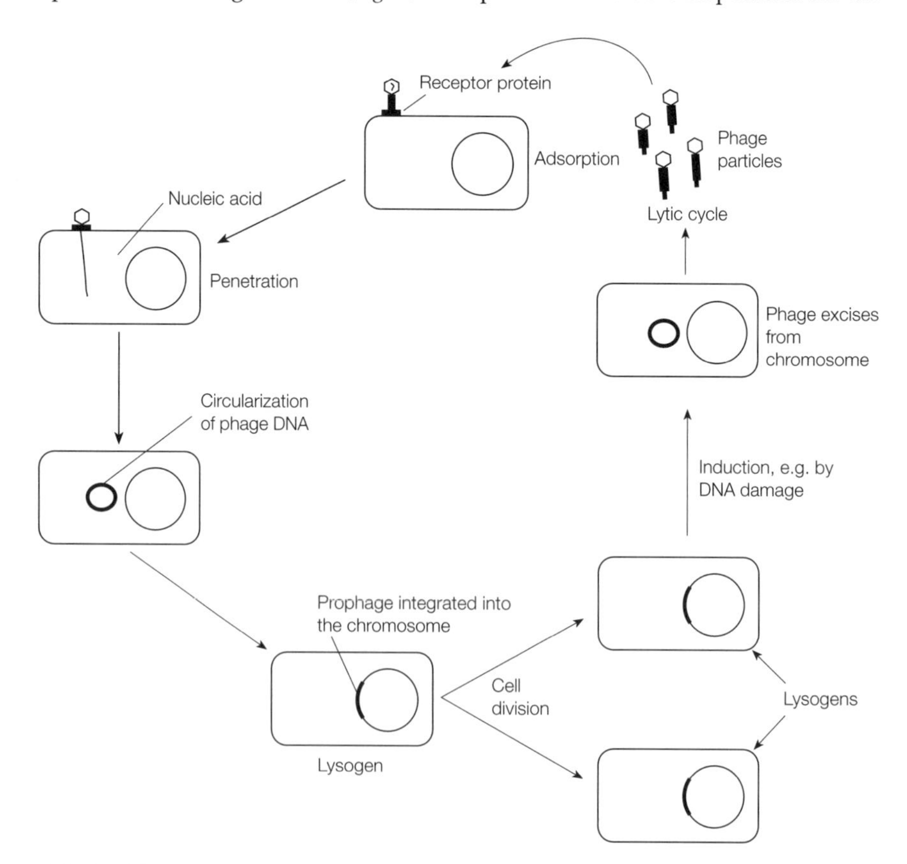

Fig. 3. The life cycle of a lysogenic phage.

phage entering into lysogeny is a repressor, produced by the *cI* gene, which prevents expression of all the phage genes necessary for phage replication and assembly. Instead, the phage enters the lysogenic state.

- The genome (which is linear) circularizes due to the presence of complementary single-stranded ends called **cos** sites.
- An integrase protein, the *int* gene product, mediates site-specific recombination (*Fig. 3* in Topic E3) between a site on the phage genome, *attP*, and a homologous site, *attB*, on the *E. coli* genome, which leads to phage DNA integration into the chromosome.

The lysogenic state can be interrupted spontaneously at any time, but especially when there is induction of the SOS response by DNA-damaging agents such as UV light. Transcription of the genes for phage production is induced. The phage is excised from the chromosome by a mechanism that is the reverse of integration mediated by the product of the *xis* gene, and λ then enters the lytic cycle.

Lysogenic phage are of particular interest because they can carry functional genes which can change the phenotype of the host. This is called **lysogenic conversion** and the properties the lysogenic phage confer often contribute to the ability of a prokaryote to survive in a host (Topic F5). Epsilon phage of *Salmonella* alter the O-antigen portion of the bacterial LPS, thus changing the antigenic properties of the bacterium. A number of Bacterial toxin genes are also carried on temperate phage including diphtheria toxin carried by phage β of *Corynebacteria diphtheriae* (Topic F6). Examples of phage conversion in this category are being discovered regularly; cholera toxin being a comparatively recent edition.

Bacteriophage λ has proved to be an extremely useful phage for gene cloning. The central region of its genome contains genes that are essential for lysogeny but not required for lytic growth. This region can therefore be replaced with up to 20 kb of foreign DNA without interfering with phage replication in the lytic cycle, making it a good system for producing large amounts of foreign DNA. This made it an excellent cloning vehicle in the past, although its use has now been superceded by other vectors.

Growth and assays for phage

Bacteriophage are grown in the laboratory in either liquid or solid culture. Prokaryotes and phage are mixed together at a ratio where the number of phage is considerably less than the number of cells. The mixture is incubated, allowing the cells to replicate and become infected with phage. The infected bacteria then lyse, releasing large numbers of phage. Consequently, the number of phage increases at a much greater rate than the cells until at a critical point the number of phage present is enough to infect all the cells in the culture. All the cells therefore lyse virtually at the same time, which can be seen as a complete reduction in the turbidity of the broth culture (see Topic D8). Alternatively, a soft agar is added to the cells and phage mixture, and this is poured on to the surface of an agar plate. The cells grow and the phage infect as described above until, eventually, the cells are all infected, causing **confluent lysis** of the bacterial lawn.

The plate method of growth is also used to measure the number of infectious phage particles in a preparation. The phage preparation is diluted, mixed with cells and, after a short time to allow adsorption of the phage on to the cells, soft agar is added and the mixture is poured on to the surface of an agar plate. The infection of a cell by a phage particle causes the production of a large number of phage, which can infect adjacent cells. When these cells lyse, yet more phage are released which infect even more cells. The uninfected cells continue to grow to form a lawn of organisms. After overnight growth, the foci where a prokaryote was originally infected by a phage is seen as a clear area of lysis called a **plaque** in a turbid lawn of bacteria. The number of plaques in the lawn indicates the number of infectious phage in the original preparation. This number is measured as plaque-forming units (pfu).

E8 Replication of bacteriophage

Key Notes

RNA phage

Most RNA phage contain single-stranded linear molecules. The RNA is a positive strand that acts as mRNA as well as the genetic material. A phage-encoded RNA replicase synthesizes a complementary copy (negative strand) of the RNA which acts as a template for the production of new RNA genomes and mRNA. Only one double-stranded RNA phage is known but little is known about how it replicates.

Single-stranded DNA phage

Single-stranded DNA phage may be icosahedral phage, such as φX174, or filamentous, such as M13. On entry into the cell, the DNA is converted into a double-stranded replicative intermediate called an RF form. Multiple copies of the RF form are produced by bidirectional replication. These then act as templates for RNA synthesis and for the synthesis of new single-stranded genomes by rolling-circle replication.

Double-stranded DNA phage

Linear DNA from T-phage is synthesized by bidirectional, semi-conservative replication. The problem of synthesis of the 5′-ends created by this mechanism is solved by the formation of concatamers of phage DNA by base-pairing between terminal direct repeats. Phage-length genomes are then cut by endonucleases. λ phage circularizes in the host and replicates by the rolling-circle method.

Related topics

DNA replication (C2)
Recombination and transposition (E3)
Bacteriophage (E7)
Transduction (E9)

RNA phage

All single-stranded RNA phage contain positive (template)-strand RNA which means that it can act as mRNA as soon as it enters the cell. They are very small phage, containing a minimum of genetic information. MS2 is a typical example; it infects *E. coli* by binding to the F-pilus, which is present on the surface of cells containing an F plasmid (Topic E6). The 3569-nucleotide genome has been sequenced and encodes four proteins: a **replicase** enzyme, a coat protein, a maturation protein and a lysis protein. In order to fit these into the genome, the proteins are overlapping (*Fig. 1*), which is a common feature of phage genomes. The enzyme that replicates the RNA is an RNA-dependent RNA polymerase which is actually made up of the phage-encoded replicase and polypeptides produced by the host. The events during replication of MS2 are shown in *Fig. 2*. The replicase makes a copy of the RNA strand called a **negative strand**. This then acts as a template for the production of new positive strands of RNA which can act as mRNA or, once enough phage components have been made, as new phage genomes. The RNA is extensively folded and packaged with the coat

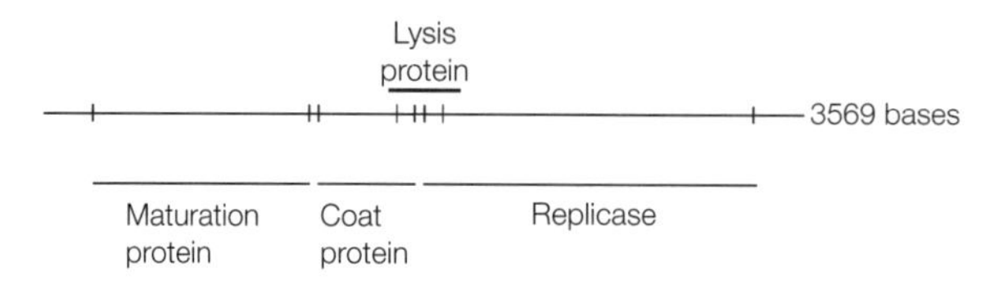

Fig. 1. Genetic map of MS2.

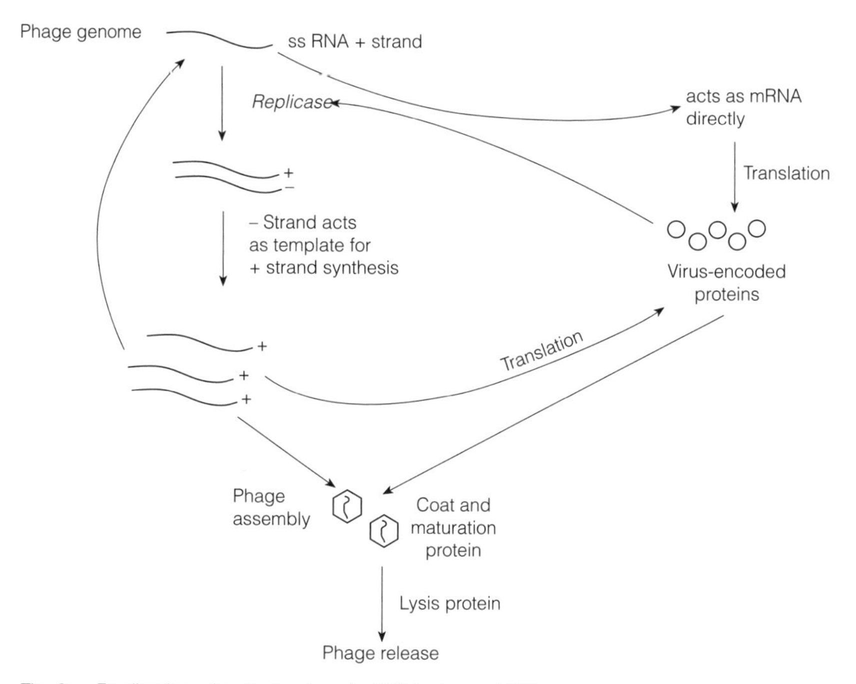

Fig. 2. Replication of a single-stranded RNA phage, MS2.

protein (180 copies) and the maturation protein (one copy). The lysis protein then causes the cell to burst, releasing the new phage particles.

Single-stranded DNA phage

Single-stranded DNA phage are typified by the icosahedral phage φX174 and the filamentous phage M13, both of which have circular genomes. Although the DNA in these phage is replicated by similar mechanisms, their life cycles are very different, as φX174 is a lytic phage whereas M13 can leave its host without killing the cell. In both cases, however, on entry into the cell the single-stranded DNA must be converted into a double-stranded form for replication to occur. This double-stranded intermediate is called the RF form, which is then replicated to produce a number of copies. These act as templates for the synthesis of mRNA and new phage proteins and also for the production of new single-stranded molecules of DNA by rolling-circle replication (see *Fig. 6*, Topic C2). The new phage-length strands of single-stranded DNA are cut off and ligated into a circle, packaged and released from the cell either by lysis (φX174) or by extrusion through the membrane (M13). Because of the nature of its replication, M13 has been used as a cloning vector for the production of single-stranded DNA to be used for DNA sequencing, as DNA can be inserted into the RF form and then subsequently purified in a single-stranded form from the progeny phage. At one time, most DNA sequencing was carried out using M13 in this way, but improved sequencing techniques using Taq polymerase have become common and can be used with double-stranded DNA plasmids.

Double-stranded DNA phage

Replication of the T-phage group, which contains linear DNA molecules, proceeds bidirectionally from a single origin of replication. The problem for linear molecules is how to replicate the last bit of DNA at the 5′-end of each strand which is left single-stranded due to the nature of DNA replication (Topic C2). This is overcome by a number of strategies. In T7 there is a direct repeat of 160 bases at the ends of the molecule (*Fig. 3*). This allows base-pairing between the single-stranded DNA portions on two separate DNA molecules to form a **dimer** twice as long as the original strand. The non-replicated portions can then be filled in by DNA polymerase and ligase. Other DNA molecules are joined until eventually a long **concatamer** is formed. A phage endonuclease cuts the

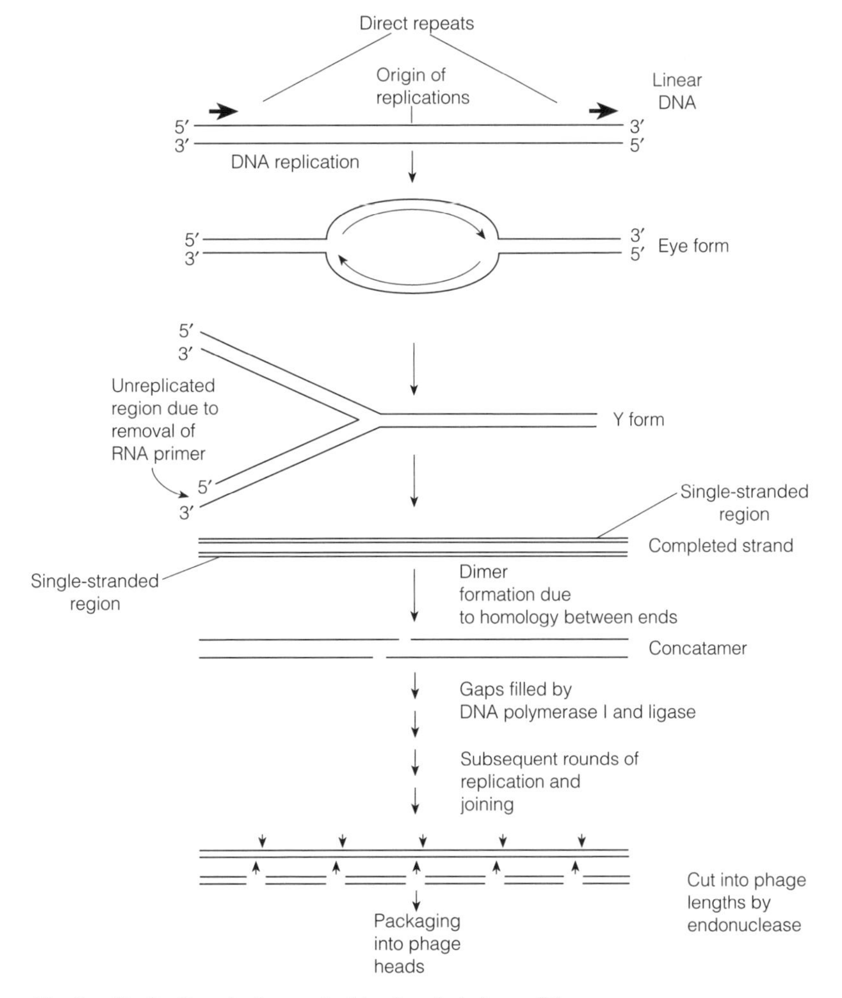

Fig. 3. Replication of a linear, double-stranded phage, T7.

concatamer at a specific site to produce phage lengths of DNA which can be packaged into heads. In the case of T4 a slightly different strategy is adopted in that the enzyme cuts the concatamer into lengths that will fit into the phage head but not necessarily at the same point in the DNA. The molecule packaged into the head has actually more than a genome's worth of DNA, and although all the genes will be present, the ends of the molecules will be different between different phage particles (*Fig. 4*). This phenomenon is called **circular permutation**.

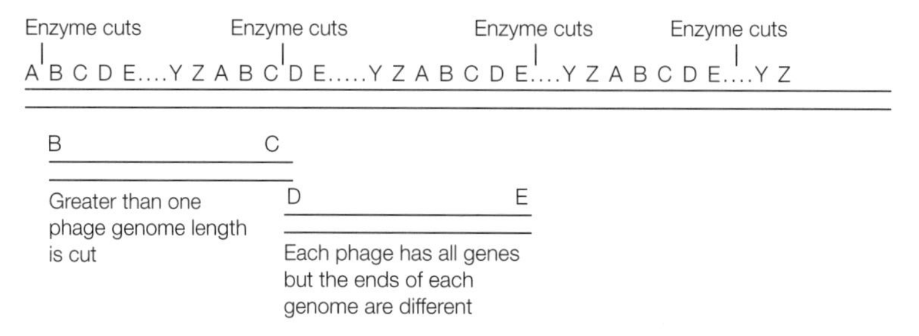

Fig. 4. The cutting of replicated T4 DNA for packaging into phage heads. Each head contains more than one genome.

In phage λ, DNA synthesis of the linear DNA proceeds via a circular intermediate formed by base-pairing between *cos* single-stranded overhangs on the DNA (Topic E7) and new double-stranded molecules are produced using the rolling-circle method of replication (see *Fig. 6*, Topic C2).

E9 TRANSDUCTION

Key Notes

Transduction — Transduction is the movement of host DNA from one bacterial cell (donor) to another cell (recipient) by a bacteriophage.

Generalized transduction — Generalized transduction is the result of a rare error during phage assembly, leading to the packaging of a piece of host genome into the phage head instead of the phage DNA. Such transducing phage are capable of infecting new hosts and injecting in the DNA but, as this DNA is not phage DNA, it will be lost unless the DNA is integrated into the host genome by recombination.

Specialized transduction — This is a feature of some lysogenic phage which spend some time integrated into sites in the host chromosome. At induction, instead of the phage genome accurately excising from the chromosome, it excises a piece of adjacent chromosome as well. Phage particles containing these genes will infect recipient cells where the whole phage may integrate into the chromosome to form a lysogen, or the chromosomal DNA may be integrated by recombination.

Related topics — Recombination and transposition (E3); Bacteriophage (E7)

Transduction

A mechanism by which DNA can be moved from one cell to another is through the agency of a phage particle; this is called **transduction**. Phage can pick up pieces of chromosomal or plasmid DNA from one bacterium, called the **donor**, and transfer it to a **recipient** bacteria where it may be integrated into the recipient genome by recombination (Topic E3). Transduction has been found in a wide range of Bacteria and phage and is thought to play a significant role in the transfer of genetic information in nature. Transduction is also an important tool for geneticists as it allows the transfer of genes between Bacteria and the mapping of Bacterial genomes; genes that are close together will be transferred to a recipient cell together. Many of the original uses of transducing phage have been replaced by gene cloning, but they are still useful tools today for the creation of bacterial strains and the analysis of genomes.

Generalized transduction

Generalized transduction is the transfer of any piece of host DNA to a recipient by a phage particle. This is due to the fact that some phage, at the point of phage assembly, will accidentally package a phage-size piece of host DNA into the phage head instead of the phage genome (*Fig. 1*). These transducing phage, on release from the host, are capable of attaching to another cell and injecting the DNA into this recipient. The DNA, however, is not phage DNA and is not capable of replicating. It will therefore only survive if it is integrated into the host chromosome by recombination (*Fig. 2*).

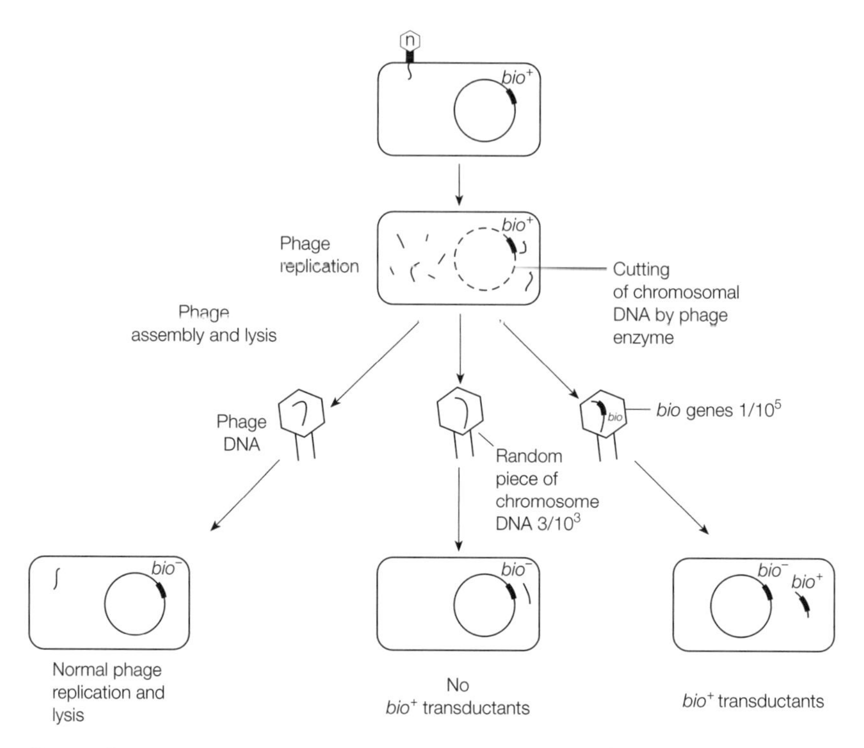

Fig. 1. Generalized transduction showing the transfer of bio genes to a recipient.

Frequently, especially if the donor and recipient are closely related, the new DNA will not make a significant difference to the recipient, but if the donor and recipient have different genotypes, a gene, or genes, of the recipient cell may be converted to that of the donor. Transduction is easy to carry out in the laboratory. A phage preparation is added to a culture of the donor strain which is allowed to incubate until complete lysis has been achieved (Topic E7). The phage **lysate** containing a few transducing phage is then used to infect a recipient culture of bacterium and this is plated out on agar plates. Bacteria that

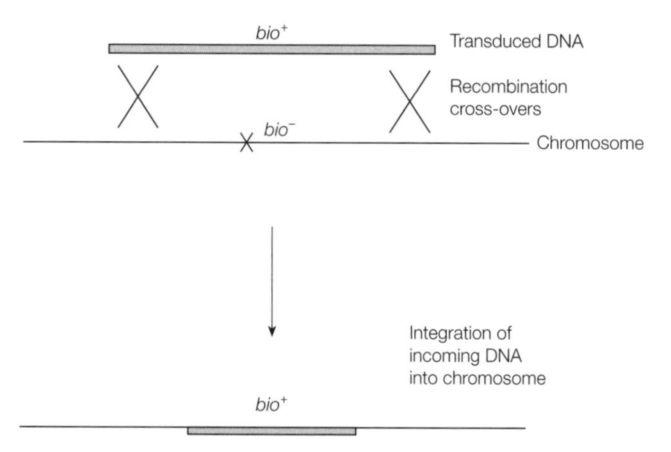

Fig. 2. Integration of transducing DNA into the host chromosome by recombination.

receive the wild-type phage will be lysed, but bacteria that receive transducing phage will be able to grow as colonies. The transfer of genes of interest may be selected by using appropriate media (Topics D6 and E1). The frequency at which a particular gene may be transferred is about one cell in every 10^6 to 10^8. Phage P22 in *Salmonella typhimurium* and phage P1 in *E. coli* are two systems frequently used in the laboratory.

Specialized transduction

Specialized transduction is a rare occurrence associated with a number of lysogenic phage that integrate into the host chromosome. Normally when a prophage is induced it excises from the chromosome in a very precise fashion to form a complete phage genome (Topic E7). Very rarely an excision event occurs which is inaccurate; a piece of chromosome adjacent to the phage is removed and a piece of the phage is left in the chromosome (*Fig. 3*). The excised phage is replicated normally, packaged into phage heads and released to infect another cell. The nature of the genes lost from the phage particle is dependent on which end of the genome is lost. In the example shown in *Fig. 3* where λ is integrated at its normal *att* site, on excision, it may pick up *gal* or *bio* genes to form λ*gal* or λp*bio*. The λ*gal* phage have lost essential genes for making mature phage so are incapable of replicating and forming plaques in a new host unless these functions are provided by a helper λ phage. The λp*bio* phage are plaque-forming, hence the 'p' designation, without any additional functions being provided, as the genes that are lost are not essential for phage replication. The fate of the DNA in the recipient is various as shown in *Fig. 4*. The transducing phage may enter the lytic cycle and produce many copies of the phage and the genes it carries or it may integrate into the host chromosome. Integration may be by site-specific recombination at the *att* site to form a lysogen (*Fig. 4a)* in which case there are now two copies of the transduced gene(s) (partial diploid), or recombination may occur between genes on the phage and in the recipient chromosome (*Fig. 4b*). In contrast to generalized transduction, the particles produced by specialized transduction are capable of moving genes at a high frequency of

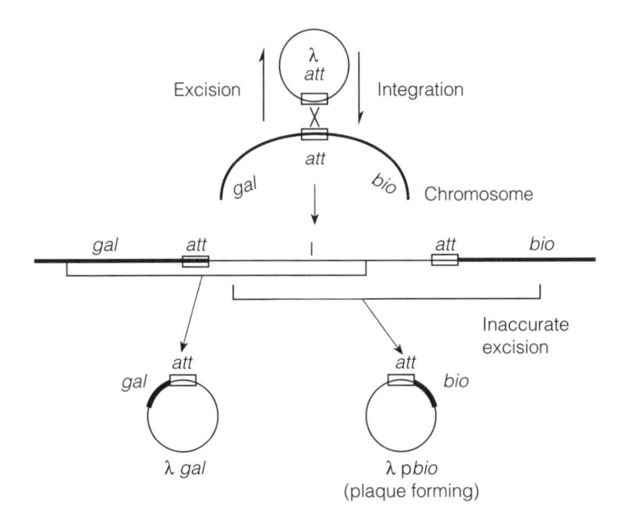

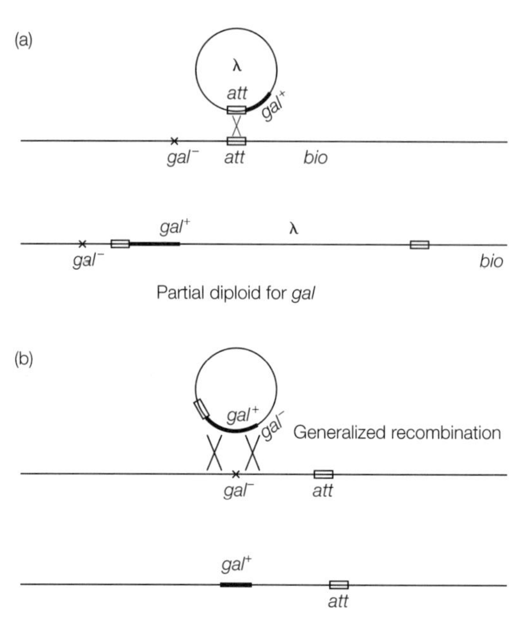

Fig. 4. Some of the fates of λ specialized transducing phage on infection of a recipient: (a) formation of λ lysogen; (b) replacement of recipient chromosomal DNA with DNA carried on λ by general recombination.

transduction in a specific manner, as all phage in a preparation carry the same piece of DNA. Using a variety of genetic techniques, specialized transducing phage have been produced for most areas of the *E. coli* chromosome. These were used in the past for: the production of large amounts of DNA for a particular gene; the transfer of genes between strains; and for the creation of partial diploid strains in order to study gene function.

E10 TRANSFORMATION

Key Notes

Overview

Transformation is the result of the uptake of free DNA from the surrounding medium. A number of bacteria such as *Bacillus*, *Streptococcus*, *Neisseria* and *Haemophilus* are capable of natural transformation.

Competence

The ability of a cell to take up DNA is dependent on the cell being in a particular state called competence. Competence is related to the presence of receptors for DNA being present on the surface of the cell. Other bacteria, such as *E. coli*, can be induced to competence by chemical treatment under cold conditions.

Uptake and fate of DNA

Depending on the species, DNA binding and transfer into the cell may be of specific or non-specific sequences. The DNA bound is normally double-stranded but in most cases it is made single-stranded as it enters the cell. *Haemophilus influenzae* takes up double-stranded DNA. The incoming DNA is integrated into the chromosome by recombination.

Related topic

Structure and organization of DNA (C1)

Overview

Our knowledge that DNA is the genetic material has come from experiments based on the fact that some species of bacteria are capable of naturally taking up DNA from their surrounding medium, and incorporating it into their genomes, by a process called **transformation**. Fred Griffith in the late 1920s first showed that a rough, non-capsulated, avirulent strain of *Streptococcus pneumoniae* could be converted into a smooth, capsulated, virulent strain by the addition of heat-killed smooth cells. Oswald Avery and his group in a series of experiments in the 1930s showed that the material carrying this transforming ability was DNA, and not the proteins of the cell. These groups were fortunate in their choice of experimental microorganism as only a few species are naturally able to take up DNA, including *Streptococcus*, *Bacillus*, *Neisseria*, *Haemophilus* and some species of archaebacteria.

Competence

Not all species within a genera will necessarily be capable of transformation nor will all cells within a population. The ability of a bacterial cell to take up DNA is associated with the development of a **competent** state in which DNA receptors on the cell surface and other transformation-specific proteins are present. For example, in *Neisseria gonorrhoeae*, only piliated cells are competent for transformation. Competence is generally dependent on growth conditions. Competence may occur in only a few cells in a culture (*B. subtilis*) or in nearly all cells (*S. pneumoniae*) depending on the genus. Competence may last for a few minutes (*S. pneumoniae*) or for several hours (*B. subtilis*).

Competence can be induced in species that are not naturally transformable (for example *E. coli*) by chemical treatments such as ice-cold $CaCl_2$ treatment followed by a brief heat shock. This competent state is not the same as natural competence.

Uptake and fate of DNA

The nature of the DNA taken up is dependent on the genera. Most bacteria will take up any DNA molecule, but *Haemophilus* requires the presence of a 10 bp specific sequence on the DNA. *Haemophilus,* a Gram-negative bacterium, binds and takes up double-stranded DNA. In contrast, the Gram-positive bacteria, *Bacillus* and *Streptococcus*, bind double-stranded DNA but degrade one strand at the same time as the other strand is imported into the cell. The incoming linear DNA is protected from degradation by competence-specific proteins and is integrated into the genome by recombination involving RecA protein (*Fig. 1*). During artificially induced competence in *E. coli,* circular plasmid DNA is preferentially taken up and in this case remains independent of the chromosome.

Transformation is probably an important mechanism by which genes are transferred naturally in the environment. Transformation has also been used in genetic mapping as genes that are close together are more likely to be taken up on the same fragment of DNA and incorporated together into the chromosome of a recipient cell.

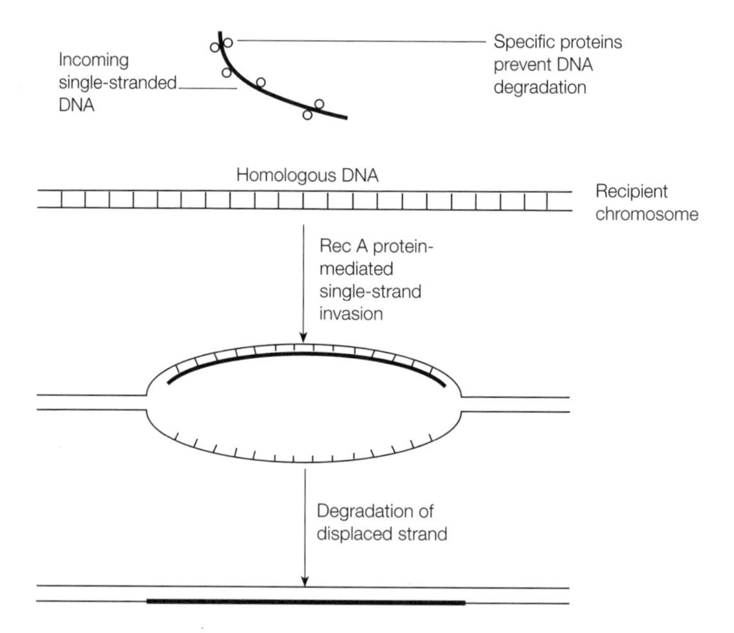

Fig. 1. Integration of single-stranded DNA into the chromosome by recombination.

F1 Prokaryotes in the environment

Key Notes

Prokaryote niche diversity

Bacteria and Archaea live in a wide variety of habitats, from temperatures of around freezing to a maximum of 115°C. Their ability to use all naturally occurring carbon compounds as growth substrates allows them to grow in most places on earth, and play a major role in the cycling of carbon, sulfur and nitrogen through the biosphere.

Cycling of elements through the biosphere

The biogeochemical cycling of elements such as carbon, sulfur and nitrogen is mainly carried out by the Bacteria and Archaea. The sulfur cycle is very complex, but the carbon and nitrogen cycles are simpler. Atmospheric and dissolved CO_2 is fixed by autotrophic microbes and plants, and the organic molecules thus generated are eventually oxidized to CO_2. Microbes also fix atmospheric nitrogen into ammonia, nitrate and nitrite, which are incorporated into organic nitrogen-containing compounds such as amino and nucleic acids. The turnover of elements allows the earth to function as a self-regulating entity.

Detection of prokaryotes in their natural habitats

For many years microbiologists relied on isolation of laboratory cultures as a method of determining which prokaryotes were present in a particular biotope. These methods are still used to some extent, but are now complemented by molecular methods, particularly fluorescence in situ hybridization (FISH) and whole genome PCR-based methods (environmental genomics).

Prokaryote commensualism

The Bacteria and Archaea almost always grow with other species (including higher plants and animals). The majority are very difficult to grow in pure culture in the laboratory. Bacteria and Archaea grow in communities in the environment, with the individual strains and species exchanging metabolites (often in competition with their peers) so that most organic compounds are ultimately converted into biomass, carbon dioxide and water.

Prokaryote niche diversity

Bacteria and Archaea can grow in low-temperature environments (for example the Antarctic bacterium *Flavobacterium frigidarium* isolated from marine sediments can grow at 4°C, as can an archaeon isolated nearby, *Methanococcoides burtonii*) as well as high-temperature ones. The highest known temperature allowing growth of a prokaryote is 95°C (*Aquifex pyrophilus*, isolated from an Icelandic marine thermal vent), while the archaeon *Pyrolobus fumarii* will grow at 113°C. Bacteria and Archaea are found in most places in the biosphere. They have become adapted to use a wide variety of compounds as sources of energy and carbon, and may be adapted to use those anthropogenically produced chemical compounds as well. The range of metabolic diversity means that microorganisms dominate nutrient-cycling processes in almost every ecosystem. This role in the cycling of elements has led Bacteria and Archaea to form very close associations with their peers as well as with plants and animals. The close nature of their commensualism, parisitism and symbiosis mean that pure cultures of many Bacteria and Archaea are extremely difficult to grow in the laboratory. These viable but non-culturable (VBNC) microorganisms can only be detected in biotopes using visualization techniques such as **fluorescence in situ hybridization** (FISH) or indirect PCR-based methods now known as **environmental genomics**.

Cycling of elements through the biosphere

James Lovelock's Gaia hypothesis portrays the earth as a self-regulating entity. The basis of this regulation is the involvement in the movement of millions of tonnes of chemical elements such as carbon, nitrogen and sulfur through the atmosphere, marine and terrestrial environments. The interconversion between solid and gaseous compounds of the elements is mostly accomplished by microorganisms, particularly the Archaea and Bacteria. The coupling of spontaneous chemical reactions with those catalyzed by microorganisms has given the study of these earth-scale transformations the name of **biogeochemistry**. An overall scheme showing the involvement of organisms in **biogeochemical cycling** is shown in *Figs. 1 and 2*.

Sulfur cycle

The biogeochemical cycling of sulfur is extremely complex, and cannot be easily described diagrammatically. The complexity of the cycle is dictated by the relatively high number of oxidation states that sulfur can have in solution and in the

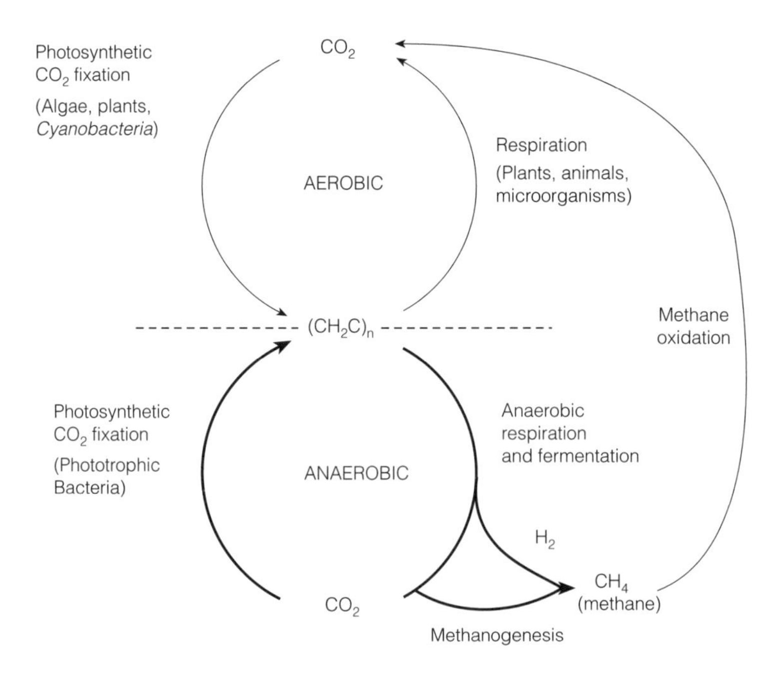

Fig. 1. Biogeochemical cycling. Transformation 1 includes precipitation, occlusion at air/water interfaces.

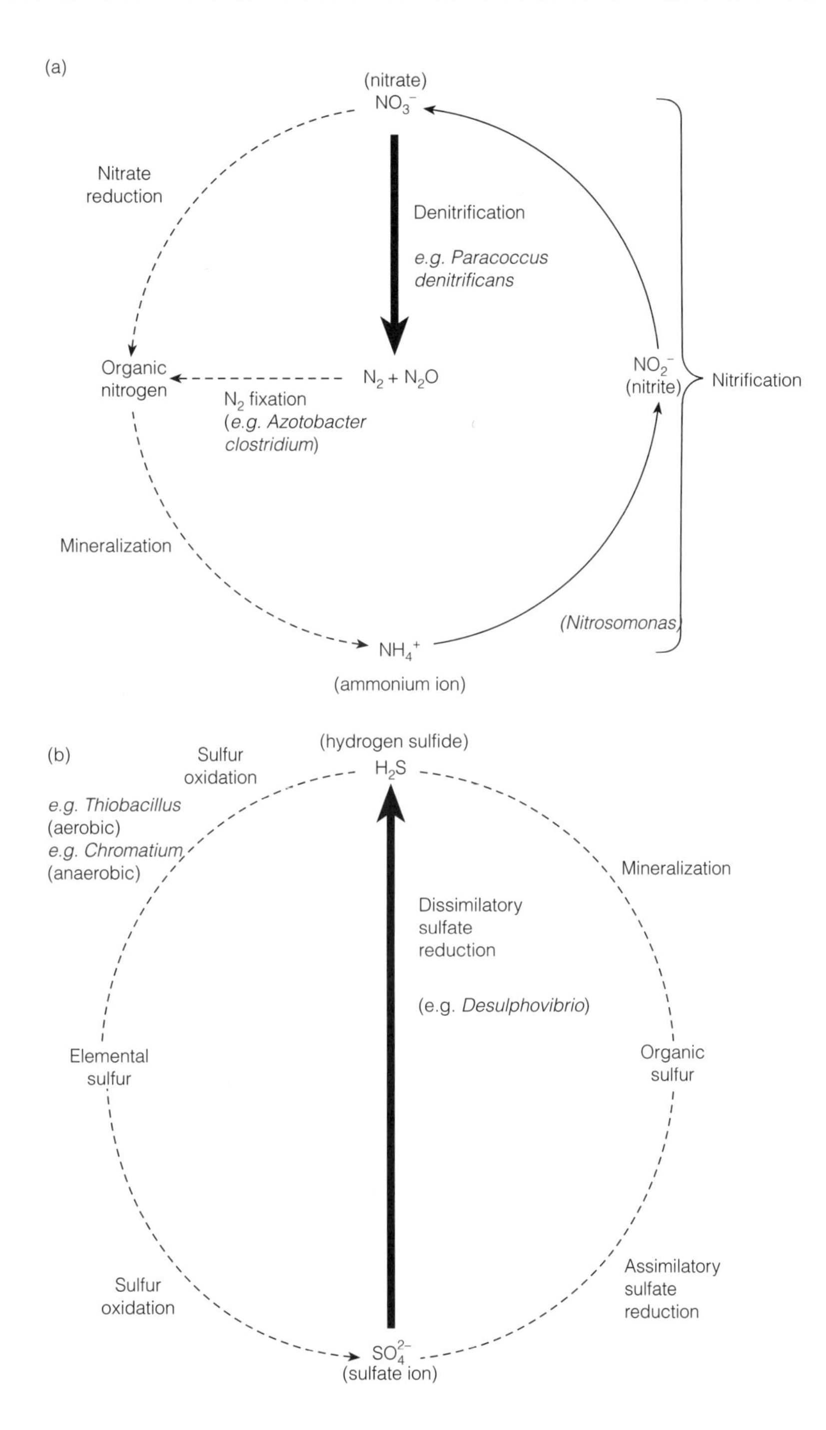

Fig. 2(a) and (b). Biogeochemical cycling. (b) Transformation 2 includes growth, fixation, secretion and decay; (c) Transformation 3 includes decay, respiration and combustion.

presence of oxygen. The most significant carbon compounds in the atmosphere is probably carbon dioxide, serving as a link between terrestrial and marine environments, but there is no direct sulfur equivalent. In prioritizing volatile atmospheric sulfur compounds, sulfur dioxide (SO_2), hydrogen sulfide (H_2S), carbonyl sulfide (COS), carbon disulfide (CS_2), dimethyl sulfide (CH_3SCH_3), dimethyl disulfide (CH_3SSCH_3) and methane thiol (CH_3SH) may be significant, depending on the geographical location examined. The nature of sulfur's chemistry may mean that non-volatile molecules, such as methane sulfonate ($CH_3SO_3^-$) and sulfate (SO_4^-) may also be present in quantity in the atmosphere as dissolved aqueous ions.

The biogeochemical cycling of sulfur serves to illustrate regulation on a global scale. The sulfur compound **dimethylsulfonium propionate** (DMSP) is produced intracellularly by marine photosynthetic algae. During and after algal blooms in the open oceans, this DMSP is released into solution on cell death, where it is metabolized by marine microorganisms to produce **dimethyl sulfide** (DMS). The generation of DMS is sufficiently high for some of the compound to enter the atmosphere as a gas above the ocean, where **photooxidation** by the suns rays leads to the breakdown of DMS into two solid forms of sulfur, sulfate and methane sulfonate. These solids act as nuclei for the condensation of water, and clouds are formed. The clouds reduce the sunlight reaching the ocean surface, and the growth of the photosynthetic algae that produced the DMSP in the first place is reduced. This regulatory cycle stops algae covering all the open ocean, though only in regions where man has not upset this cycle by dumping of sewerage or other compounds that the algae might feed on.

Sulfur cycling also occurs through terrestrial ecosystems, where a combination of **sulfate-reducing bacteria** (such as *Desulfobacter* and *Desulfovibrio* species) and **sulfate-oxidizing bacteria** (such as *Thiobacillus* species) interconvert elemental sulfur, hydrogen sulfide, sulfate, thiosulfate and polythionates. Sulfur is also assimilated by all organisms into the amino acids cysteine and methionine and so ultimately into proteins. Sulfur is also assimilated for the prosthetic protein groups known as iron-sulfur clusters.

Carbon cycle

Man's geologically recent release of greenhouse gases such as carbon dioxide (CO_2) into the atmosphere is generally held to be responsible for global warming. In the absence of man, carbon dioxide serves as an atmospheric link between carbon released in the marine and terrestrial environments, with methane playing a secondary role. Outside the atmosphere, carbon forms a wide range of compounds from simple metal cyanides to complex macromolecules such as starch, lignin, cellulose and nucleic acids. Microorganisms take part in the ultimate conversion of all of these into carbon dioxide or methane, and these processes may be broadly divided into the aerobic and anaerobic (*Fig. 2a*). Carbon compounds are broken down aerobically either by respiration or anaerobically by fermentation. The release of carbon dioxide is balanced by the fixation of CO_2 photosynthetically by plants and microorganisms, or by the aerobic chemolithotrophic growth of bacteria such as *Paracoccus denitrificans*.

Nitrogen cycle

The nitrogen cycle is the easiest to define in terms of the number of participating major nitrogenous compounds. Proteins, prosthetic groups such as chlorophyl, nucleic acids and other macromolecules can be regarded as fixed nitrogen, present in the marine or terrestrial environments. Ammonia, nitrate and nitrite ions may participate either in the recycling of fixed nitrogen, or in the generation of atmospheric molecular nitrogen (*Fig. 2b*). The bulk of molecular nitrogen is

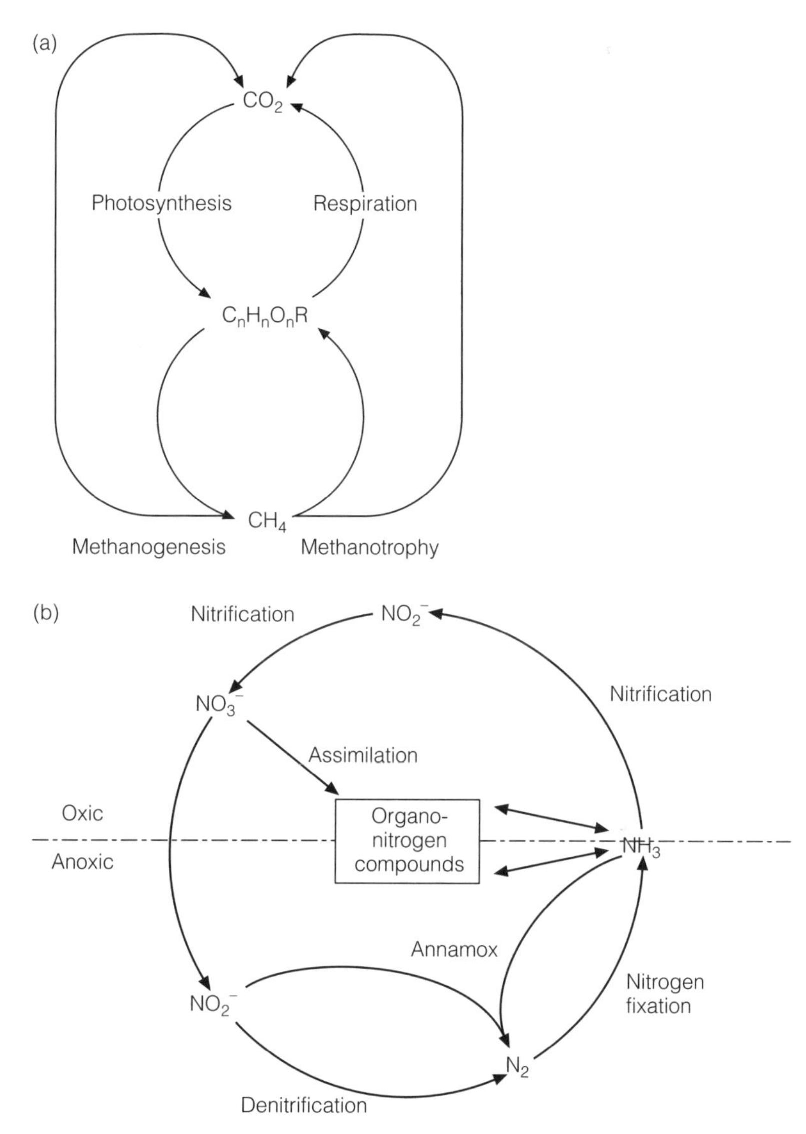

Fig 2. The biogeochemical cycling of the elements. (a) The carbon cycle; (b) The nitrogen cycle.

reduced by prokaryote nitrogen fixation, while very small amounts are reduced by lightning strikes. The biologically mediated processes have been called **mineralization**, **nitrification**, **denitrification**, **nitrogen fixation** and **nitrate reduction**.

Mineralization of fixed nitrogen is performed by many microorganisms during the degradation of organic matter, and results in the generation of free ammonia. This ammonia may be reassimilated, or converted into nitrate by the process known as nitrification. Well characterized nitrifiers include *Nitrosomonas europea* and are a two-step process via nitrite (*Fig. 2b*).

The conversion of nitrate to molecular nitrogen (denitrification) is performed by a taxonomically undefined group of organisms including the Bacteria

Paracoccus denitrificans and *Pseudomonas aeruginosa* as well as the Archaea *Haloarcula marismortui* and *Pyrobaculum aerophilum*. Recent evidence suggests that some fungi can also denitrify. Nitrate is converted sequentially into nitrite, nitric oxide and nitrous oxide before nitrogen is released from the organism. Each stage of this process is energetically beneficial to the organism, and is generally coupled to the electron transport chain. For this reason it is also called **nitrate respiration**.

Both Archaea such as *Methanosarcina barkeri* and Bacteria such as *Rhizobium* species are also capable of nitrogen fixation or **diazotrophy**. However, the rhizobia (bacteria classified as *Rhizobium* and *Bradyrhizobium*) have the unique ability to form nodules on the roots of certain plants, allowing nitrogen to be fixed almost directly from atmosphere to plant. The majority of diazotrophs are free living, and employ various methods to protect the obligately anaerobic enzyme nitrogenase from the effects of molecular oxygen.

Detection of prokaryotes in their natural habitats

The isolation and laboratory cultivation of a particular prokaryote from an epitope was at one time the sole evidence that that species was present. However, many studies have shown that there are many more bacteria in the environment than can be cultured using laboratory media, and nucleic acid analysis has complemented classical bacterial ecology. Individual bacterial cells can be labeled with a dye attached to an oligonucleotide which hybridizes to ribosomal RNA. The organisms to which the oligonucleotide bind will then show up under a fluorescence microscope. Careful choice of oligonucleotide can mean that these fluorescent probes can be species or even strain specific, allowing the microbial ecologist to examine and even count various sorts of microorganism in a particular niche. Various developments of this **fluorescence in situ hybridization** technique, or FISH, have allowed the differentiation between metabolically active and inactive microorganisms.

The development of PCR means that the total ribosomal RNA (RNA should only be produced by active cells) in a bacterial community can be examined. By sequencing a library of ribosomal cDNA from a particular epitope, the ecologist has some idea of the diversity of active bacteria. Extending this idea of amplification of DNA to other genes has developed into the field of environmental genomics.

Both FISH and PCR of total genomic DNA suggest that there is a far greater diversity of Bacteria and Archaea than first thought, to the extent that the idea of discrete species barriers is beginning to break down. Although these new organisms are often held to be viable but non-culturable (VBNC), new techniques and laboratory practices are finally yielding pure and mixed cultures suitable for physiological and biochemical studies.

Prokaryote commensualism

Bacteria are classically studied as pure cultures in the laboratory. These cultures are regarded as clonal and genetically stable. However, the situation in the environment is somewhat different: Bacteria and Archaea rarely grow as pure cultures, more often forming associations with organisms around them. These associations can range from the symbiotic (see *Table 1*) to the pathogenic. The diversity of life in a particular niche often makes **functional ecology** (the study of which organisms are responsible for which activities) challenging. Many ecosystems have evolved to include many Bacteria and Archaea exchanging and competing for metabolic intermediates to a state where no single chemical compound is completely metabolized by a single organism.

Table 1. Examples of associations between prokaryotes and higher organisms (see also Topic I4)

Bacterial/Archaeal species	Partner	Description
Rhizobium leguminosarum	Peas and other legumes	Allows plant to fix nitrogen directly from the atmosphere
Various methanogenic Archaea	Ruminant animals	The presence of Archaea in the gut results in the digestion of the cellulose content of plants
Symbiodinium species	Various coral species	Photosynthetic metabolism of the bacteria provide nutrients for the coral
Aeromonas veronii	Various leech species	Allows the digestion of blood
Vibrio fischeri	Squid	Confers a bioluminescent property on the squid

F2 Prokaryotes in industry

Key Notes

Biotechnology

Biotechnology is the use of living organisms in technology and industry. Prokaryotes have been exploited for many years in the manufacture of food and other useful products. For large-scale applications, prokaryotes are grown in fermenters using either batch, fed-batch or continuous-culture processes.

Gene technology and biotechnology

The efficient production of some proteins and chemicals can only be carried out by mutant or recombinant microorganisms. Gene technology provides a means of providing virus-free human proteins, as well as for the overexpression of useful enzymes from extremeophiles.

Fermented foods

Prokaryotes are used in the manufacture of a surprising diversity of processed food, e.g. cheese, yoghurt and soy sauce. The role of bacteria is predominantly to convert carbohydrates and alcohols into organic acids, lowering the pH and increasing the storage life and palatability of the food. Starter cultures are used to reduce the duration of the fermentation process.

Biocatalysis

Microorganisms are a convenient source of enzymes. The regiospecific and stereospecific properties of enzymes can be exploited to produce enantiomerically pure preparations of some chemicals. These bacterial enzymes may either be used directly, or held in immobilized cells for ease of use. The use of prokaryote enzymes can significantly reduce the cost and increase the yield of reactions an organic chemist would find difficult to perform.

Bioremediation

The biodeterioration of man-made compounds by microbes can be a problem (e.g. in the case of paints and plastics) or a benefit when it comes to cleaning up the environment. Bioremediation is the use of microbes to treat waste water and soil where unacceptable levels of pollutants have accumulated. Even highly toxic compounds such as polychlorinated biphenyls (PCBs) can be metabolized by some bacteria.

Biotechnology

Prokaryotes have been used to make a wide variety of products for many thousands of years, bur lately this has been called biotechnology. In order to produce products or intermediates, bacteria are grown in large vats (fermenters) protected from contamination and changes in pH, temperature and dissolved oxygen concentration. Industrial fermenters vary in size from a few litres to several thousand. Modern biotechnology relies heavily on developments in chemical engineering to regulate and monitor the processes occurring in these fermenters, and this regulation makes the growth of bacteria under these conditions somewhat different from that found in ordinary shake flasks. The optimization of fermenters requires a thorough understanding of the kinetics of prokaryote growth.

Industrial fermenters allow the growth of prokaryotes by one of three processes: batch, fed-batch and continuous fermentation. A batch fermenter is sterilized, inoculated and allowed to grow before the entire culture volume is harvested. A fed batch system follows a similar regimen, but instead only some of the culture is harvested after growth, and fresh medium is added to restart the process. Continuous culture uses a constant in-flow of medium offset by the exit of spent medium, product and biomass and is not used so much in the production of products for human consumption because of perceived problems with culture sterility. Most developments in fermentation systems focus on the provision or exclusion of oxygen from the vessel, depending on the process required. The amount of dissolved oxygen in a culture can be increased by a variety of methods, from the simple mechanical stirred vessel to the air-lift fermenter.

Biotechnology now plays an important role in many industries, including the food-processing industry, providing both enzymes (e.g. invertase for the manufacture of glucose syrup) and biomolecules (e.g. sodium glutamate).

Gene technology and biotechnology

The exploitation of biological potential for industrial processes currently relies on metabolic diversity of microorganisms to provide biological solutions to common problems. Improvement of biotechnological processes would seem an attractive start for the modern molecular biologist, but public opinion and government regulation of the use of genetically modified organisms has restricted the widespread use of gene technology.

However, the production of many drugs and fine chemical products relies on **recombinant** or **mutant microorganisms** to produce enzymes (see below) and many human proteins have been cloned in *Escherichia coli* (*Table 1*) so that significant quantities can be produced for therapeutic use without the danger of viral contamination. As more enzymes are studied with a view to industrial application, those from the extremeophiles seem more attractive. Although bacteria growing at high temperature or extremes of pH are often easier to keep sterile on a large scale, protein yields are often low. Cloning into well studied prokaryotes such as *E. coli* or *Bacillus subtilis* allows much more controlled and efficient **heterologous overexpression** of useful enzymes such as amylases and lipases from extreme halophiles for washing powder.

Fermented foods

Today's industrial fermented foods have their origins in much older processes. Foods that use bacteria in their production include soy sauce (*Pediococcus*

Table 1. A selection of human proteins cloned in E. coli *and their therapeutic use*

Protein	Function	Therapeutic use
Urokinase	Plasminogen activator	Anticoagulant
Serum albumin	Major blood protein	Synthetic plasma constituent
Factor VIII, Factor X	Blood-clotting	Prevention of bleeding in hemophiliacs
Interferons	Can cause cells to become resistant to some viruses	Antiviral therapy
Growth hormone releasing factor (HGH)	Permits the action of growth hormone in the body	Growth promotion, recovery from physical stress
Erythropoietin (EPO)	Stimulates production of red blood cells	Replacement of cells after chemotherapy; treatment of anemia

species), cheese, yoghurt (*Lactobacillus, Enterococcus, Bifidiobacterium* species as well as many others), sauerkraut (*Lactobacillus, Streptococcus, Leuconostoc* species) and the traditional manufacture of vinegar (*Acinetobacter* species). In general, prokaryotes are used to acidify foodstuff to allow storage, and a table of foods made by microorganisms is shown in *Table 2*.

Table 2. What are microbes used for in the food industry?

Foods	Flavorings
Fermented meat	Vinegar
Soy sauce	Nucleotides
Cheeses/milk/yoghurt	Amino acids
Mushrooms and edible fungi	Vitamins
Baker's yeast	
Coffee	
Pickles, sauerkraut, olives	
Organic acids	**Starter cultures**
Citric acid	Dairy industry (cheese etc., silage)
Itaconic acid	Leguminous crops

Bacterial food fermentations are generally complex and involve a succession of different organisms to render the final product palatable. The process rarely involves bacteria alone, but the production of fermented cabbage (sauerkraut) uses bacteria with progressively higher tolerance to acid. *Streptococcus faecalis* and *Leuconostoc mesenteroides* initiate the conversion of plant sugars into organic acids, and are succeeded by *Lactobacillus brevis* and finally *Lactobacillus plantarum*. A similar process is used to preserve the human foods cucumbers (as gerkhins) and olives, while in agriculture *Enterococcus, Pediococcus* and *Lactobacillus* species successively acidify cut grass to produce silage for winter feeding of cattle. Many of these processes will occur because of endemic prokaryotes without the aid of man, but the use of **starter cultures** grown up in the laboratory increases the speed of the fermentation.

Biocatalysis

Microorganisms can be used as a convenient source of enzymes, since these same proteins provide them with the means to grow on a variety of complex organic compounds. The enzymes can be used in their purified form (either immobilized or in solution) but are sometimes more stable when whole cells are used. Bacterial cells may be rendered non-viable by **immobilization** or **permeabilization** but still retain catalytic activity. This can alleviate the need to provide cofactors such as ATP, which can add significantly to costs. The reactions the enzymes catalyze are often **regiospecific** (attacking a single group on a molecule but leaving others of the same chemical composition) and **stereospecific** (attacking one enantiomer such as D-glucose, but not the corresponding stereoisomer). The specificity of enzymes has allowed industrial chemists to perform reactions that would be impossible by normal synthetic routes, but mostly **biotransformations** are cheaper to perform and have a higher yield.

Examples of biotransformations are provided in *Table 3*, but one of the most economically significant biotransformations is the production of acrylamide (a polymer used in many chemical processes as well as in the cosmetics industry). It is possible to use a copper catalyst to convert acrylonitrile into acrylamide, but this reaction must be performed at 100°C, after which the catalyst must be regenerated and the unreacted highly toxic acrylonitrile must also be rigorously separated from the product. However, these problems are avoided by the use of immobilized *Pseudomonas chlorophis* performing the biotransformation in a bioreactor. The reaction can be run at 10°C, so heating costs are reduced and the bacterial enzyme responsible (nitrile hydratase) converts over 99.9% of the acrylonitrile into acrylamide. Around half a million tonnes of acrylamide are produced annually by this process.

Table 3. Examples of industrial production of organic compounds by prokaryotes

Prokaryote source	Chemical	Major application
Acetobacter	Acetic acid	Solvent, starting compound for many synthetic reactions
Clostridium	Isopropanol	Solvent, antifreeze
Clostridium	Acetone	Solvent, starting compound for many synthetic reactions
Bacillus	Acrylic acid	Precursor for acrylonitrile and other polymers
Bacillus	Propylene glycol	Solvent, antifreeze, antifungal compound

Although biotransformations have been used in the bulk chemical industry, the main application is in the production of **fine chemicals** such as antibiotic derivatives. The cost savings can be dramatic: cortisone was first synthesized as a 31-step organic synthesis starting from 615 kg of deoxycholic acid. This yielded 1 kg of cortisone, which was sold as an anti-inflammatory drug in the 1940s at around $200 per gram. Use of enzymes from the fungus *Aspergillus niger* in some of the steps reduced the cost to $6 per gram in 1952. The use of mycobacterial enzymes allowed plant sterols to be used as much cheaper-starting compounds so that by 1980 the price of cortisone had dropped to 46 cents in the United States, about a quarter of one percent of the original cost.

Bioremediation

The metabolic diversity of microbes has allowed their commercial exploitation in the field of waste water and soil clean-up. The **biodeterioration** of paints, plastics and other man-made compounds by microbes is often seen as a problem, but where unacceptable levels of these compounds accumulate, this property can be a benefit. Although the technology is still in its infancy, compounds previously considered **recalcitrant** are now being subjected to treatment by a variety of microbes. Pilot studies have shown the efficacy of the use of bacteria against compounds such as trichloroethylene (**TCE**) and polychlorinated biphenyls (**PCBs**). The use of microbes is not without problems: the **biosurfactants** of *Pseudomonas* species have been used to allow endemic soil bacteria **bioavailability** to emulsified crude oil after tanker spillages. However, the emulsified oil proved much more mobile than the crude slick, and a bigger problem was created as the hydrocarbons moved deep into gravel and towards potable water sources. Experimental waste water treatment plants, where polluted medium is much more contained, have been far more successful than *in situ* approaches.

F3 Bacterial disease – an overview

Key Notes

Pathogens

Infectious disease is the change in the structure or function of a host caused by a microorganism. Microorganisms that are able to cause disease in their host are described as pathogens. Virulence is a measurement of a pathogen's disease-causing capacity. The features of microorganisms that confer the ability to infect a host and avoid the immune system are called virulence factors. The microbes that are normally found colonizing a host without causing damage are called commensals.

The nature of microbial disease

Microbial diseases may be chronic, such as tuberculosis, or acute, typified by Staphylococcal food poisoning. The symptoms of the disease depend on the site of infection, the ability of the organism to combat the immune system and the toxic products it produces. Pathogens such as *Streptococcus pyogenes* are capable of causing many different diseases from impetigo to rheumatic fever. Others may show a range of different symptoms during the course of a disease, such as *Treponema pallidum*, the causative agent of syphilis.

Koch's postulates

Koch's postulates are a number of criteria that have been used in the past to prove that a bacterium is responsible for a particular disease.

1. The microorganisms should be found in all cases and at all sites of the disease.
2. The microorganisms should be isolated from the infected person and maintained in pure culture.
3. The pure cultured microbe should cause symptoms of the disease on inoculation into a susceptible individual.
4. The microorganisms should be reisolated from the intentionally infected host.

Related topics

Human defense mechanisms (F4)
Entry and colonization of human hosts (F5)
Detrimental effects of fungi in their environment (H5)
Detrimental effects of Chlorophyta and Protista (I5)
Virus infection (J8)

Pathogens

Although millions of microorganisms, from a wide range of species, live in and on the human body, their presence is normally beneficial to the host and, in some cases, even essential. These microbes are referred to as the **normal flora** or **commensals**. The healthy human body has a wide range of defense mechanisms, the **immune system**, with which it can protect itself from microbes capable of causing damage **(disease).** Microbes that cause disease are called **pathogens**. **Infectious disease** is considered to be a change in the normal structure or function of the host's body as a result of a microbial or parasitic infection. This change to the host, manifested as a set of **symptoms,** may be due to the effect of microbial products such as toxins or the result of the host's immune reaction to the presence of the microbe (Topic F4). Pain, fever, redness and swelling are common symptoms of bacterial disease. There are many definitions associated with the microbial ability to cause disease, which are listed in *Table 1*.

For a bacterium to be pathogenic it must be able to:

- spread between hosts,
- invade and remain within the host,
- acquire nutrients,
- avoid or damage the host immune system,
- become established in the host.

The features of bacteria that allow them to do this are called **virulence factors** (see Topic F5). The genes for many virulence factors are carried on plasmids (Topic E5) or phage (Topic E7), but not all virulence factors are plasmid borne (F5).

The nature of the host is an important factor in pathogenicity. Bacteria that are capable of infecting normal, healthy hosts may be considered as true pathogens. However, if the host is damaged, other **opportunistic pathogens** may be able to infect and cause damage. *Pseudomonas aeruginosa*, for example, is unable to infect healthy skin but will infect and cause serious damage to burnt skin. People who are immunocompromised as a result of genetic defects, medical treatment or infection with certain viruses, such as HIV, are particularly susceptible to infection and may be infected by many normally harmless microbes.

The nature of bacterial disease

Bacillus anthracis was the first bacterium proven to be the causative agent for a disease, anthrax, in the 1870s. This work was done by Robert Koch. Subsequently, the **etiologies** (causes) of many common and sometimes fatal diseases were shown to be a result of bacterial infection. Bacterial diseases range from **chronic** infections such as tuberculosis caused by *Mycobacteria tuberculosis*, and other related mycobacterial species, which are capable of survival within macrophages (Topic F5), to **acute** infections, such as Staphylococcal food poisoning, caused by production of an enterotoxin that induces vomiting and diarrhea (Topic F6).

Table 1. Some definitions used in describing pathogenicity

Pathogens	Bacteria that are capable of causing disease
Colonization	The establishment and multiplication of bacteria on a site in the host
Infection	The establishment and multiplication of a pathogen on a site in the host
Commensals or resident microflora	Bacteria that are normally found colonizing sites in the body
Asymptomatic carriage	Infection with disease-causing bacteria without the production of any symptoms
Systemic	Infection where bacteria are spread throughout the body
Sepsis	The presence of bacteria or their products in the blood or tissues
Septicemia	The multiplication of bacteria in the blood
Toxemia	Disease caused by the presence of a toxin in the blood

The type of disease caused by a bacterium may depend on: the site at which it infects (skin, respiratory tract, gastrointestinal tract, genito-urinary tract, tissues or blood stream), which normally reflects the mechanism by which it is transmitted (Topic F5); its ability to combat the immune system (Topic F4); and the toxic products that it produces.

In many cases of Bacterial disease the symptoms are due to the production of very powerful toxins capable of causing extensive cell damage at distances from the site of infection. Examples of such diseases are: diphtheria, an infection of the upper respiratory tract by *Corynebacteria diphtheriae*; tetanus, an infection of deep wounds, where anerobic conditions prevail, by *Clostridium tetani*; and cholera, an infection of the intestinal tract by *Vibrio cholera*. Other diseases such as gonorrhea (a sexually transmitted disease caused by *Neisseria gonorrhoeae*) and Legionnaires disease (a form of pneumonia caused by inhalation of an airborne bacterium, *Legionella pneumophila*) result from the ability of these microbes to survive within cells (Topic F5) and avoid the immune defense mechanisms. Normally, a number of different virulence mechanisms will be associated with the ability of a particular microbe to cause disease.

Some Bacteria are capable of causing just one kind of disease such as cholera or whooping cough (caused by *Bordatella pertussis*). Other bacteria may cause a range of different diseases, depending on the site at which they act, the virulence factors they carry, and the stage of the disease process. An example is *Streptococcus pyogenes*. This pathogen is a major cause of sore throats, but, if an erythrogenic exotoxin is produced by *S. pyogenes*, the disease caused is the more serious one of scarlet fever, characterized by a diffuse rash. In a small number of cases *S. pyogenes* infections may result in rheumatic fever, a creeping arthritis, which may also be associated with heart valve damage, or acute glomulerulonephritis (kidney damage). These diseases are thought to be the result of autoimmune responses to antigens on the bacteria which are similar to host antigens. *S. pyogenes* is also capable of causing a skin disease, impetigo. A number of new strains of the organism cause a potentially fatal septicemia as a result of a toxin named streptococcal pyogenic exotoxin. Another example of the variety of symptoms caused by one bacterium is shown by the spirochaetes, *Borrelia burgdorfei* (causes Lyme's disease, transmitted by deer ticks) and *Treponema pallidum* (syphilis). These have three stages of disease if the original infection is not treated, reflecting the movement of the microbe in the body with time.

This section has attempted to describe just a few of the many bacterial diseases that exist. Others will be discussed in the context of virulence factors in later sections.

Koch's postulates

For some infections it is very easy to identify the causative agent and establish how it causes disease. The symptoms of cholera, for example, can be clearly explained by the presence of the organism *Vibrio cholera* in the small intestine and the production of a toxin that causes the local secretion of copious amounts of water (Topic F6). For other diseases, the cause and effect relationship may not be as clear. For example, *Helicobacter pylori* has only recently been accepted as the cause of many gastric ulcers as it was believed for a long time that bacteria could not survive in the stomach.

Traditionally, the relationship between a bacterium and a disease has been established if the bacterium fulfills a number of criteria laid down by Robert Koch in the late 1800s:

1. The microorganism should be found in all cases and at all sites of the disease.
2. The microorganism should be isolated from the infected person and maintained in pure culture.
3. The pure cultured microbe should cause symptoms of the disease on inoculation into a susceptible individual.
4. The bacteria should be reisolated from the intentionally infected host.

In practice, in most cases, it has proved difficult to fulfill all of these criteria. Pure culture of Bacteria in the laboratory has often been a problem. Although we are fairly certain that *Mycobacterium leprae* is the cause of leprosy, it has yet to be cultured in laboratory medium, and has so far been cultured only in an animal model. In the case of an intestinal disease called Whipple's disease, the bacterial agent responsible has only been shown using modern polymerase chain reaction (PCR)-based DNA detection methods. Criteria 3 and 4 of Koch's postulates are also difficult to meet if no animal model for the disease is available or there is a lack of human volunteers.

A modern-day modification of these postulates is now being used to establish if a particular virulence factor is necessary for the production of an infection. Potential virulence factors are first identified, then removed by mutation or replaced by cloning and the microbe is then tested for virulence. In this way a clear understanding of the disease process of a number of microorganisms has been obtained.

F4 HUMAN DEFENSE MECHANISMS

Key Notes

Overview

The host immune defense system consists of two components: a non-specific constitutive defense aimed at preventing infection and an inducible response which responds to the presence of a specific microbe or cellular component in the host.

Defense of surfaces

The body is protected by a number of defense mechanisms that are present at all times and act immediately on contact with any foreign body. These include physical barriers and mechanical removal, chemicals such as lysozyme and lactoferrin, the normal microbial flora on mucosal membrane surfaces, the protein complement system and phagocytosis.

Phagocytosis

Phagocytic polymorphonucleocytes (PMNs) and macrophages ingest and destroy foreign molecules or cells. Their activity is enhanced (opsonized) by the presence of complement proteins or antibodies on the surface of the microbe.

Complement

Complement is a set of proteins circulating in the blood stream which, on activation by the presence of microbes or antibody–antigen complexes, attract phagocytes, enhance phagocytic ability and form a membrane–attack complex (MAC), which can kill Gram-negative bacteria.

The inflammatory response

Infection results in a complex series of events, mediated by proteins, called the inflammatory response, which helps to prevent the spread of the microbes and aids in their destruction. This results in the classic symptoms of infection: pain, redness, swelling and fever.

Specific induced immune response

Antibodies (humoral immunity), produced by B lymphocytes, and receptors on T lymphocytes (cell-mediated immunity) recognize and bind to specific foreign molecules (antigens), resulting in a variety of responses depending on the nature of the target. On a first encounter, the response takes between 5 and 7 days to reach full potential, but there is also a memory component in this system, which allows a much more rapid response on a second exposure.

Antibodies

Antibodies are proteins called immunoglobulins consisting of two light and two heavy chains. One end of the molecule carries the antigen-binding site and at the opposite end is the Fc portion which mediates complement activation and enhances phagocytosis.

Cell-mediated immunity

T cells have a major role in the stimulation of B cells to produce antibodies and in the killing of host cells that have intracellular microorganisms.

Related topics	Bacterial disease – an overview (F3) Entry and colonization of human hosts (F5)	Bacterial toxins and human disease (F6)

Overview

The function of the immune system of the host is, initially, to prevent invasion of the host by microorganisms (bacteria, fungi, protozoa and viruses). However, if infection occurs, the system must recognize the presence of a foreign invader and act to neutralize or eradicate it. There are two components in this system:

- a non-specific, constitutive (innate) set of defenses which act immediately against most microbes;
- a specific, inducible (adaptive) reaction, mediated by antibodies (produced by B-lymphocytes) and T-lymphocytes cells, which is initiated by the presence of a specific organism or cellular component (**antigen**).

Defense of surfaces

Bacteria can enter the body through any site that is open to the environment. All these sites are therefore protected by a number of non-specific defense mechanisms (*Table 1*). The main external barrier, the **skin**, is a dry, thick layer of dead, keratinized cells virtually impermeable to all microorganisms. Natural openings in the skin such as pores, hair follicles and sweat glands are protected by the secretion of toxic chemicals such as fatty acids and **lysozyme**. Generally, the only way bacteria can penetrate this surface is through wounds. With the exception of the eye, all other surfaces to which bacteria can gain access, such as the gastrointestinal tract, the respiratory tract and the genito-urinary tract are covered in **mucosal membranes**. The cells in these membranes are protected by thick layers of **mucus**, a mixture of proteins and polysaccharides, which prevents penetration of bacteria into the cell surface. Ciliated cells may also

Table 1. Components of the non-specific immune response

Defense	Site	Function
Mechanical barriers	Skin, mucosal membranes, mucin	Prevents penetration of microbes
Flushing	Tears, urine, mucus, saliva, liquid movement through gut, coughing and sneezing	Removal of microorganisms and prevention of attachment
pH	Fatty acids on skin, stomach acidity, vaginal acidity	Prevention of microbial growth
Lysozyme	Most sites and tissues	Breaks down Bacterial peptidoglycan
Lactoperoxidase	Mucosal membranes and lysosomes	Kills bacteria by producing toxic superoxide radicals
Lactoferrin/transferrin	Mucosal surfaces, blood and tissues	Chelates iron essential for microbial growth
Commensal microflora	Skin and most mucosal surfaces, exception the lungs	Competes for colonization sites Colicins may inhibit bacterial growth
Phagocytosis in tissues	PMNs in blood/tissue; macrophages and lungs	Engulfs bacteria into phagosomes in which they are destroyed
Complement proteins	Blood/tissue. Activated by presence of microorganisms or antigen–antibody interactions	Attracts phagocytes. Enhances phagocytosis (opsonization) Kills Gram-negative Bacteria

present, which help in the removal of bacteria by the constant movement of the mucus. Similarly, all mucous membranes are constantly flushed by fluids which:

- prevent bacteria becoming established at one site by moving them through the system;
- contain antibacterial compounds which kill or inhibit bacterial growth.

Finally, a major part of the non-specific defense system of the host is the complex set of residential commensals present on the skin and on most mucosal membranes. These microorganisms compete with the invading pathogenic bacteria for nutrients and attachment sites. They may also contribute to the host defense against disease by the production of antibacterial compounds, such as **colicins.** A good example of the role of commensals has been shown in mature, premenopausal women where lactic acid produced by *Lactobacillus* species keeps the pH of the vagina around five inhibiting the growth of potential pathogens.

Phagocytosis

Circulating in the blood and tissues are cells called **phagocytes** that are capable of ingesting and killing bacteria. Two main types of cell are important in the control of bacterial disease: short-lived **polymorphonuclear leukocytes** (PMNs, neutrophils) found mainly in the bloodstream, and **macrophages** found in the tissues and in the lungs. Macrophages can live for weeks and have a number of complex roles in the immune system. These include:

- phagocytosis and destruction of foreign material;
- the secretion of proteins called **cytokines** to stimulate the **inflammatory response** which acts to prevent bacterial multiplication and spread;
- presentation of antigens on their surface in a way that induces the specific immune responses B- and T-lymphocytes.

Macrophages are first produced from stem-cells as monocytes which circulate in the bloodstream. On attraction to a site of infection, or entry into a tissue, they differentiate into macrophages. The steps of ingestion and killing by phagocytic cells are shown in *Fig. 1*. The phagocyte ingests the bacterium, by endocytosis, into a membrane-surrounded sac called a **phagosome**. This in turn fuses with lysosomes that contain a range of antibacterial compounds which kill and destroy the bacterium. Phagocytosis is enhanced by the presence of antibody and complement (see later) on the surface of the bacterium. This is called **opsonization**.

Complement

As well as lysozyme, iron-chelating agents and other antibacterial agents (*Table 1*), serum contains a number of proteins that make up the **complement** system. These proteins on their own are not toxic but are activated by the presence of components of microorganisms in the host. There are two pathways of activation: the classical pathway which is initiated by **antibody–antigen complexes** and the alternate pathway which is triggered by **microbial surface molecules** such as cell walls. Activation causes a cascade of proteolytic reactions which leads to the production of a number of products that activate the immune system:

- C3b binds to bacteria enhancing their phagocytosis;
- C3a and C5a trigger the inflammatory response and activate phagocytes;
- a membrane attack complex which forms pores in the membranes of Gram-negative bacteria, leading to cell death.

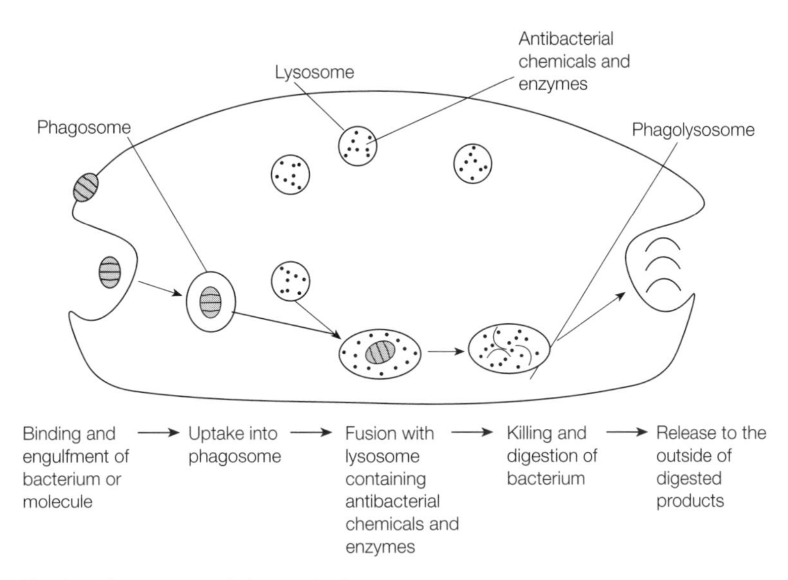

Fig. 1. The process of phagocytosis.

The inflammatory response

The infiltration of phagocytes to the site of infection, cytokines released by macrophages, the complement cascade and tissue damage result in a complicated series of events (the inflammatory response) that leads to the classic symptoms of **inflammation:** heat, redness, swelling and pain. The function of this is to prevent the multiplication and spread of the microbes and aid in their eventual destruction.

Specific induced immune response

The presence of foreign molecules **antigens**, which might be protein or carbohydrate, in the host can induce a highly efficient specific defense response. The induction process is complicated, involving a number of different cell types, leading to the presentation of the foreign antigen on the surface of an antigen-presenting cell (for example, a macrophage). Depending on the nature of the antigen and its presentation, the immune response may be in the form of **antibodies** (produced by B-lymphocytes), sometimes referred to as humoral immunity, or it may be **cell-mediated** by T-lymphocytes. The antibodies and the receptors on the surface of T-lymphocytes recognize and bind to antigens, resulting in a variety of responses depending on the nature of the target. This type of immune response takes between 5 and 7 days to reach full potential on a first encounter with a pathogen. However, an important feature of this response is the production of a **memory** component, which allows a much more rapid response on second exposure to the antigen.

Antibodies

Antibodies are proteins called immunoglobulins (Igs). They are made up of two heavy and two light chains, which create two functionally important areas, as shown in *Fig. 2*. There are two sites that recognize and bind the antigen (called the **Fab** regions) and the **Fc** portion at the other end of the molecule which mediates the function of the antibody. The result of a specific antibody–antigen interaction depends on the nature of the antigen and antibody and may include:

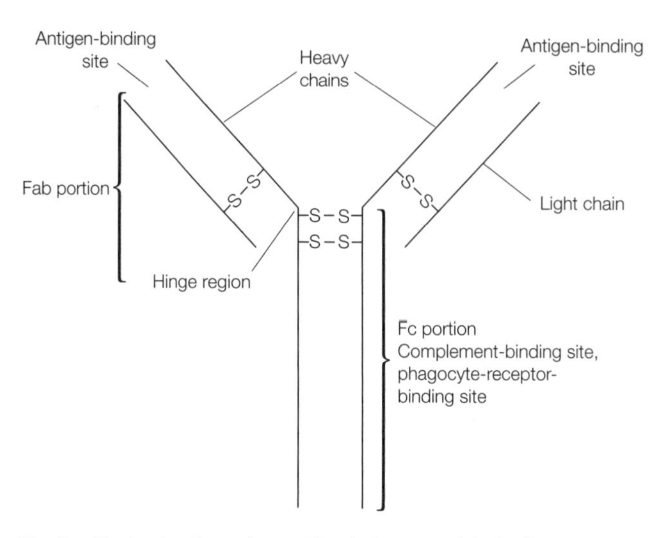

Fig. 2. Basic structure of an antibody: immunoglobulin G.

- enhanced phagocytosis;
- complement activation;
- neutralization of toxins;
- inactivation of proteins;
- inhibition of binding of toxins or bacteria to surfaces.

However, the final aim is to neutralize and remove the foreign molecule or infective agent from the host. There are several types of antibody molecule, which differ in structure and function. The normal circulating antibodies responsible for the immune response to bacteria are IgG and IgM. The antibody that protects mucosal membranes is secretory IgA.

Cell-mediated immunity

T-lymphocytes (T-cells) carry antibody-like receptors on their surface and are induced to function by the presence of a foreign antigen in the blood. T-cells play several roles in the specific immune defense system, depending on the nature of the antigen involved in the initial activation of the response:

- T-helper cells are important for the stimulation of B cells to produce antibodies in response to protein antigens.
- Cytotoxic T-cells kill cells that display foreign antigens, such as cells infected with *Mycobacterium tuberculosis*.
- Suppressor T-cells are involved in down-regulating the immune response.

F5 Entry and colonization of human hosts

Key Notes

Overview

In order to set up an infection a bacterium must reach a site at which it can survive, accumulate nutrients to allow it to replicate, and overcome the host immune defense mechanisms.

Transmission between hosts

Bacteria are transmitted by aerosols, in water or food, by direct contact or via animal vectors. The sites in the host that they reach depend on the transfer mechanism and the microbe itself. Some microbes may remain on mucosal surfaces, others penetrate through into the tissues and others are injected directly into the tissues or bloodstream.

Colonization of surfaces

Adhesin molecules on the surface of cells bind to cell surfaces and prevent the mechanical flushing of the bacteria from the host. Adhesins may be polysaccharide or protein, pili or capsules. Motility and mucinases, which break down mucus, also assist in colonization.

Invasion of cells

Bacterial proteins called invasins induce non-phagocytic cells to take up bacteria. Some bacteria use this mechanism to penetrate through epithelial layers and to spread within the host.

Acquisition of nutrients

In order to survive within the host, bacteria must be able to acquire nutrients. Toxins that lyse cells and extracellular enzymes assist the bacteria to do this. Iron-chelating proteins are also important virulence factors as they allow the bacteria to compete with the host for iron.

Spread of bacteria

Extracellular enzymes and toxins which destroy host tissue allow the dissemination of the microbe in the host.

Avoidance of host defense mechanisms

Many mechanisms contribute to the ability of Bacterial pathogens to withstand the onslaught of the host defense systems. Some particularly successful strategies include: destruction of phagocytes and complement; survival of bacteria within phagocytes; disguises that prevent the host from recognizing the bacteria as foreign.

Organization of virulence genes

Pathogenicity islands consist of groups of virulence genes located together on the chromosome. Evidence suggests that these genes have been acquired from an outside source in a single event.

Related topics

Prokaryote cell structure (D2)
Bacterial cell envelope and cell wall synthesis (D3)
Human defense mechanisms (F4)
Bacterial toxins and human disease (F6)

Overview

What has a Bacterium to do to cause infection (*Fig. 1*)? It must reach an appropriate site in the host where it can survive and multiply. To do this it must overcome the host immune defense mechanisms as described in Topic F4. The properties or **virulence factors** that a bacterium requires to infect therefore depend on two factors:

- the mechanism by which the bacteria enter the host;
- the nature of the site of colonization, whether it is a surface such as the mucosal membranes or within the tissues of the host.

Bacteria that survive on mucosal surfaces, such as *Vibrio cholerae* in the gut, will encounter a different range of host defense mechanisms to bacteria, like *Yersinia pestis*, that are capable of survival in the bloodstream or tissues of the host.

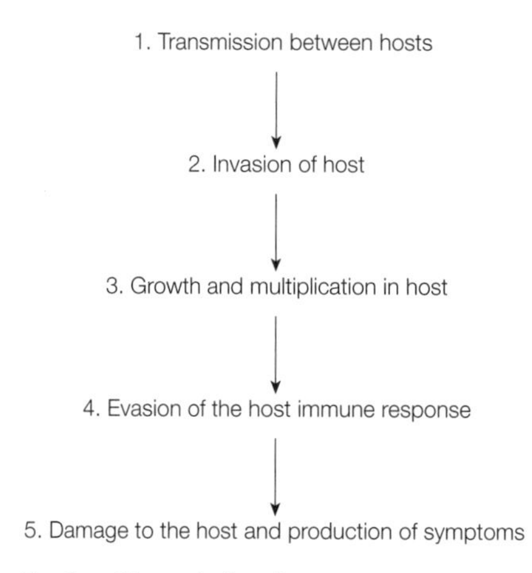

Fig. 1. Stages in the disease process.

Transmission between hosts

Organisms may be transmitted between hosts in a number of ways, as shown in *Table 1*. The route of transfer depends on the nature of the bacteria. Bacteria that can survive in the environment may be spread in food, water and air whereas others, such as *Neisseria gonorrhoeae*, cannot survive outside the human host and must be spread by direct contact. The nature of transmission also dictates the final site of infection: bacteria that are spread in the air or aerosols will gain access to the naso-pharynx and the lungs; those spread by food or water will gain access to the gut. Many bacteria, such as *V. cholerae*, will remain on the mucosal membranes and may only cause damage by the production of extracellular toxins or enzymes (Topic F6). Other bacteria such as *Salmonella typhi* enter into the host tissues by penetrating through the mucous membranes. Some bacteria enter directly into the host bloodstream, often through animal bites, an example is *Y. pestis* carried by the rat flea. Finally, many bacteria may enter the host through wounds and burns.

Colonization of surfaces

In order to survive as a pathogen or commensal, the immediate priority for most bacteria is to remain in the host. Nearly all surfaces in the body are protected by some form of flushing movement that removes any invading bacteria that cannot stick to the surface of the epithelial cells. Most pathogenic bacteria carry adhesins on the surface of their cells to allow the attachment of the bacterium to

Table 1. Some examples of how bacteria are transmitted between hosts

Nature of transmission	Organism	Disease	Site of infection
Air/aerosols	*Corynebacteria diphtheriae*	Diphtheria	Throat but toxin passes into bloodstream
	Legionella pneumophila	Legionnaire's disease	Alveolar macrophages in the lung
	Bordatella pertussis	Whooping cough	Ciliated epithelial cells in the upper respiratory tract
Food/water-borne	*Vibrio cholera*	Cholera	Gut
	Salmonella typhi	Typhoid	Gut → invasion through mucosa → systemic infection
	Shigella dysenteriae	Dysentry	Gut → invasion of mucosa
Direct contact	*Neisseria gonorrhoeae*	Gonorrhea	Genito-urinary tract → invasion of epithelial cells → systemic infection (1% of cases)
	Mycobacteria leprae	Leprosy	Skin
Vector-borne	*Borrelia* sp. (vector – deer tick)	Lyme's disease	Local to bite → systemic → neurons
	Yersinia pestis (vector – flea)	Bubonic plague	Blood and lymph glands

host cell surfaces. Many adhesins are proteins and are often in the form of pili (fimbriae), long cylindrical-shaped structures which stick up from the surface of the cell (*Fig. 2*; see also Topic D2). The structure of the pili is thought to help overcome the electrostatic repulsion resulting from the negative charge on both types of cell. However, membrane proteins and other surface proteins can also act as adhesins (*Table 2*), sometimes in the absence of pili. The receptors to which the adhesins bind on the surface of the host cells are often specific glycoproteins or glycolipids which may not be present in all cell types. For this reason, microbes infecting different animals or different sites may require different adhesins. A good example of this are the enterotoxigenic strains of *E. coli*, which

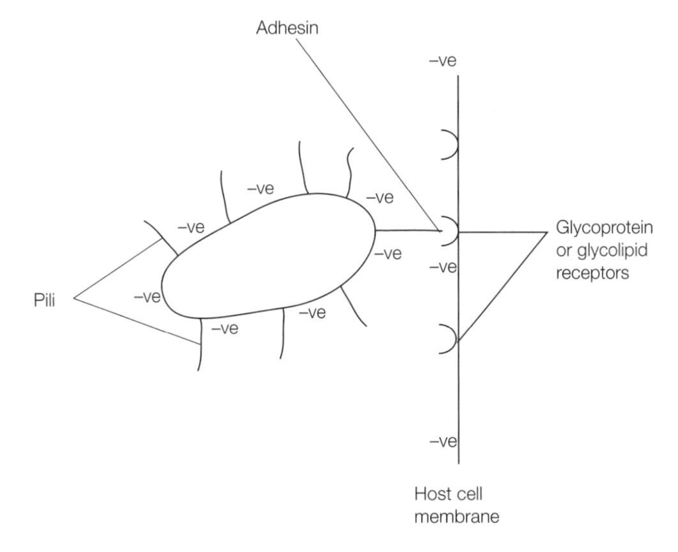

Fig. 2. The role of pili in adherence to cell membranes.

Table 2. Examples of the nature of some adhesins and the sites to which they bind

Adhesin	Bacterium	Binding site
K88 pili	*E. coli*	Pig intestinal mucosa
CFA/I, CFA/II pili	*E. coli*	Human intestinal mucosa
PAP pili	*E. coli*	Human urinary tract
Outer membrane protein II	*N. gonorrhoeae*	Urogenital tract
Protein F surface protein	*Strep. pyogenes*	Fibronectin
Exopolysaccharides	*Strep. mutans*	Teeth

cause a diarrheal disease by the production of toxins that induce secretion of fluid from the epithelial cells (Topic F6) after colonization of the gut. *E. coli* strains pathogenic to pigs carry a piliated adhesin, K88, that binds to pig epithelium but not human. Disease in humans is caused by strains that carry different types of adhesin such as CFA/I.

Exopolysaccharides may also act as adhesins, an important example being those produced by *Streptococcus mutans* in the formation of dental plaque. Exopolysaccharides have also been implicated in the formation of microcolonies of *Pseudomonas aeruginosa* in the lungs of cystic fibrosis patients.

In addition to the problem of adhesion, most surfaces are covered with mucus so bacteria must first penetrate this to reach the epithelial cell surface. A number of virulence factors may contribute to the ability of bacteria to overcome this barrier (*Table 3*):

- motility caused by the presence of flagella will help the bacteria to move through the mucus;
- mucinases break down mucus.
- ciliostatic proteins can inhibit the movement of cilia on the surface of cells in the naso-pharynx, therefore reducing movement.

Invasion of cells

Some bacteria, such as *Shigella dysenteriae, Listeria monocytogenes, Y. entercolitica* and *S. typhi*, can penetrate through epithelial layers to set up either localized or systemic infections. Bacteria such as *Salmonella* sp. may penetrate through

Table 3. Virulence factors that assist the bacterium to colonize a host

Virulence factors	Function
Flagella and chemotaxis	Movement to an appropriate site
Adhesins	Attachment to cell surfaces; prevents bacteria from being washed out
Mucinases	Break down mucus, help bacteria gain access to cell surface
Ciliostatic proteins	Inhibit action of cilia, preventing removal of bacteria from surface
Enzymes such as DNases, hyaluronidases, proteases, collagenases	Nutrient acquisition; dissemination of the microbe
Toxins	Nutrient acquisition; dissemination of the microbe; destruction of cells of the immune system
Siderophores	Iron scavenging
Invasins	Trigger non-phagocytic cells to take up bacteria

specialist M cells in the gut. These are naturally phagocytic cells which have a role in the gut-associated immune system. Other microorganisms, for example *Shigella* sp., that penetrate through the epithelium have the ability to induce naturally non-phagocytic cells to become phagocytic. Proteins produced by the bacteria called **invasins** bind to receptors in the host cell surface, **integrins**, and induce rearrangements of actin molecules, part of the cell cytoskeleton, under the membrane. The bacteria are then engulfed into the cells by an invagination of the host cell membrane. Subsequently they may pass through the cell in these membrane vesicles (*S. typhi*) or escape from the vesicle into cytoplasm (*Sh. dysenteriae*). The nature of the subsequent disease depends on other virulence factors present on the bacteria. *Sh. dysenteriae* spreads to adjacent cells and causes a localized infection with extensive tissue damage due to multiplication of the bacteria in the cells and to the production of a toxin (Topic F6). In contrast, *S. typhi*, which causes typhoid fever, enters the bloodstream and spreads throughout the body. The disease symptoms of high fever and anorexia are as a result of the presence of the bacteria in the blood and the effect of LPS on the immune system (Topic F6). Recently a number of virulence genes called **Type III secretion systems** have been identified in the proteobacteria. The products of these genes specialize in the direct export of virulence factors into host cells to subvert the normal functioning of these cells to benefit the invading bacteria.

Acquisition of nutrients

In order to replicate, bacteria must be able to acquire nutrients from the host. Bacteria produce many toxins (Topic F6) and enzymes that lyse cells, causing varying amounts of damage ranging from local tissue damage to the death of the host. There is much debate as to the role of these compounds and why they are produced by bacteria, but it may be simply that they are virulence factors which allow the release of nutrients from the cell. Another group of virulence factors that contribute to bacterial growth is siderophores, iron-chelating agents, that can remove iron from transferrin or lactoferrin (*Table 3*). Iron is essential for bacterial growth, and mechanisms that keep the free concentration of iron below the level necessary for bacterial growth are an important aspect of the host defense system (Topic F4).

Spread of bacteria

The spread of bacteria from the sites of infection is aided by the production of a large number of extracellular enzymes, some of which are listed in *Table 3*. Toxins and enzymes that break down cells (Topic F6) may also play a role in the dissemination of the organism.

Avoidance of host defense mechanisms

A large number of virulence factors have been found that help the bacteria to avoid the host defense mechanisms. A list of different types is given in *Table 4* but this is, by no means, all of them. As the initial line of defense for the host is to destroy the bacteria by the action of phagocytes (potentiated by components of the complement system; Topic F4), avoidance of phagocytosis and complement killing are important features of successful pathogens. Phagocytosis may be avoided by:

- killing the phagocyte (cytolytic toxins),
- inhibiting their movement (toxins that inhibit cellular function)
- inhibiting phagocytosis (capsules).

One of the most interesting strategies that has evolved in a number of bacteria is the ability to survive within the phagocytes. This can lead to serious problems as

Table 4. Examples of virulence factors that protect bacteria from the host's immune response

Target	Virulence factors	Function
Phagocytes	Cytolytic toxins	Destruction of phagocytic cells
	Toxins that inhibit cellular function	Inhibition of chemotaxis/mobilization of phagocytes
	Capsules and cell surface proteins such as Protein M of *Strep. pyogenes*	Inhibition of phagocytosis
	Mycobacterial cell-envelope sulfatides	Prevention of phagosome–lysosome fusion
	Listeriolysin O (*L. monocytogenes*)	Breaks down the phagosome membrane allowing the bacteria to escape into the cytoplasm
	Catalase, superoxide dismutase, cell-envelope components	Protection of cell from intracellular killing by phagocytes
Complement	Proteases	Destruction of components of complement
	Capsules and long LPS side chains	Prevent activation of complement or hinder access of complement to surface
Antibodies	IgA proteases	Inactivation of IgA on mucosal surfaces
	Protein A (*Staph. aureus*)	Binding of Fc portion of IgG disguises bacterium by a coat of IgG
	Capsules containing sialic acid	Sialic acid is a host polysaccharide therefore not recognized as foreign
	Antigenic variation	Evasion of the antibody response

bacteria such as *Mycobacterium tuberculosis* and *L. monocytogenes* can persist for a long time in the host without being eradicated. Three different mechanisms can help the bacteria in this survival:

- prevention of fusion of the phagosome with the lysosome;
- escape from the phagosome into the cytoplasm before fusion with the lysosome;
- production by the bacteria of enzymes and other proteins that overcome the effect of the intracellular killing components of the phagosome.

Bacteria interact with complement in a number of ways in order to destroy or minimize its antibacterial activity:

- proteases may destroy the individual components of the proteolytic cascade;
- potential complement-activating components may be hidden by the presence of capsules or long O-antigen side chains on LPS (Topic D3);
- bacteria shed molecules that bind to complement, therefore acting as decoys.

Bacteria have evolved a number of strategies to avoid the effect of the host's specific immune response. Bacteria, such as *N. gonorrhoeae*, that survive on mucosal surfaces often produce specific proteases which destroy IgA, the secretory immunoglobulin. Other bacteria try to prevent the induction of the specific immune reaction by looking like part of the host. They can do this by either producing polysaccharide capsules that resemble host polysaccharides, such as sialic acid (*E. coli*), or by coating themselves with host proteins such as fibronectin. Protein A produced by *Staphylococcus aureus* is a very interesting example of this phenomenon in that it binds to the Fc portion of immunoglobulins (Topic F4) therefore creating a coat of antibodies on its surface but with the antigen-binding part sticking up. This is seen as normal host protein by the host defense system and, therefore, acts as a disguise. Finally, antigenic variation is a very neat way of avoiding the specific immune response. In this case, the bacteria change the antigenicity of some component of their cell surface such as pili or outer membrane proteins. This means that the host's specific antibody response, which is against a component that is no longer there, is ineffective so the bacteria survive.

Organization of virulence genes

It has long been observed that virulence factor genes are often associated with mobile genetic elements. Toxin genes have been found on plasmids (enterotoxin of *E. coli;* Topic E5) and bacteriophage (diphtheria toxin; Topic E7), adhesin molecules are often found on plasmids, as are colicin genes. This suggests that pathogenic bacteria may have gradually acquired a collection of genes by horizontal transmission which allow them to infect a host and cause disease. Recently it has been shown that virulence genes are often grouped together on the chromosome in so called 'pathogenicity islands'. These pieces of DNA have characteristics that are different from the rest of the bacterial DNA suggesting that they have been acquired in a single step from an external source. In addition, insertion sequences and repetitive DNA have been found associated with these elements implying that transposition (Topic E3) may have been involved in their acquisition. This block horizontal transmission of virulence genes is a matter of concern as it implies the evolution of new and even more virulent bacterial pathogens in the future.

F6 Bacterial toxins and human disease

Key Notes

Endotoxins

Endotoxin is the lipid A portion of lipopolysaccharide (LPS) found on the surface of Gram-negative bacteria. It causes septic shock by induction of the host immune system when present in the bloodstream.

Exotoxins

The word exotoxin may be used to describe any extracellular protein produced by bacteria which induces a toxic effect on host cells. Exotoxins may be split into three groups based on their sites of action: membrane-damaging toxins, toxins that act on sites inside the cell, and superantigens, toxins that over-induce the immune system.

Membrane-damaging toxins

Damage to the host cell membranes may be as a result of enzyme activity that destroys components of the cell membrane, such as sphingomyelinases and phospholipases, or by holes made in the membrane by pore-forming proteins such as *Staphylococcus aureus* α-toxin.

Intracellularly acting toxins

Some of the most toxic compounds known to man such as botulinum toxin and diphtheria toxin are bi-partite toxins consisting of two components: a binding (B) part that binds to receptors on the host cell surface and an active (A) component that is the actual toxic part. Once inside the cell many of the known bi-partite toxins act by ADP-ribosylating a molecule in the host cell; for example, the eukaryotic elongation factor EF-2 (diphtheria toxin) and the Gs protein, which regulates the activity of adenylate cyclase in eukaryotic membranes (cholera toxin). Other activities associated with toxins include ribonuclease (Shiga toxin) and metallopeptidases (botulinal and tetanus toxin).

Related topics

Bacterial cell envelope and cell wall synthesis (D3)
Human host defense mechanisms (F4)
Entry and colonization of human hosts (F5)

Endotoxins

Traditionally, bacterial toxins have been split into two groups on the basis of whether they are associated with the bacterial cell, **endotoxins**, or released into the medium by growing bacteria, **exotoxins**. Endotoxin activity is due to the lipid A portion of LPS, which is an integral part of the outer membrane of Gram-negative bacteria (Topic D3). The endotoxin is normally released from Gram-negative cells in the bloodstream after lysis as a result of the action of the host immune system or by the action of some antibiotics. Once released, the LPS activates the complement cascade by the alternative pathway (Topic F4) and induces the release of cytokines and other defense mediators from macrophages and other cells. These bioactive compounds are a normal part of the host defense mechanisms and help to rid the body of infection, but when induced by LPS, the amounts of these are much higher than normal and cause toxicity to the host itself. This is called **septic** shock which has the symptoms of fever, circulatory collapse and organ failure, and eventually death. Components of Gram-positive cells, including lipoteichoic acids and peptidoglycan, can also induce septic shock.

Exotoxins

Traditionally, the word exotoxin has been used to describe bacterial proteins produced by both Gram-negative and Gram-positive bacteria that are toxic to host cells. Generally, the word exotoxin is shortened to toxin, when describing bacterial proteins, as not all exotoxins are necessarily secreted into the surrounding medium but may be held on the surface of the cell or in the periplasmic space. The exotoxins are generally heat labile; there are a number of heat-stable toxins such as the *E. coli* ST enterotoxin. They are water soluble, so can act at a site distant from the original site of infection. In some cases, such as botulism, there is no need for colonization of the host by the bacterium at all, the toxin alone is enough to cause the disease (toxemia).

The nomenclature of toxins is varied:

- Some toxins, especially those that are the clear cause of disease symptoms, are named after the bacteria that produces them such as cholera toxin (produced by *Vibrio cholerae*) or diphtheria toxin (*Corynebacterium diphtheriae*).
- Others are named after their sites of action such as neurotoxins (acting on nerve cells), enterotoxins (acting on the gut) or leukotoxins (active against white blood cells). Toxins that are active against a number of cell types may just be described as cytotoxins.
- Some toxins are just given a letter or number such as *Pseudomonas aeruginosa* toxin A or *Staphylococcus aureus* α-toxin.
- Toxins may be named on the basis of their activities such as phospholipase C (produced by *Listeria monocytogenes*) and adenylate cyclase produced by *Bordatella pertussis*.

Some bacteria, especially the Gram-positives, may produce a number of different toxins, others may only produce one. Toxin production is generally very carefully controlled at the level of transcription in bacteria (Topic C5) and, in some cases, only a subset of bacteria within a strain will actually secrete a toxin. Most toxins are either membrane-damaging toxins or toxins that act inside the cell. A small third group, called **superantigens,** interact with T-helper cells, causing over-induction of the immune system and symptoms similar to shock.

The role of toxins in the disease process is a matter of some debate. However, as discussed in Topic F5 they may play a number of roles including:

- the release of nutrients from cells;
- destruction of cells associated with the immune defense;
- escape from the phagolysosome by intracellular bacteria.

Membrane-damaging toxins

These cytolytic toxins cause lysis by destroying the integrity of the plasma membranes of the host cell. Usually they are active against a wide range of cell types and can be divided into two groups on the basis of their mechanisms of action: enzymic action and pore forming. A large number of different Gram-positive bacteria produce enzymes such as phospholipases and sphingomyelinases, which break down the lipid components on the membrane. This destabilizes the membrane, eventually leading to cell lysis. The second group of toxins includes the thiol-activated cytolysins, streptolysin O (produced by *Streptococcus pyogenes)* and listeriolysin O (produced by *L. monocytogenes)* and *Staph. aureus* α-toxin. These toxins form pores in the host cell membranes which results in the leakage of nutrients and essential ions from the cell and eventually cell lysis.

Intracellularly acting toxins

The most potent group of toxins known to man are the bacterial exotoxins that act on intracellular targets. These include proteins such as botulinum toxin which inhibits the release of neurotransmitters in the peripheral nervous system, cholera toxin that acts in the gut to induce watery diarrhea and diphtheria toxin which inhibits protein synthesis (*Table 1*). Although these toxins cause a very disparate set of symptoms at different sites in the body they do tend to have a common **bipartite** structure. One part of the molecule is responsible for binding to specific receptors in the host cell membrane, called the B-portion while a separate part of the molecule, the A-portion carries the toxic enzymic activity. These toxins are, therefore, frequently called **A–B type toxins**. The structure of two of these toxins is shown in *Fig. 1*. Diphtheria toxin is one of the simplest of the A–B toxins consisting of just one A and one B portion. These are synthesized together as a single molecule, then cleaved to give two peptides joined by a disulfide bridge. In contrast, cholera toxin consists of five B subunits surrounding an active A_1–A_2 molecule, which is synthesized separately.

The B portion is responsible for the cell specificity of most of these toxins. It will bind only to cells that have the correct receptor. Hence, in the case of botulinum toxin, the receptor glycoprotein or glycolipid is only found in the peripheral neurones whereas diphtheria toxin binds to growth factor precursor found on many cell types. After binding to a receptor, the toxin–receptor complex is internalized in the cell either by endocytosis, followed by translocation of the A portion from the endocytic vacuole into the cytoplasm, or by translocation of the A portion across the host plasma membrane, directly into the cytoplasm.

In many cases the enzymic activity of the A–B-type toxins is ADP-ribosylation of a target site in the cell. The A-portion catalyzes the removal of the ADP-ribosyl group from NAD and the attachment of that group to a cellular protein. The target proteins and the consequences of the ADP-ribosylation are various, as shown in *Table 1*, and the varied nature of these explains why toxins cause a wide range of different symptoms. Diphtheria toxin, for example, inhibits protein synthesis, with one molecule of the toxin enough to kill the cell. In

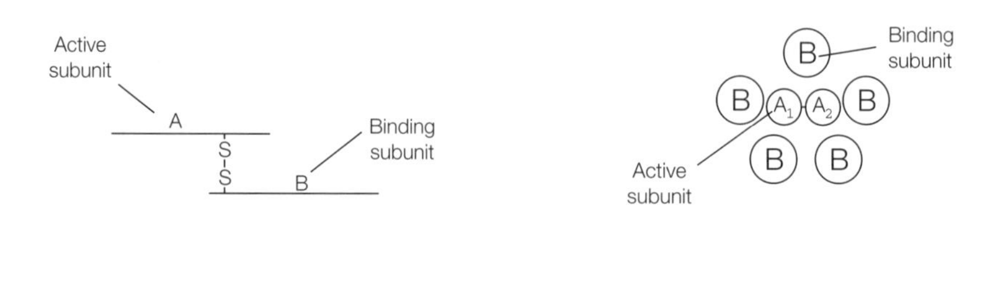

Fig. 1. Structure of A–B type exotoxins.

Table 1. Some examples of bacterial exotoxins that act inside the cell

Bacteria	Toxin	Site of action	Mode of action	Symptom/role in disease
Vibrio cholerae *E. coli*	Cholera toxin LT toxin	Intestine	ADP-ribosylation of Gs (stimulatory) protein which regulates adenylate cyclase	Water secretion into the intestine resulting in watery diarrhea
Corynebacterium diphtheriae *Pseudomonas aeruginosa*	Diphtheria toxin Exotoxin A	Many cell types	ADP-ribosylation of EF-2 leads to inhibition of protein synthesis	Cell death, general organ damage
Bordatella pertussis	Pertussis toxin	Many cell types	ADP-ribosylation of Gi (inhibitory) protein which regulates adenylate cyclase	Adherence to cells, cell damage, fluid secretion
Shigella dysenteriae *E. coli*	Shiga toxin Vero (Shiga-like) toxin	Many cell types	RNA glycosidase enzyme modifies 28S rRNA in 60S ribosome subunit. Inhibits protein synthesis	Cell death
Clostridum tetani	Tetanus toxin	Neurons in central nervous system	Metallopeptidase inhibits release of neurotransmitters	Spastic paralysis
Cl. botulinum	Botulinal toxin	Peripheral neurons	Metallopeptidase inhibits release of neurotransmitters	Flaccid paralysis

contrast, cholera toxin modifies Gs protein, a regulatory protein of adenylate cyclase, causing it to be permanently switched to the production of cAMP. In the gastrointestinal tract, where the toxin acts, the most significant effect of the rise in cAMP is the production of an ion imbalance, leading to massive water loss from the cell. This is seen as copious watery diarrhea, causing dehydration and death, if not treated.

Other enzymic activities are also associated with the A portion of A–B type toxins. Shiga toxin, produced by *Shigella dysenteriae*, is a ribonuclease which inactivates the 60S subunit of the ribosome. The neurotoxins, botulinum toxin, which causes flaccid paralysis, and tetanus toxin, which induces spastic paralysis, are both zinc-requiring endopeptidases. The different symptoms caused by each toxin are probably as a result of the different cell specificities of the toxins. Botulinum toxin acts on the peripheral nervous system whereas tetanus toxin interferes primarily with the central nervous system.

F7 Control of bacterial infection

Key Notes

Disease in the population

The most effective tools in the control of bacterial diseases have been public health policies, which have removed pathogenic bacteria from water and food and reduced their populations in the environment. Similarly, better housing and nutrition have raised the health of individuals, thus making them more resistant to infection.

Vaccination

Whole microbes (live, attenuated or killed), fractions of microbes or inactivated microbial products, such as toxoids, may be used to induce a specific immune response in a host. This provides protection from infection with that microbe in the long term. Vaccines are not available for all bacterial infections.

Antibiotics

Antibiotics are molecules, produced by microbes, which inhibit the growth, or kill, other microbes. The antibiotics normally used to treat bacterial infections are those that are selectively toxic to bacterial cells and do not harm the host. These antibiotics target sites such as peptidoglycan, ribosomes and nucleic acid synthesis which are significantly different between prokaryotic and eukaryotic cells. The effectiveness of antibiotic therapy is endangered by the development of bacterial antibiotic resistance and, particularly, by the spread of antibiotic-resistance genes between bacteria.

Related topics

Translation (C7)
Bacterial cell envelope and cell wall synthesis (D3)
Plasmids (E5)
Human defense mechanisms (F4)
Virus vaccines (J10)
Antiviral chemotherapy (J11)

Disease in the population

Deaths from bacterial diseases have become a rare occurrence in the developed world in the last 50 years as a result of a number of measures designed to reduce the spread of (and even eradicate) pathogenic bacteria in our environment and to improve the health of the population. These are shown in *Table 1*. The importance of sanitation, good housing and nutrition in reducing the incidence of bacterial disease cannot be emphasized enough. Even before the introduction of vaccines and antibiotics, the reduction of diseases such as tuberculosis had occurred as a result of public health measures, designed to prevent the spread of the causative microorganism, *Mycobacterium tuberculosis*, between individuals. In the developing countries, where there is a lack of sanitation and a large incidence of malnutrition, bacterial diseases are still a major cause of death.

Vaccination

Vaccination has proved to be a useful way of protecting the individual, and the population, from a number of bacterial diseases. Vaccination, or active immu-

Table 1. Strategies used to control bacterial infections

Prevention of the spread of bacteria within the population	Improvements in water treatment and sewage disposal prevents the spread of water-borne diseases such as cholera Introduction of food hygiene regulations stops the spread of food-borne diseases such as typhoid Reduction in overcrowding prevents the spread of aerosol-borne diseases such as tuberculosis Control of the vector (e.g. rats or mosquitos) prevents the spread of animal-borne diseases Vaccination reduces the incidence of the organism in the environment Education improves the general standard of hygiene in the population
Improvement in the health of the individual	Better nutrition improves an individual's resistance to disease Vaccination protects an individual against infection with specific bacteria Antibiotics can cure an individual of a bacterial infection

nization, is the artificial introduction of antigens from a microbe into an individual, in a controlled way, leading to the stimulation of the immune system without the symptoms of the full-blown disease. This results in the production of memory cells within the host (see Topic F4) so that, on a second encounter with the microbe, the immune system can generate a rapid antibody response: thereby preventing infection. Vaccines may be whole cells, cellular fractions (e.g. cell-surface components such as capsule material) or inactivated bacterial toxins (toxoids). Both live and dead bacteria are used in vaccines. Live vaccines are more useful as they mimic more effectively the natural disease process, causing long-lasting immunity. However, the bacteria must be **attenuated** in some way to ensure that they do not cause significant disease in the host. Examples of some of the bacterial vaccines used today are given in *Table 2*. Although vaccines have been particularly useful in the control of diseases such as diphtheria, tetanus (both are toxoid vaccines) and tuberculosis (a live, attenuated vaccine), there are many bacteria for which it has proven difficult to produce a safe, effective vaccine. Finally, passive immunization, the introduction of antibodies to a particular microbe, is used occasionally to protect against infection.

Antibiotics

Molecules that inhibit the growth of, or kill, bacteria are called **antibacterial agents**. If these compounds are isolated from microbes they are called **antibiotics**. There are many millions of antibiotics, produced mainly by soil bacteria

Table 2. Examples of vaccines available for protection against bacterial diseases

Disease	Vaccine components
Diphtheria	Inactivated toxin (toxoid)
Tetanus	Inactivated toxin (toxoid)
Tuberculosis	Attenuated *Mycobacterium bovis* (BCG)
Whooping cough	Subcellular fractions and pertussis toxoid
Haemophilus influenzae meningitis	Capsular polysaccharide – linked to a protein carrier
Meningococcal meningitis	Capsular polysaccharides
Typhoid	Killed cells of *Salmonella typhi* Live oral attenuated strain Ty21A Polysaccharides

and fungi, that are active against bacteria but only a few can be used to control human bacterial disease. These are the few that, although they are toxic to bacteria, have no significant effect on the human host. The reason for this so-called **selective toxicity** is that the sites at which these antibiotics act are either, unique to bacteria such as peptidoglycans (Topic D3) or, very different between prokaryotes and eukaryotes such as ribosomes (Topic C7) and nucleic acid synthesis (Topics C2 and C4). *Table 3* shows the sites in the bacterial cell at which a number of clinically important antibiotics act.

Some of the factors that affect antibiotic therapy are listed below:

- The antibiotic must reach the site of infection in the host.
- The antibiotic has to reach its target site in the cell. This is easier for antibiotics, such as penicillin, which act on peptidoglycan than for those, like tetracycline, which must penetrate through the plasma membrane to reach their target sites, the ribosomes.
- Gram-negative bacteria are often intrinsically resistant to the action of antibiotics due to the presence of the outer membrane which acts as an additional barrier for the antibiotic to cross and protects the peptidoglycan (Topics D2 and D3).
- Broad-spectrum antibiotics are effective against a wide range of different Gram-positive and Gram-negative bacteria whereas other antibiotics may have only a narrow range.
- All the pathogenic bacteria must be eradicated from the host by either inhibiting the growth of the microbes (**bacteriostatic** antibiotics), which can then be removed by the immune systems, or by killing them directly (**bactericidal** antibiotics).

Antibiotics have proved to be of great benefit to humankind and have ensured that people no longer need die from diseases such as streptococcal scarlet fever, or as a result of wound infections. However, there is a problem. Bacteria can become resistant to the action of antibiotics. The mechanisms by which they do this include: the production of enzymes that break down the antibiotic, reduction in permeability to the antibiotic and alterations to the target site as shown in *Table 4*.

Antibiotic resistance may arise by mutation (Topic E1) but more often the genes for antibiotic resistance are transferred between bacteria by conjugation, transduction and transformation (Topics E6, E9 and E10). Antibiotic resistance, carried by plasmids, has caused particular concern as these plasmids may carry

Table 3. Target sites for antibiotics in bacterial cells

Target site	Mode of action	Example
Cell wall (peptidoglycan biosynthesis,Topic D4)	Inhibition of cross-linking	Penicillins and cephalosporins (β-lactam antibiotics)
	Inhibition of polymerization	Glycopeptide antibiotics (e.g. vancomycin)
Protein synthesis (Topic C7)	Inhibition of translocation of ribosome	Aminoglycoside antibiotics (e.g. Gentamicin)
	Inhibition of binding of aminoacyl tRNAs	Tetracycline
Nucleic acid synthesis (Topic C2)	Inhibition of tetrahydrofolic acid synthesis	Sulfonamides and trimethoprim
	Inhibition of DNA gyrase	Quinolone antibiotics (e.g. ciprofloxacin)

Table 4. Common mechanisms of antibiotic resistance in bacteria

Mechanism of resistance	Examples	
Antibiotic inactivation	β-Lactamase	Penicillin resistance
	Chloramphenicol acetyl transferase	Chloramphenicol resistance (CAT)
	Aminoglycoside modifying enzymes	Aminoglycoside resistance
Reduction in permeability	Reduced uptake	Natural resistance of many Gram-negative bacteria due to presence of outer membrane
	Antibiotic efflux	Tetracycline resistance
Alteration of target site	Change in target site so that it is no longer sensitive to drug	Sulfonamide resistance
	New target site produced that is not sensitive to the antibiotic	Methicillin resistance
	Overproduction of target site	Trimethoprim resistance

genes that confer resistance to many different antibiotics at the same time (Topic E5). Multiply-resistant bacteria are therefore becoming a problem, particularly in the hospital environment, and the fear is that it will not be long before there is a bacterial strain that is untreatable by all known antibiotics.

G1 TAXONOMY

Key Notes

Current taxonomic status of the eukaryotic microbes

The objective of current taxonomic schemes is to create monophyletic groups of microorganisms which are assumed to have a single ancestor. This is achieved by studying characters of an individual and comparing them with those of other, closely related, organisms. Using such information phylogenetic trees can be constructed that can indicate probable evolutionary sequences of the eukaryotic microbes.

The protista

Previously, members of this group were grouped into two form groups, the algae and protozoa. A more modern approach is to group both colorless and pigmented species together in a monophyletic taxonomy.

Amitochondrial protists

At the base of the phylogenetic tree is a group of amitochondrial protests, i.e. microbes that lack mitochondria and live anaerobically. It is from this group that other lineages of eukaryotic microorganisms have evolved.

Mitochondrial protists

Mitochondrial protists possess mitochondria with either discoid, tubular or lamellar cristae, depending on their position on the phylogenetic tree. They are aerobic organisms. They can be free-living, symbiotic or parasitic on other species.

Fungi and chlorophyta

Fungi and chlorophyta have lamellar cristae in their mitochondria and they share much of their cellular structure with higher plants and animals.

Sequence of evolutionary events

A number of phenomena prevent us from knowing what the exact events of evolution were. Both primitive and derived features can be found in some protists, and loss of features, parallel evolution and transfer of genetic material can all contribute to the difficulty in creating a definitive tree.

Related topics

Prokaryote taxonomy (D1)
Fungal structure and growth (H1)
Chlorophytan and Protistan taxonomy and structure (I1)
Virus taxonomy (J2)

Current taxonomic status of the eukaryotic microbes

Establishing relationships within the different members of the fungi, chlorophyta and protistan microbes relies on studying **characters**, which are features or attributes of an individual organism that can be used to compare it with another organism. These features can be morphological, anatomical, ultrastructural, biochemical or based on sequences of nucleic acids. The objective of such a study is to create **monophyletic** groups which are assumed to have a single ancestor, usually extinct; a similar approach is used in the creation of classification systems for prokaryotes (Topic D1). A **cladistic** approach would not assume features from an ancestor, but would merely define a monophyletic group on the basis of shared characters. New information based on the presence and type of mitochondria, and the DNA sequencing of ribosomal RNA, place the fungal, chlorophytan and protistan members of the eukaryotic microbes into a complex **phylogenetic tree** (*Fig. 1*).

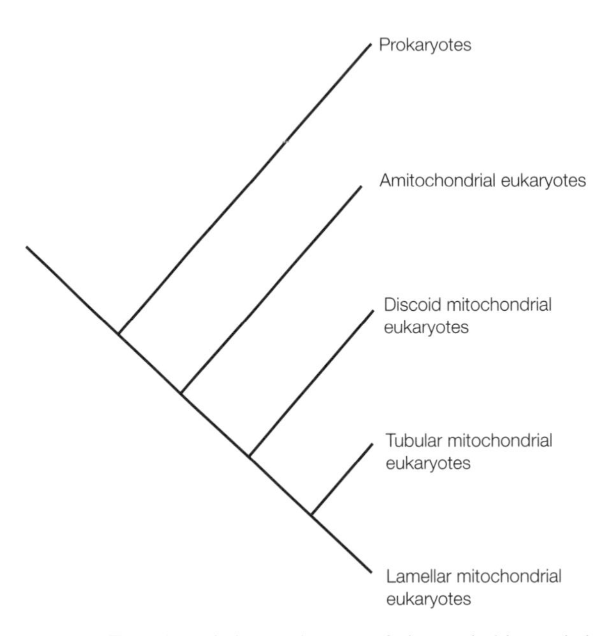

Fig. 1. Tentative phylogenetic tree of the probable evolutionary sequence of protistan microbes.

The protista

The protista are a paraphyletic group of organisms that are not animals, true fungi or green plants. There are approximately 200 000 named species. Traditionally they were classified into non-monophyletic, adaptive groups called the flagellates, algae and protozoa. There are about 80 different patterns of organization, clustered into 60 lineages.

In this book we have followed a classification based on current information published on the Tree of Life website.

Amitochondrial protists

At the base of the tree are amitochondrial protists, which separated from other lineages before mitochondria were acquired (see Topic I1). These are anaerobic organisms that may inhabit sediments or animal intestines or they can be animal parasites.These organisms most strongly resemble the colorless, **anaerobic** cell from which, by symbiotic acquisition of mitochondria and in some cases chloroplasts, the rest of the lineages developed.

Mitochondrial protists

The next group in the protistan lineage are the **euglenozoa**, those organisms that have mitochondria with **discoid christae**. These organisms are **aerobes**. The euglenids, ameboflagellates and acrasid slime molds share these characteristics (see Topic I1). Many members of this group have two flagellae. Some are free-living while others are parasites within animals and man.

Organisms having mitochondria with **tubular** cristae, the **alveolata** appear next on the phylogenetic tree, and within this group we see the myxomycete and dictyosteloid slime molds, dinoflagellates, ciliates, apicomplexans, diatoms, oomycetes, radiolaria and heliozoans.

The fungi and chlorophyta

At the top of the tree are organisms that have mitochondria with **lamellar** cristae, the chlorophyta, chytrids, higher fungi, plants and animals (see Topics H1 and I1).

Sequence of evolutionary events

Exact orders of evolution are difficult to determine, because of morphological inconsistencies within groups. For instance, dinoflagellates have a number of less specialized features, but they also have more derived features including being biflagellate and having scales, a characteristic which brings them close to the alveolata (see Topic I1).

Even sequencing of nucleic acid to provide phylogenetic **gene trees** may not answer all the questions. For instance, sequencing the small subunit nuclear-encoded rRNA reveals that one can follow the evolution of an individual gene, but not necessarily the whole organism, as different regions of rRNA genes evolve at different rates. Phylogenetic patterns may also be confused because of transfer of genetic information between lineages as a consequence of **endosymbiosis** and other mechanisms. **Parallel** evolution within these organisms is also likely to occur, and superficial similarities within distinct lineages may arise because of loss of features, which is likely to be a significant factor within these apparently simple organisms.

G2 EUKARYOTIC CELL STRUCTURE

Key Notes

Eukaryotes

Eukaryote cells have complex, membrane-bound, subcellular organelles which compartmentalize cell functions. The distinguishing feature of a eukaryotic cell is the nucleus.

Plasma membrane

The plasma membrane is a semi-permeable barrier between the outside and inside of the cell, and it is involved in cell–cell recognition, endo- and exocytosis and adhesion to surfaces. Transport systems in the membrane allow it to import materials selectively into the cell.

Cytoplasm

The cytoplasm is 70–85% water, but also contains proteins, sugars and salts in solution. The organelles of the cell are suspended in the cytoplasm. Both fungal and photosynthetic protista have single membrane-bound vacuoles in their cells.

Cytoskeleton

The cytoskeleton of the cell is made up of microtubules, intermediate filaments and microfilaments, which maintain the shape of the cell and carry out numerous functions such as motility and the transport of organelles.

Nucleus and ribosomes

The nucleus is a double membrane-bound organelle which contains the chromosomal DNA of the cell. Inside the nucleus is the nucleolus, which is the site of ribosomal RNA synthesis. Ribosomes are made of two subunits of RNA plus proteins, and they are the site of DNA translation and protein synthesis.

Endoplasmic reticulum

The endoplasmic reticulum (ER) is a complex of membrane tubes and plates which is continuous in places with the nuclear membrane. The ER can be smooth or it may be termed rough where ribosomes are attached to it. The main function of this organelle is the synthesis and transport of proteins and lipids.

Golgi body

The Golgi body is a series of flattened, membrane-bound fenestrated sacs and vesicles. Vesicles secreted from the ER fuse with the *cis*-Golgi, and their contents are then further processed by resident biochemical processes. Processed materials are then secreted from the *trans*-Golgi in vesicles which fuse with other organelles or with the plasma membrane.

Lysosomes and peroxisomes

Lysosomes and peroxisomes are membrane-bound sacs secreted from the Golgi. Lysosomes contain acid hydrolases involved in intracellular digestion. Peroxisomes contain amino and fatty acid-degrading enzymes and the enzyme catalase, which detoxifies hydrogen peroxide released by degradative processes.

Mitochondria

Mitochondria are the site of respiration and oxidative phosphorylation in aerobic organisms. They are bound by a double membrane, the inner one being in-folded to form plates or tubes called cristae. ATP production is located in particles attached to the cristae.

Hydrogenosomes

Hydrogenosomes are organelles found in some amitochondrial, anaerobic groups. Their function is energy production. They contain enzymes of electron transport which use terminal electron acceptors that generate hydrogen.

Glycosomes

Glycosomes contain the enzymes of glycolysis and are found only in the apicomplexa.

Chloroplasts

Chloroplasts are double membrane-bound organelles which contain the photosynthetic pigment chlorophyll. Within the chloroplast are stacks of flattened sacs called thylakoids where the photosynthetic systems are located.

Cell walls

Cell walls are found in the photosynthetic protista (cellulose-based) and the fungi (chitin-based). They delimit the outside of the cell from the environment and are important in maintaining cell rigidity and controlling excess of water influx due to osmosis.

Flagella

Flagella are microtubule-containing extensions of the cell membrane. They provide the cell with motility by their flexuous bending, which is controlled by the microtubule motor protein dynein.

Cilia

Cilia have the same internal structure as flagellae but they are smaller and more numerous. Co-ordinated movement, which can be seen as wave-like beating, is required for motility.

Contractile vacuoles

Contractile vacuoles are found in free-living freshwater protozoa. They expel water absorbed into the cell by osmosis through pores in the cell surface.

Related topics

The microbial world (A1)
Heterotrophic pathways (B1)
Electron transport, oxidative phosphorylation and β-oxidation of fatty acids (B2)
Autotrophic reactions (B3)
DNA replication (C2)
Transcription (C4)
Translation (C7)
Prokaryote cell structure (D2)
Bacterial cell envelope and cell wall synthesis (D3)
Taxonomy (G1)
Cell division and ploidy (G3)
Chlorophytan and protistan taxonomy and structure (I1)

Eukaryotes

Eukaryotic cells are compartmentalized by **membranes**. The cell contains several different types of membrane-bound organelle in which different biochemical and physiological processes can occur in a regulated way (*Fig. 1*). Membranes also transport information, metabolic intermediates and end-products from the site of biosynthesis to the site of use.

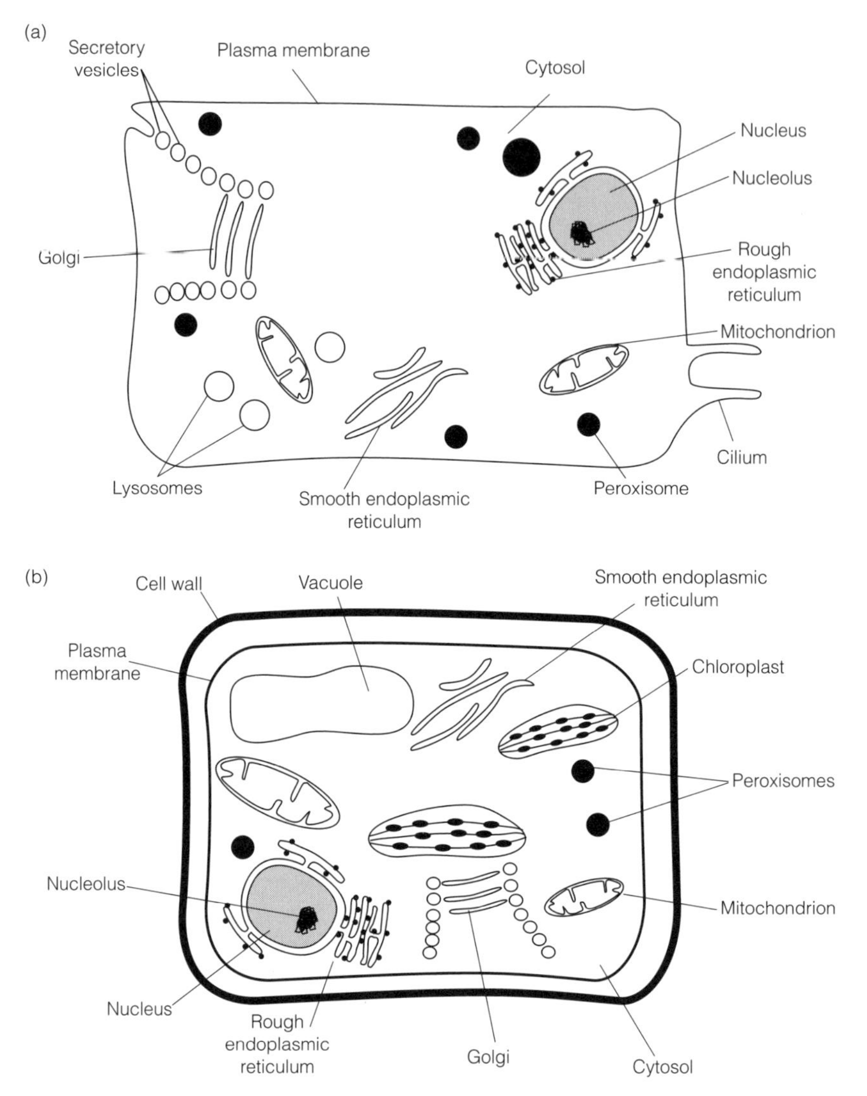

Fig. 1. Eukaryotic cell structure: (a) structure of a typical animal cell (b) structure of a typical plant cell.

Plasma membrane

The plasma membrane of eukaryotes is a **semi-permeable** barrier that forms the boundary between the outside and the inside of the cell. It is similar to that of the prokaryotes (see Topic D3), except that it contains **sterols**, flat molecules that give the membrane a greater rigidity and which stabilize the eukaryotic cell. There are transport systems in the membrane that selectively import materials into the cell, and it is also involved in **endo**- and **exo-cytosis**, where food particles are engulfed and waste products are expelled from the cell in membrane-bound vesicles. The plasma membrane is involved in key interactive processes between cells, like **cell–cell recognition** systems, as well as adhesion of the cell to solid surfaces.

Cytoplasm

The cytoplasm is a dilute solution (70–85% water) of proteins, sugars and salts in which all other organelles are suspended. It has **sol–gel** properties, which means it can be liquid or semi-solid depending on its molecular organization. Vacuoles act as storage sites for nutrients and waste products in a very weak solution. The high water content of the vacuole maintains a high cell turgor pressure.

Cytoskeleton

The eukaryotic cell is further stabilized by a cytoskeleton made up of **tubulin** containing **microtubules** (25 nm diameter), and **microfilaments** (4–7 nm diameter) and **intermediate fibers** (8–10 nm diameter). The cytoskeleton is a dynamic structure, providing support for the cell but also the machinery for ameboid movement, cytoplasmic streaming and nuclear and cell division (see Topic G3).

Nucleus and ribosomes

The nucleus contains the DNA of the microbe. In eukaryotic microbes the nucleus usually contains more than one **chromosome**, in which the DNA is protected by **histone** proteins. In diploid organisms these chromosomes are paired (see Topic G3). However, in some protista there are two distinct types of nuclei within one cell, and they may be polyploid. The larger of the two nuclei are termed **macronuclei** and this nucleus is associated with cellular function. The smaller **micronucleus** functions in controlling reproduction.

The nucleus is surrounded by the **nuclear membrane**, a double membrane which is perforated by many pores where the two membranes fuse. It is via these pores that the nucleus remains in constant control of the rest of the cell machinery via **mRNA** and **ribosomes**. In places the nuclear membrane is also continuous with the endoplasmic reticulum. Within the nucleus is the **nucleolus**, an RNA-rich area where rRNA is synthesized (see Topics C4 and C7 for DNA transcription and translation). Eukaryotic ribosomes are essentially very similar to those of prokaryotes (see Topic C7) but they are slightly larger, their two subunits are of 60S and 40S, making a dimer of 80S. Their function is the same as that of prokaryotes (see Topic C7).

Endoplasmic reticulum

The outer membrane of the nucleus is in places continuous with a complex, three-dimensional array of membrane tubes and sheets, the endoplasmic reticulum (ER). Tubular ER can be studded with ribosomes, and described as **rough ER** (RER), where **ribosomal translation** and **protein modification** takes place (see Topics C2 and C4). These proteins can either be secreted into the lumen of the ER or inserted into the membrane. Plates of **smooth ER** are associated with **lipid synthesis** and **protein** and **lipid transport** across cells.

Golgi

The Golgi is composed of stacks of a flattened series of membrane-bound sacs or **cisternae**, surrounded by a complex of tubes and vesicles. There is a definite **polarity** across the stack, the ***cis*** or forming face receiving vesicles from the ER, the contents of these vesicles then being processed by the Golgi, to be budded from the sides or the ***trans*** (maturing) face of the organelle. The Golgi apparatus processes and packages materials for secretion into other subcellular organelles or from the cell membrane. Golgi in fungi are less well developed than in algae, and tend to have fewer or single cisternae. They are sometimes termed **dictyosomes.**

Lysosomes and peroxisomes

The Golgi body generates these single-membrane-bound organelles which contain enzymes (**acid hydrolases** in the lysosome, **aminases, amidases** and

lipases plus **catalase** in the peroxisomes) needed in the digestion of many different macromolecules. The internal pH of the lysosome is **acidic** (pH 3.5–5) to enable the enzymes to work at the optimum pH, and this pH is maintained by proton pumps present on the membrane. The breakdown of amino and fatty acids by the peroxisomes generates **hydrogen peroxide**, a potentially cytotoxic by-product. The enzyme **catalase**, also present in the peroxisome, degrades the peroxide into water and oxygen, protecting the cell.

Mitochondria

Mitochondria are double-membrane-bound organelles where the processes of **respiration** and **oxidative phosphorylation** occur (see Topics B1 and B2). They are approximately 2–3 mm long and 1 mm in diameter. Their numbers in a cell vary. They contain a small, circular DNA molecule which encodes some of the mitochondrial proteins, and 70S ribosomes (see Topics C4 and C7). The inner membranes of the mitochondria contain an **ATP/ADP transporter** that moves the ATP, which is synthesized in the organelle, outwards into the cytoplasm. ATP production is located on particles attached to the cristae, the inner infolded mitochondrial membrane (*Fig. 2*). Their structures differ slightly between the three protistan groups. Not all protista have mitochondria (see Topic I2) and in these cells metabolism is essentially anaerobic (see Topic I4). In the aerobic, very primitive eukaryotes' mitochondrial **cristae** are discoid (see Topic G1). Mitochondria of the fungi are large and highly lobed with flat, plate-like cristae, while those of the chlorophyta have much more inflated cristae.

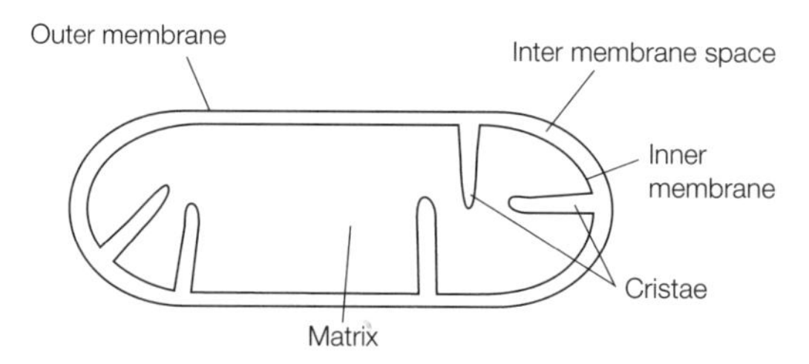

Fig. 2. Structure of a mitochondrion.

Hydrogenosomes

Hydrogenosomes are unique organelles found in anaerobic protista that lack mitochondria. They are membrane-bound organelles containing **electron-transport pathways** in which **hydrogenase** enzymes transfer electrons to terminal electron acceptors which generate molecular hydrogen (see Topic I4) and ATP.

Glycosomes

Glycosomes are unique to the protistan group Apicomplexa. This organelle is surrounded by a single unit membrane and contains the **enzymes of glycolysis** (see Topic B1).

Chloroplasts

Chloroplasts are **chlorophyll**-containing organelles that can use light energy to fix carbon dioxide into carbohydrates (**photosynthesis**) (see Topics B3 and I2). They are bound by double membranes and contain flattened membrane sacs called **thylakoids** where the light reaction of photosynthesis takes place (*Fig. 3*). In photosynthetic protista and the chlorophyta these organelles are large, almost filling the cell. The **pyrenoid** is a proteinaceous region within the chloroplast where polysaccharide biosynthesis takes place.

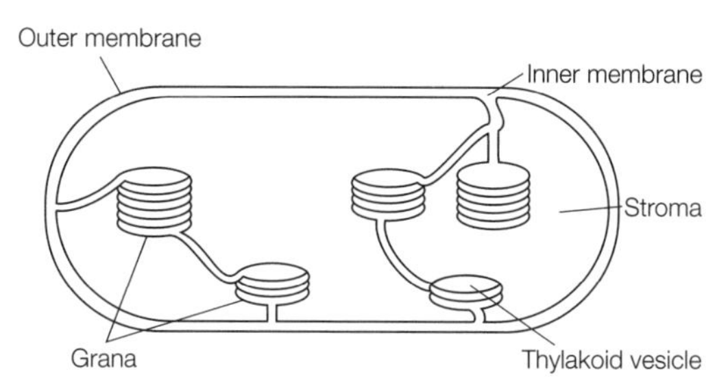

Fig. 3. Structure of a chloroplast.

Cell walls

The protoplasts of fungal and photosynthetic protistan cells are in most cases surrounded by rigid cell walls. In the fungi the cell wall is composed of a microcrystalline polymer of **chitin** (repeating units of β1–4 linked NAG) and amorphous β-glucans, while in the chlorophyta and some other photosynthetic protista, cell walls are composed of **cellulose** (repeating units of β1–4 linked glucose) and **hemicelluloses** (see Topics H1 and I1).

Flagella

Flagella are membrane-bound extensions of the cell, which contain microtubules (see Topic I2). The microtubules are arranged as a bundle of nine doublets around the periphery of the flagellum, with a pair of single microtubules running within them. This structure is called the **axoneme**. Flagella provide cells with motility, because they flex and bend when supplied with ATP. Each outer pair of microtubules has arms projecting towards a neighboring doublet (*Fig. 4*) and a spoke extending to the inner pair of microtubules. Microtubules are formed from **tubulin,** a self-assembling protein**.** Tubulin is composed of two subunits α and β arranged in a helical fashion. The projecting arms between outer subunits are made up of the protein, **dyenin**. This protein is involved in converting the energy released from ATP hydrolysis into mechanical energy for flagellar movement. Movement is produced by the interaction of the dyenin arms with one of the microtubules of adjacent doublets. A basal body (**kinetosome**) anchors the flagellum within the cytoplasm.

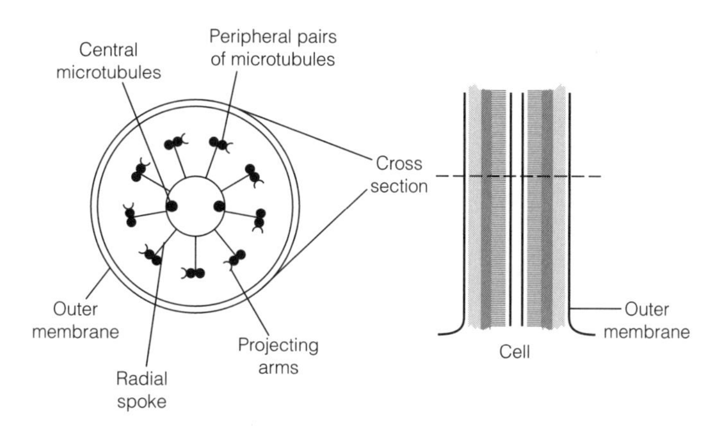

Fig. 4. General structure of a flagellum or cilium.

Cilia

Cilia have the same internal structure as flagellae but they are typically shorter in length. They are usually present on a cell in great numbers and propel the cell by co-ordinated beating seen as waves over the surface of the organism.

Contractile vacuoles

Contractile vacuoles are found in free-living freshwater non-photosynthetic protistans. Their function is to regulate osmotic pressure within the cell by expelling water from a central vacuole through a pore in the outer surface. The simplest contractile vacuoles consist of a vacuole that can form anywhere in the cell, to a fixed structure that is surrounded by bands of microtubules and surrounded by collecting canals that collect fluid from the cytoplasm (*Fig. 5*).

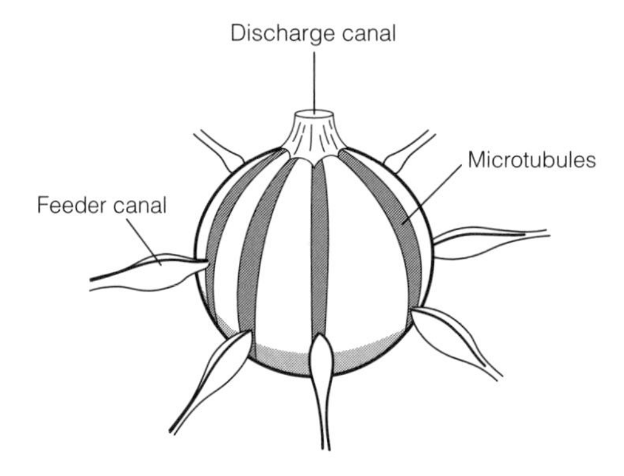

Fig. 5. Diagram of a contractile vacuole.

G3 CELL DIVISION AND PLOIDY

Key Notes

Cell cycle

The cell cycle describes all the events that occur in a cell from the end of one cell division to the end of the next. There are four phases in the eukaryotic cell cycle, which include cell growth (G1), DNA synthesis (S), a second gap or growth phase (G2) and finally nuclear division (N).

Mitosis and asexual cell division

Mitosis is nuclear division which results in progeny nuclei that are identical to the parent. It is usually followed by cytokinesis, cell division, which produces cells that have the same phenotype as the parent. In some eukaryotic microbes multiple nuclear divisions may occur without cytokinesis, giving rise to large, multinucleate cells. There are four stages of mitosis, prophase, metaphase, anaphase and telophase. During prophase chromosomes duplicate to form chromatids, joined at the centromeres. Centromeres are attached to spindle microtubules. During metaphase, chromatids are arranged across centre of the cell, and during anaphase the microtubules pull the sister apart to the poles of the dividing cell. During telophase the microtubules disappear and the nuclear envelopes fuse.

Meiosis and sexual cell division

Meiosis is a nuclear division where there is a halving of chromosome numbers from a diploid to a haploid state. In organisms with an extended diploid phase of their life cycle, meiosis produces haploid gametes, and it is immediately followed by gamete fusion and formation of a new diploid organism. In organisms with an extended haploid stage, diploid formation is immediately followed by meiosis and produces the new haploid organism. There are eight stages to meiosis; prophase 1, metaphase 1, anaphase 1 and telophase 1, and prophase 2, metaphase 2, anaphase 2 and telophase 2. During prophase 1, homologous chromosomes associate and duplicate. At this point there may be recombination. During metaphase 1, homologous chromosomes assemble on the spindle across the cell, and they are separated during anaphase 1 and telophase 1. Prophase and metaphase 2 are transient and the chromatids assemble across the cell to be separated during anaphase 2. During telophase 2 cell division usually occurs.

Chromosomes

Eukaryotic microorganisms package the large amount of DNA they contain into chromosomes. Chromosomes contain a single, linear double strand of DNA tightly bound with histone proteins. There are usually between four and eight chromosomes per cell.

Histones

Histones are basic proteins that bind to DNA to condense and fold it. They are vital to the structure of DNA and have been highly conserved during evolution.

Related topics	Structure and organization of DNA (C1) DNA replication (C2) Prokaryote growth and cell cycle (D7)	Recombination and transposition (E3) Reproduction in fungi (H3) Life cycles in the Chlorophyta and Protista (I3)

Cell cycle

The cell cycle consists of an **interphase** during which growth of the cell (**G_1**) occurs, the cell increases in volume to maximum size and there is synthesis of cytoplasmic constituents and RNA (*Fig. 1*). Synthesis of DNA occurs next, during the **S** phase, as chromosomes are duplicated in preparation for nuclear division. During the final part of the cell cycle a second gap or growth phase occurs (**G_2**), when specific cell division-related proteins are synthesized. Nuclear division (**N**) then follows to complete the cell cycle (cf. the bacterial cell cycle, D7).

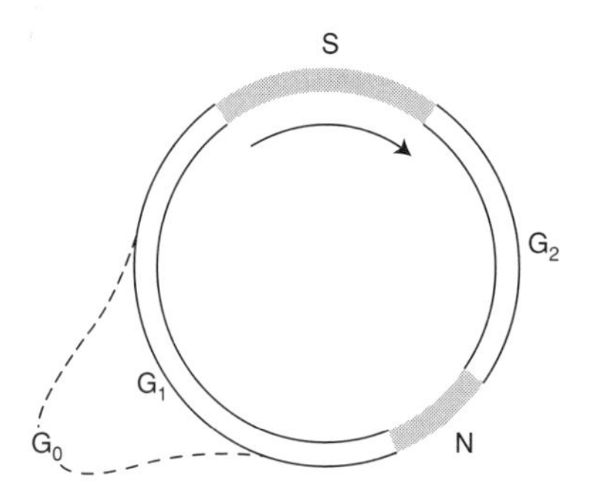

Fig 1. The eukaryotic cell cycle. The S phase is typically 6–8 h long, G2 is a phase in which the cell prepares for mitosis and lasts for 2–6 h, and nuclear division (N) takes only about 1 h. The length of G1 is very variable and depends on the cell type. Cells can enter G0, a quiescent phase, instead of continuing with the cell cycle. From Hames, B.D. et al., Instant Notes in Biochemistry. *© BIOS Scientific Publishers Limited, 1997.*

Mitosis and asexual cell division

Asexual cell division in unicellular eukaryotic organisms is synonymous with growth. Division is usually by **binary fission** after a single nuclear division. The parent cell divides, usually longitudinally, into two even-sized identical progeny cells. Division can be by **multiple fission** (see Topic I5), where many nuclear divisions occur, producing either a large multinucleate **coenocyte** or many uninucleate progeny cells. All cells produced from mitosis are genetically identical to their parent.

In both cases **somatic** cell division is preceded by the mitotic division of the cell nucleus. In mitosis the replicated DNA from the S phase of the life cycle is separated equally into two progeny cells. The events of mitosis can be separated into four stages for convenience, but each flows into the other as a continuous process.

The first phase in mitosis is the **prophase**. In this phase microtubules form from the **microtubule organizing centers** (**MTOCs**). In fungi these structures are known as **spindle pole bodies** (**SPBs**), located close to the nuclear envelope. In the motile species of protista the MTOC is called a **centriole** and it becomes surrounded by microtubules in a process termed **aster formation** (*Fig. 2a*). The MTOCs begin to move towards opposite poles of the nucleus, and spindle microtubules appear between them. Single chromosomes, which have been

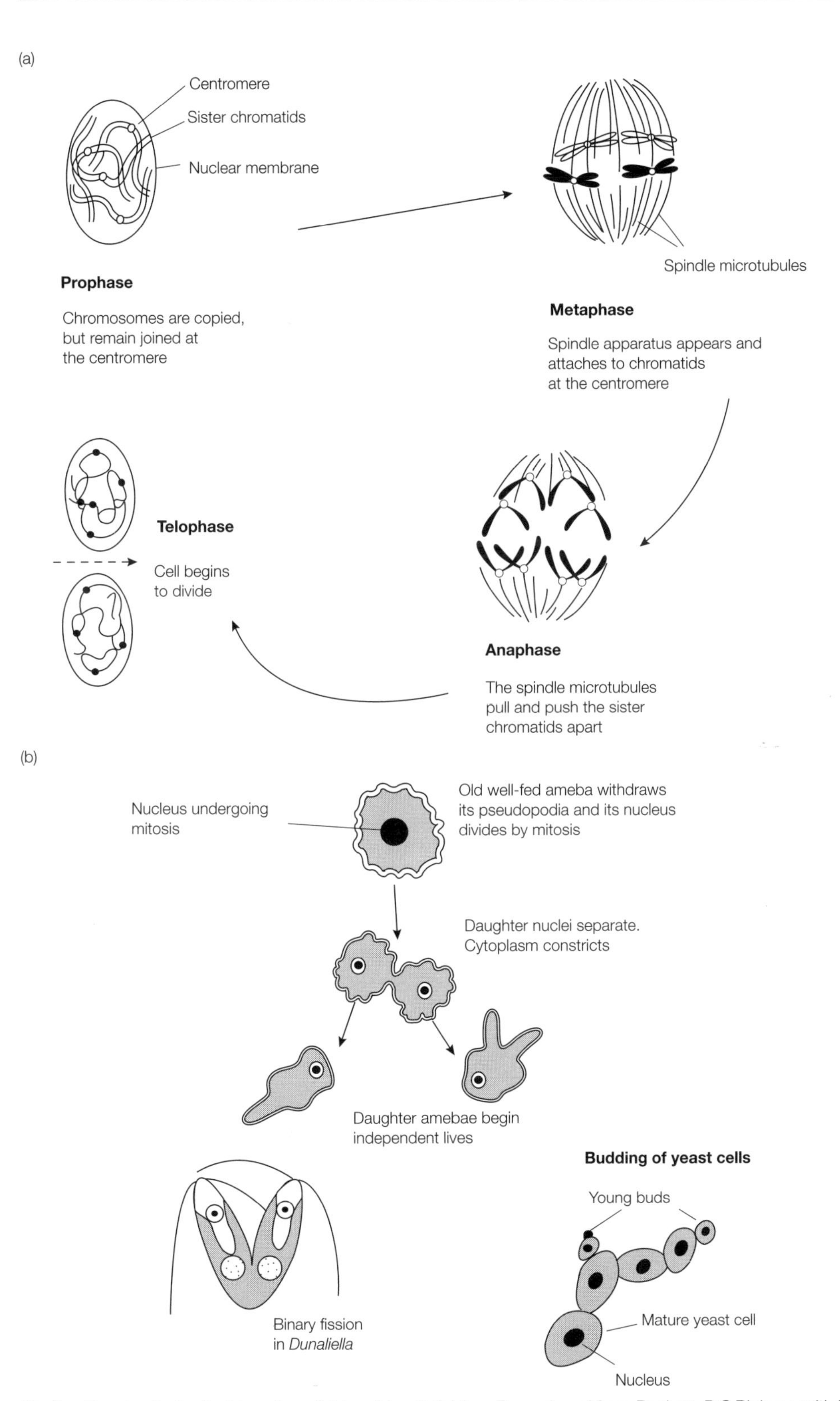

Fig 2. Events of mitosis: (a) nuclear division, (b) cell division. Reproduced from Beckett, B.S Biology, *with kind permission.*

duplicated to form **chromatids**, are joined together at their **centromeres**. These centromeres are also attached to spindle microtubules. By late prophase–early **metaphase**, MTOCs are opposite each other and the spindle is complete, with chromosomes aligned across the center in a **metaphase plate**. In many species of fungi, chromosomes remain extremely indistinct during mitosis, and they do not appear to assemble across a metaphase plate.

The nuclear envelope may disappear between the prophase and the beginning of the metaphase in some eukaryotic microbes, but in some protista the nuclear envelope remains intact throughout the process, the microtubules of the spindle penetrating through it.

In the next phase of mitosis, the **anaphase**, pairs of chromatids that were held together at the centromere begin to separate simultaneously, and spindle microtubules begin to pull them towards the two poles of the cell. By the end of the anaphase the chromatids have been pulled close to the MTOCs. In many species of the fungi there is an asynchronous chromosome separation during the anaphase.

In the final phase of mitosis, the **telophase**, the aster microtubules disappear, and the nuclear envelope reforms if it has disintegrated. In the two progeny nuclei the MTOC duplicates, and **cell division** commences with the division of the cytoplasm by an invaginating plasma membrane or the formation of a **cell plate** by Golgi-derived vesicles across the midline between the two nuclei (*Fig. 2b*). In some protista separation can be by **budding**, producing a progeny cell that is much smaller than the parent.

Meiosis and sexual cell division

Most protista are **haploid** for most of their life cycle. They have only one set of chromosomes. The **diploid** stage (two sets of chromosomes) is often very transient and found only in resting structures such as spores. The life cycle ends with meiosis, which returns the new cell to its haploid state. In organisms with a dominant diploid vegetative phase, meiosis occurs just before cell division, producing haploid gametes which then fuse to reform the diploid.

At the end of the interphase and before meiosis begins, the duplication of chromosomes to chromatids occurs just as it does in mitosis. However, as the cell is diploid, **homologous** pairs of chromosomes associate during **prophase 1** (*Fig. 3*), and at this point it is possible for genetic recombination to occur (cf. Topic E3).

During metaphase 1 chromosomes assemble across the metaphase plate, and in anaphase 1 homologous chromosomes are separated. Telophase 1 is very transient, and the chromosomes rapidly move into the prophase and metaphase 2, where pairs of chromatids assemble across the metaphase plate. The chromatids separate from each other at anaphase 2, and in telophase 2 cytoplasm begins to separate around the four progeny nuclei, each containing a haploid complement of chromosomes.

In some circumstances **multiple sets** of chromosomes can exist in a cell and this is termed **polyploidy**. Nuclear division in polyploid organisms is complex and often results in the loss of single chromosomes, leading to odd numbers of chromosomes in some progeny cells. This is termed **aneuploidy**. Polyploid and aneuploid cells are usually unable to participate in meiosis because of their odd chromosome numbers.

Chromosomes

Compared with the prokaryotes, eukaryotic microbes contain much more DNA, and therefore have had to evolve structures which pack, store and present DNA

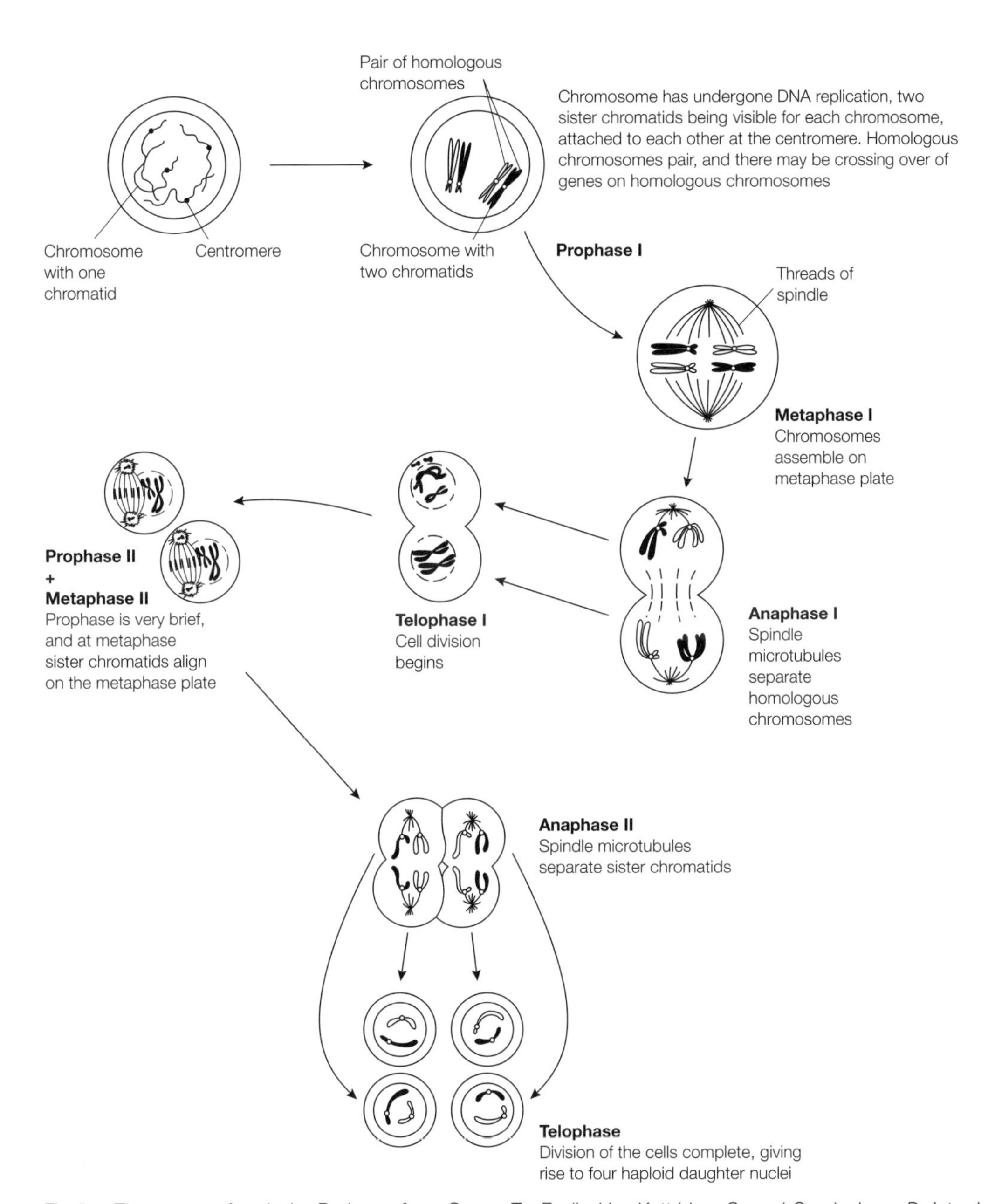

Fig 3. The events of meiosis. Redrawn from Gross, T., Faull, J.L., Kettridge, S. and Sproingham, D. Introductory Microbiology *1995, Stanley Thornes Ltd., Cheltenham, UK.*

during different parts of the cell cycle (see Topic C1). There are usually between four and eight chromosomes per cell, but there can be more or less. Chromosomes contain a single linear double strand of DNA, tightly condensed with **histone** proteins. The combined DNA and histone is often termed **chromatin**. The histone proteins bind to the DNA and create three levels of folding. During the prophase, chromatin is very condensed but, in the interphase, chromatin is described as dispersed.

Histones

Histones are very basic proteins because they contain many basic amino acids (lysine and arginine), with positively charged side chains. These side chains associate with the negatively charged phosphate groups on the DNA molecule. The histones are vital to the structure of DNA and have been **highly conserved** during evolution (see Topic C1).

H1 FUNGAL STRUCTURE AND GROWTH

Key Notes

Fungal structure

Fungi are heterotrophic, eukaryotic organisms with a filamentous, tubular structure, a single branch of which is called a hypha. A network of hyphae is called a mycelium. Hyphae are bound by firm, chitin-containing walls and contain most eukaryotic organelles. Not all fungi are multicellular, some are single-celled and are termed yeasts.

Fungal taxonomy

There are four phyla within the fungi, divided from each other on the basis of differences in their mechanisms of sexual reproduction. The four phyla are the Zygomycota, Chytridiomycota, Ascomycota and Basidiomycota. A fifth group exists which contains fungi where sexual reproduction is not known but where asexual reproduction is seen. These fungi are placed in the phylum Deuteromycota.

Fungal wall structure and growth

Fungal walls are formed from semi-crystalline chitin microfibrils embedded in an amorphous matrix of β-glucan. In the Ascomycota and Basidiomycota, hyphae grow by tip growth followed by septation. In the Chytridiomycota and Zygomycota fungal hyphae grow by tip growth but remain aseptate.

Colonial growth

Colonial growth is characterized by the radial extension of mycelium over and through a substrate, creating a circular or spherical fungal colony.

Kinetics of growth

Fungal growth can be measured by measuring mycelial mass changes with time under excess of nutrient conditions. From this information the specific growth rate can be calculated. After a lag phase, a brief period of exponential growth follows as hyphal tips are initiated. As the new hypha extends, it grows at a linear rate until nutrient depletion causes a retardation phase, followed by a stationary phase.

Hyphal growth unit

Hyphal growth may also be measured by microscopy and by counting the total numbers of hyphal tips, and dividing that number by the total length of mycelium in the colony, the average length of hypha required to support a growing tip can be calculated. This is termed the hyphal growth unit.

Peripheral growth zone

The peripheral growth zone is the region of mycelium behind the tip, which permits radial extension at a rate equal to the specific growth rate.

Related topics

Prokaryote cell structure (D2)
Growth in the laboratory (D6)
Prokaryote growth and cell cycle (D7)
Techniques used to study microorganisms (D8)
Taxonomy (G1)
Eukaryotic cell structure (G2)
Reproduction in fungi (H3)
Chlorophytan and Protistan taxonomy and structure (I1)

Fungal structure

Fungi are filamentous, non-photosynthetic, eukaryotic microorganisms that have a **heterotrophic** nutrition (see Topic B1). Their basic cellular unit is described as a **hypha** (*Fig. 1*). This is a tubular compartment which is surrounded by a rigid, **chitin**-containing wall. The hypha extends by tip growth, and multiplies by branching, creating a fine network called a **mycelium**. Hyphae contain nuclei, mitochondria, ribosomes, Golgi and membrane-bound vesicles within a plasma–membrane bound cytoplasm (see Topic G2). The subcellular structures are supported and organized by microtubules and endoplasmic reticulum. The cytoplasmic contents of the hypha tend to be concentrated towards the growing tip. Older parts of the hypha are heavily vacuolated and may be separated from the younger areas by cross walls called **septae**. Not all fungi are multicellular, some are **unicellular** and are termed **yeasts**. These grow by binary fission or budding.

Fungal taxonomy

In the past the fungi were a **polyphyletic** group which contained microorganisms that had very different ancestors. Current thinking now prefers a **monophyletic** classification where all groups within a phylum are descendants of one ancestor (see Topic G1). Fungi are currently divided into four major phyla on the basis of their morphology and sexual reproduction.

In the **Zygomycota** and the **Chytridiomycota** the vegetative mycelium is non-septate, and complete septa are only found in reproductive structures. Asexual reproduction is by the formation of **sporangia**, and sexual reproduction by the formation of non-motile **zygospores** or motile **zoospores** respectively.

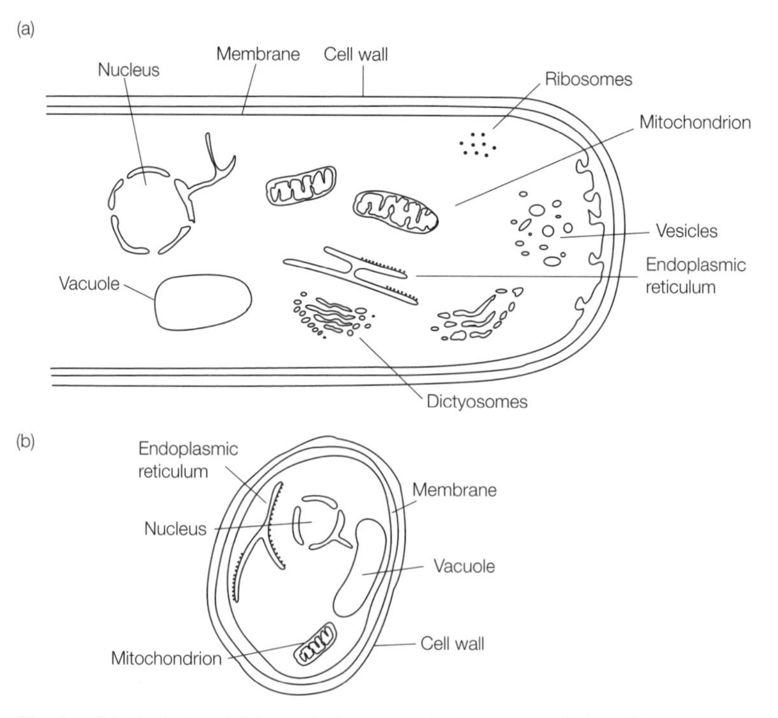

Fig. 1. (a) Hyphal and (b) yeast structures. Redrawn fron Grove, S.M., Bracker, C.E. and Morre, B.J. American Journal of Botany, *vol. 57, pp. 245–266, with kind permission from the Botanical Society of America.*

The other two phyla have a more complex mycelium with elaborate, perforate septa. They are divided into the **Ascomycota** and the **Basidiomycota**. Members of the Ascomycota produce asexual **conidiospores** and sexual **ascospores** in sac-shaped cells called **asci**. Fungi from the **Basidiomycota** rarely produce asexual spores, and produce their sexual spores from club-shaped **basidia** in complex fruit bodies.

A fifth group exists in the higher fungi which contains all forms that are not associated with a sexual reproductive stage, and these are termed the **Deuteromycota**.

Within each of the major phyla are several classes of fungi, indicated by names ending with *-etes*, for example Basidiomycetes, and within classes can be sub-classes or orders, indicated by names ending in *-ales*, within which there are genera and then species (see Topic D1). The important differences between fungi used to distinguish taxonomic groups are summarized in *Table 1*.

Table 1. Features of the main groups of fungi

Group	Perforate septae present or absent	Asexual sporulation	Sexual sporulation
Zygomycota	Absent	Non-motile sporangiospores	Zygospore
Chytridiomycota	Absent	Motile zoospores	Oospore
Ascomycota	Present	Conidiospores	Ascospores
Basidiomycota	Present	Rare	Basidiospores
Deuteromycota	Present	Conidiospores	None

Fungal wall structure and growth

Fungal walls are rigid structures formed from layers of semi-crystalline chitin **microfibrils** that are embedded in an amorphous matix of **β-glucan**. Some protein may also be present. Growth occurs at the hyphal tip by the fusion of membrane-bound vesicles containing wall-softening enzymes, cell-wall monomers and cell-wall polymerizing enzymes derived from the Golgi with the hyphal tip membrane (*Fig. 2*). The fungal wall is softened, extended by turgor pressure and then rigidified.

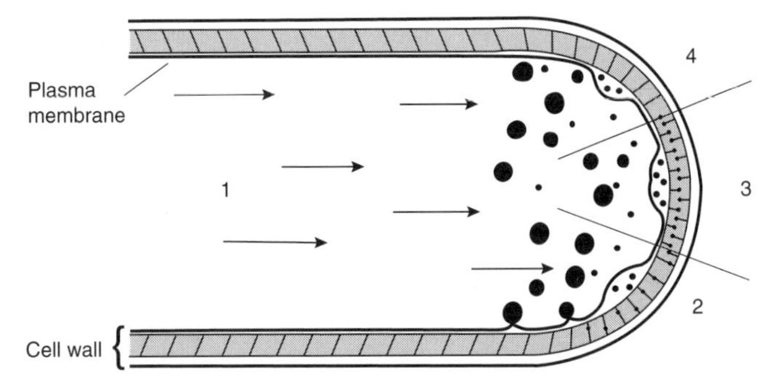

Fig. 2. Hyphal tip growth. Step 1, vesicles migrate to the apical regions of the hyphae. Step 2, wall lysing enzymes break fibrils in the existing wall, and turgor pressure causes the wall to expand. Step 3, amorphous wall polymers and precursors pass through the fibrillar layer. Step 4, wall synthesizing enzymes rebuild the wall fibrils. Redrawn from Isaac, S, Fungal–plant interactions *1991 with kind permission of Kluwer Academic publishers.*

Septae are cross walls that form within the mycelium. Growth in the Zygomycota and Chytridiomycota is not accompanied by septum formation, and the mycelium is cenocytic. Septae only occur in these groups to delimit reproductive structures from the parent mycelium, and they are complete (see Topic H3). In the Ascomycota and Basidiomycota growth of the mycelium is accompanied by the formation of incomplete septae. Septae in the ascomycetes are perforate, and covered by endoplasmic reticulum membranes to limit movement of large organelles such as nuclei from compartment to compartment. This structure is called the **dolipore** septum. In dikaryotic basidiomycetes, septum formation is co-ordinated with divisions of the two mating-type nuclei, maintaining the dikaryotic state by the formation of **clamp connections**. These septae resemble crozier formation in the formation of asci (see Topic H3).

Colonial growth

Hyphal tip growth allows fungi to extend into new regions from a point source or **inoculum**. Older parts of the hyphae are often emptied of contents as the cytoplasm is taken forwards with the growing tip. This creates the radiating colonial pattern seen on agar plates (*Fig. 3*), in ringworm infections of skin and fairy rings in grass lawns.

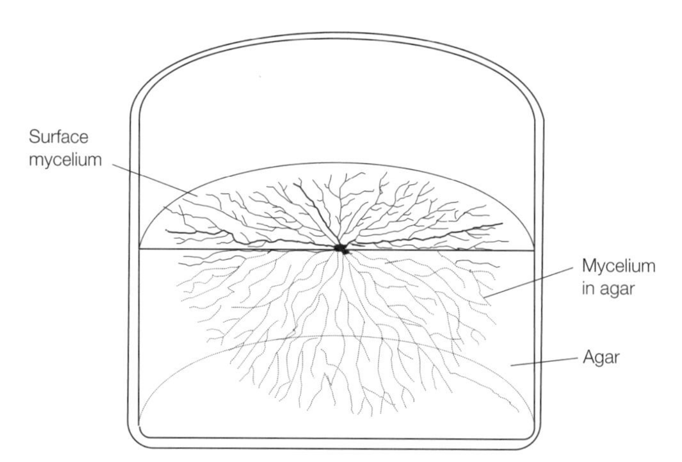

Fig. 3. Colonial growth patterns of filamentous fungi. Reproduced from Ingold, C.T. and Hudson, H.J. The Biology of Fungi *1993, with kind permission from Kluwer Academic publishers.*

Kinetics of growth

When fungi are filamentous their growth rate cannot be established by cell counting using a hemocytometer or by turbidometric measurements (which can be used to measure bacterial and yeast growth) (see Topics D6 and D8). However, by measuring mass (M) changes with time (t) under excess nutrient conditions the **specific growth rate** (m) for the culture can be calculated using the formula:

$$\frac{\mathrm{d}M}{\mathrm{d}t} = \mathrm{m}M$$

Fungal growth in a given medium follows the growth phases of lag, acceleration, exponential, linear, retardation, stationary and decline (see Topic D7, Fig. 3).

Exponential growth occurs only for a brief period as hyphae branches are initiated, and then the new hypha extends at a linear rate into uncolonized regions of substrate. Only hyphal tips contribute to extension growth. However, older hyphae can grow aerially or differentiate to produce sporing structures.

Hyphal growth unit

It is also possible to observe hyphal growth by microscopy, measuring tip growth and branching rates of mycelium. From these data the hyphal growth unit and the **peripheral growth zone** can be calculated. The hyphal growth unit (G), which is the average length of hypha that is required to support tip growth, is defined as the **ratio** between the total length of mycelium and the total number of tips:

$$\frac{\text{Total number of tips}}{\text{Total length of mycelium}} = \text{Hyphal growth unit}$$

In most fungi the hyphal growth unit includes the tip compartment plus two or three sub-apical compartments. The ratio increases exponentially from germination, but stabilizes to give a constant figure for a particular strain under any given set of environmental conditions.

Peripheral growth zone

The peripheral growth zone is the region of mycelium behind the tip, needed to support maximum growth of the hyphal tip. This zone permits radial extension at a rate equal to that of the **specific growth rate** of unicells in liquid culture. The mean rate of hyphal extension (E) is a function of the hyphal growth unit and the specific growth rate

$$E = G\mu$$

The **radial extension rate** (K_r) is a function of the peripheral growth zone (v) and the specific growth rate (μ):

$$K_r = \omega\mu$$

H2 FUNGAL NUTRITION

Key Notes

Carbon nutrition

Fungi require organic carbon compounds to satisfy their carbon and energy requirements. They obtain this carbon by saprotrophy, symbiosis or parasitism. The carbon must be available in a soluble form in order to cross the rigid cell wall, or must be broken down by enzymes secreted by the fungal cells.

Carbon metabolism

Fungi normally utilize glycolysis and aerobic metabolism of carbohydrates. Some can use fermentative pathways under reduced oxygen levels. A few fungi are truly anaerobic.

Nitrogen nutrition

Fungi cannot fix gaseous nitrogen but can utilize nitrate, ammonia and some amino acids as nitrogen sources.

Macro-, micro-nutrients and growth factors

Most macro- and micro-nutrients that fungi require are present in excess in their environments. Phosphorus and iron may be in short supply, and fungi have specific mechanisms to obtain these nutrients. Some fungi may require external supplies of some vitamins, sterols and growth factors.

Water, pH and temperature

Fungi require water for nutrient uptake and are therefore restricted to damp environments. They occupy acidic environments between pH 4 and 6, and by their activity further acidify it. Most fungi are mesophilic, growing between 5° and 40°C, but some can tolerate high or low temperatures.

Secondary metabolism

Secondary metabolites, derived from many different metabolic pathways, are produced by fungi when vegetative growth becomes restricted by nutrient depletion or stress. Such compounds may offer a competitive advantage to the producer.

Related topics

Heterotrophic pathways (B1)
Autotrophic reactions (B3)
Biosynthetic pathways (B4)
Growth in the laboratory (D6)
Control of bacterial infection (F7)
Eukaryotic cell structure (G2)
Chlorophytan and Protistan taxonomy and structure (I1)
Chlorophytan and Protistan nutrition and metabolism (I2)

Carbon nutrition

Fungi are heterotrophic for carbon (see Topic B1). They need organic compounds to satisfy energy and carbon requirements. There are three main modes of nutrition: **saprotrophy**, where fungi utilize dead plant, animal or microbial remains; **parasitism**, where fungi utilize living tissues of plants and animals to the detriment of the host; and **symbiosis**, where fungi live with living tissues to the benefit of the host.

Carbohydrates must enter hyphae in a soluble form because the rigid cell wall prevents endocytosis. Soluble sugars cross the fungal wall by diffusion, followed by active uptake across the fungal membrane (see Topic G2). This type of nutrition is seen in the symbiotic and some parasitic fungi. For the saprophytic fungi most carbon in the environment is not in a soluble form but is present as a complex polymer like cellulose, chitin or lignin. These polymers have to be broken down enzymically before they can be utilized. Fungi release **degradative enzymes** into their environments. Different classes of enzyme can be produced, including the **cellulases**, **chitinases**, **proteases** and multi-component **lignin-degrading enzymes**, depending on the type of substrate the fungus is growing on. Regulation of these enzymes is by **substrate induction** and **end-product inhibition** (see Topic B4).

Carbon metabolism

Once within the hypha, carbon and energy metabolism is by the processes of **glycolysis** and the **carboxylic acid cycle** (see Topic B1). Fungi are usually aerobic, but some species, for example the yeasts, are capable of living in low oxygen-tension environments and utilizing **fermentative** pathways of metabolism (see Topic B4). Recently, truly **anaerobic** fungi have been discovered within animal rumen and in anaerobic sewage-sludge digesters.

Nitrogen nutrition

Fungi are heterotrophic for nitrogen (see Topic B4). They cannot fix gaseous nitrogen, but they can utilize nitrate, ammonia and some amino acids by direct uptake across the hyphal membrane. Complex nitrogen sources, such as peptides and proteins, can be utilized after extracellular proteases have degraded them into amino acids.

Macro-, micro-nutrients and growth factors

Phosphorus, potassium, magnesium, calcium and sulfur are all macronutrients required by fungi. All but phosphorus are usually available to excess in the fungal environment. Phosphorus can sometimes be in short supply, particularly in soils, and fungi have the ability to produce extracellular **phosphatase** enzymes which allow them to access otherwise unavailable phosphate stores.

Micronutrients include copper, manganese, sodium, zinc and molybdenum, all of which are usually available to excess in the environment (see Topic D6). Iron is relatively insoluble and therefore not easily assimilated, but fungi can produce **siderophores** or organic acids, which can **chelate** or alter iron solubility and improve its availability.

Some fungi may require pre-formed vitamins, for example, thiamin and biotin (see Topic D6). Other requirements can be for sterols, riboflavin, nicotinic acid and folic acid.

Water, pH and temperature

Fungi require water for nutrient uptake and they are therefore restricted to fairly moist environments such as host tissue if they are parasites or symbionts, or soils and damp substrates if they are saprophytes. Desiccation causes death unless the fungus is specialized, as they are in the lichens (see Topic H4). Some fungi are wholly aquatic.

Fungi tend to occupy acidic environments, and by their metabolic activity (respiration and organic acid secretion) tend to further acidify it. They grow optimally at pH 4–6.

Most fungi are **mesophilic**, growing between 5° and 40°C. Some are **psychrophilic** and are able to grow at under 5°C, others are **thermotolerant** or **thermophilic** and can grow at over 50°C.

Secondary metabolites

Nutrient depletion, competition or other types of metabolic stress which limit fungal growth promote the formation and secretion of secondary metabolites. These compounds can be produced by many different metabolic pathways and include compounds termed **antibiotics** (active against bacteria, protista and other fungi) (see Topic F7), **plant hormones** (gibberellic acid and indoleacetic acid (IAA)) and **cytotoxic** and **cytostimulatory** compounds.

H3 Reproduction in fungi

Key Notes

Life cycles

All fungi undergo a period of vegetative growth during which their mycelium exploits a substrate. This stage is followed by asexual and/or sexual reproduction, which differs in each of the four phyla.

Reproduction in the Chytridiomycota

Asexual reproduction in the Chytridiomycota is usually by the formation of motile, uniflagellate zoospores within spore-containing structures called sporangia. Sexual reproduction is by the formation of diploid oospores following the fusion of two haploid cells. These may undergo meiosis, or there may be a period of vegetative growth in the diploid state.

Reproduction in the Zygomycota

Fungi in the Zygomycota reproduce asexually by the formation of non-motile sporangiospores within sporangia elevated on aerial hyphae. Sexual reproduction is by the formation of diploid zygospores, within which meiosis occurs.

Reproduction in the Ascomycota

Fungi in the Ascomycota reproduce asexually by the formation of conidiospores from hyphal tips. Sexual reproduction is by the fusion of hyphae rapidly followed by meiosis and the production of ascospores.

Reproduction in the Basidiomycota

Basidiomycete fungi rarely reproduce asexually. Sexual reproduction is by the formation of basidiospores on the gills or pores of large fruit bodies we know as mushrooms and toadstools.

Fungal spores

Spores allow fungi to spread, to maintain genetic diversity and to survive adverse conditions. Spores produced asexually are generally adapted for dispersal while those produced sexually are often adapted for survival.

Spore discharge

Spores may be discharged from parent mycelium by passive or active means. Passive mechanisms include using wind and water as dispersants; active mechanisms use explosive principles.

Air spora

Spores in the atmosphere can affect human, animal and plant health. They can cause allergies and spread plant disease.

Related topics

Cell division and ploidy (G3)
Fungal structure and growth (H1)
Fungal nutrition (H2)
Beneficial effects of fungi in their environment (H4)
Detrimental effects of fungi in their environment (H5)

Life cycles

Each of the four fungal groups is characterized by differences in their life cycles. All fungi are characterized by having a period of vegetative growth where their biomass increases. The length of time and the amount of biomass needed before sporulation can occur varies. Almost all fungi reproduce by the production of

spores, but a few have lost all sporing structures and are referred to as ***mycelia sterilia***. Different types of spore are produced in different parts of the life cycle.

Reproduction in Chytridiomycota

Fungi in the Chytridiomycota are quite distinct from other fungi as they have extremely simple thalli and motile zoospores. Some species within this group can be so simple that they consist of a single vegetative cell within (**endobiotic**) or upon (**epibiotic**) a host cell, the whole of which is converted into a **sporangium**, a structure containing spores. These types are termed **holocarpic** forms.

Other members of this group have a more complex morphology, and have **rhizoids** and a simple mycelium. Asexual reproduction in the chytridiomycota is by the production of motile **zoospores** in sporangia that are delimited from the vegetative mycelium by complete septae. The zoospores have a single, posterior flagellum. Sexual reproduction occurs in some members of the chytridiomycota by the production of **diploid** spores after either somatic fusion of haploid cells, either two different mating-type mycelia, fusion of two motile **gametes**, or fusion of one motile gamete with a **non-motile egg** (*Fig. 1*). The resulting spore may undergo meiosis to produce a haploid mycelium or it may germinate to produce a diploid vegetative mycelium, which can undergo asexual reproduction by the production of diploid zoospores. The diploid mycelium can also produce resting sporangia in which meiosis occurs, generating haploid zoospores that germinate to produce haploid vegetative mycelium.

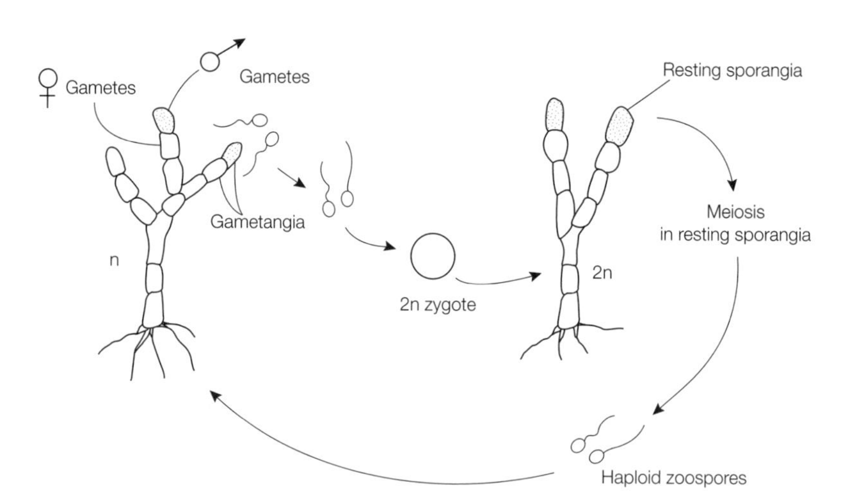

Fig. 1. Chytrid life cycle.

Reproduction in Zygomycota

In the Zygomycota, asexual reproduction begins with the production of aerial hyphae. The tip of an aerial hypha, now called a **sporangiophore**, is separated from the vegetative hyphae by a complete septum called a **columella** (*Fig. 2*). The cytoplasmic contents of the tip differentiate into a sporangium containing many asexual spores. The spores contain haploid nuclei derived from repeated mitotic divisions of a nucleus from the vegetative mycelium (see Topic G3). Dispersal of the spores is by wind or water.

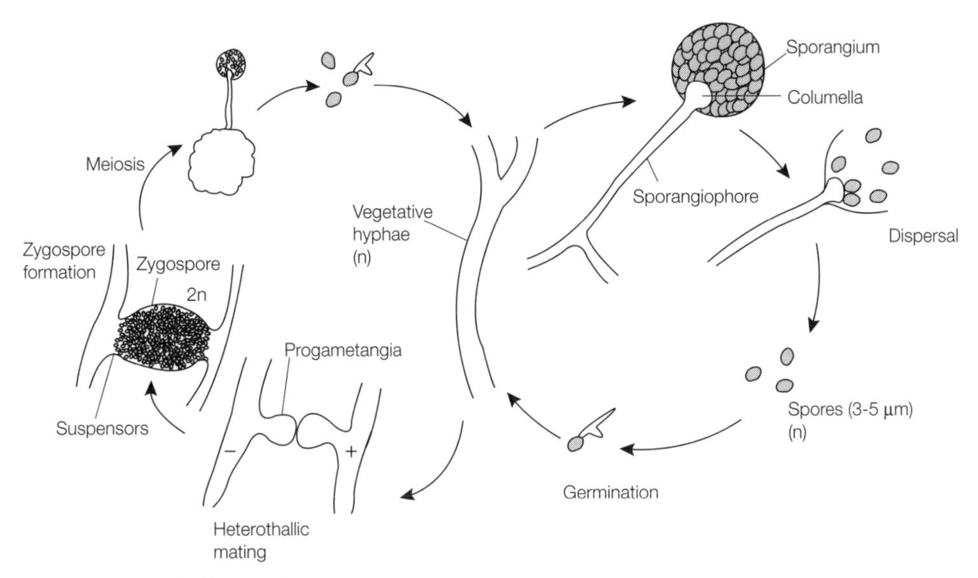

Fig. 2. Life cycle of a typical zygomycete.

In sexual reproduction, two nuclei of different mating types fuse together within a specialized cell called a **zygospore** (*Fig. 2*). In some species the different mating-type nuclei may be within one mycelium (**homothalism**). In other species, two mycelia with different mating-type nuclei must fuse (**heterothalism**). In both cases, fusion occurs between modified hyphal tips called **progametangia**, which once fused are termed the zygospore. Within the developing zygospore meiosis occurs; usually three of the nuclear products degenerate, leaving only one nuclear type present in the germinating mycelium (see Topic G3).

Reproduction in Ascomycota

The vegetative stage of the Ascomycota life cycle is accompanied or followed by asexual sporulation by the production of single spores called conidia from the tips of aerial hyphae called conidiophores (*Fig. 3*). The spores can be delimited by a complete transverse wall formation followed by spore differentiation (*Fig. 3a*) termed **thallic** spore formation, or more usually by the extrusion of the wall from the hyphal tip, termed **blastic** spore formation (*Fig. 3b*). These spores can be single-celled and contain one haploid nucleus, or they can be multicellular and contain several haploid nuclei produced by mitosis (see Topic G3).

Spores can be produced from single, unprotected conidiophores or they can be produced from **aggregations** that are large enough to be seen with the naked eye (*Fig. 3c*). The conidiophores can aggregate into stalked structures where the spores produced are exposed at the top (**synnema** or **coremia**). Alternatively, varying amounts of sterile fungal tissue can protect the conidia, as in the flask-shaped **pycnidia**. Some species produce conidia in plant tissue, and the conidial aggregations erupt through the plant epidermis as a cup-shaped **acervulus** or a cushion-shaped **sporodochium**.

Sexual reproduction in this group occurs after **somatic** fusion of different mating-type mycelia. A transient diploid phase is rapidly followed by the formation of **ascospores** within sac-shaped **asci** differentiated from modified

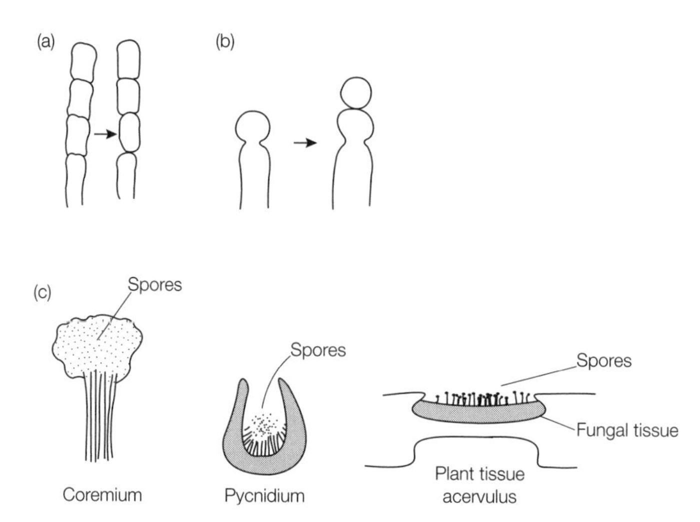

Fig. 3. Asexual reproduction in the Acsomycetes (a) thallic spore formation; (b) blastic spore formation; (c) aggregations of conidiophores. Redrawn from Ingold, C.T. and Hudson, H.J., The Biology of Fungi, *1993, with kind permission from Kluwer Academic Publishers.*

hyphal tips. In the initial stages of ascal development hooked hyphal tips form, called **croziers** or **shepherds' crooks** because of their shape. They have distinctive septae at their base which insure that two different mating-type nuclei are maintained in the terminal cell. Formation of the septae is coordinated with nuclear division (*Fig. 4*). In yeasts all these events occur within one cell, after fusion of two mating-type cells, the whole cell being converted into an ascus.

In more complex Ascomyceta many asci form together, creating a fertile tissue called a **hymenium**. In some groups the hymenium can be supported or even enclosed by large amounts of vegetative mycelium. The whole structure is called a **fruit body** or **sporocarp** and is used as a major taxonomic feature (*Fig. 5*). They can become large enough to be seen with the naked eye. Flask-shaped sexual reproductive bodies are called **perithecia**, cup-shaped bodies are called **apothecia** and closed bodies are called **cleistothecia**. These structures

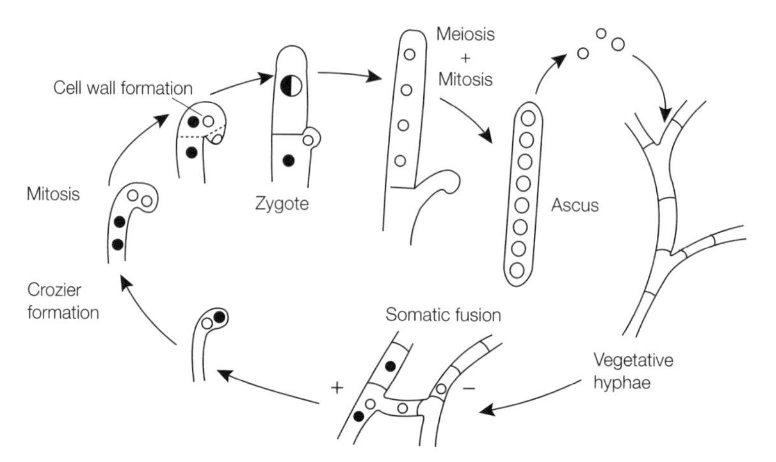

Fig. 4. Sexual reproduction in the Ascomycetes.

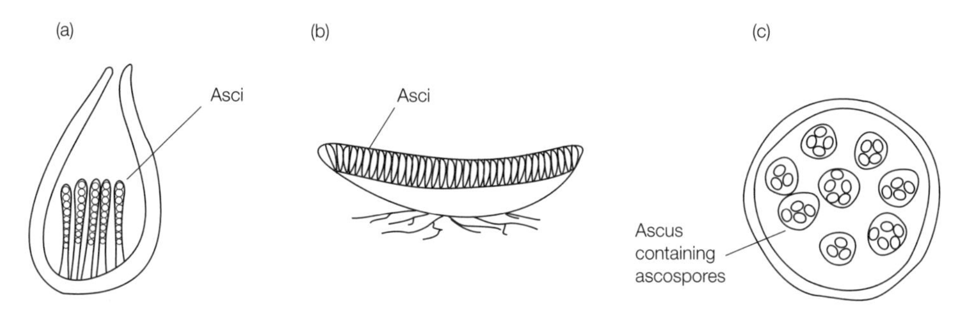

Fig. 5. Structures of sexual sporocarps in the Ascomycetes: (a) perithecium; (b) apothecium; (c) cleistothecium. Redrawn from Ingold, C.T. and Hudson, H.J., The Biology of Fungi, *with kind permission from Kluwer Academic Publishers.*

have evolved to protect the asci and assist in spore dispersal, but the hymenium itself is unaffected by the presence of water.

Reproduction in Basidiomycota

This group of fungi are characterized by the most complex and large structures found in the fungi. They are also distinctive in that they very rarely produce asexual spores. Much of the life cycle is spent as vegetative mycelium, exploiting complex substrates. A preliminary requisite for the onset of sexual reproduction is the acquisition of two mating types of nuclei by the fusion of compatible hyphae. Single representatives of the two mating-type nuclei are held within every hyphal compartment for extended periods of time. This is termed a **dikaryotic** state, and its maintenance requires elaborate septum formation during growth and nuclear division (see Topic H1).

Onset of sexual-spore formation is triggered by environmental conditions and begins with the formation of a **fruit body primordium**. Dikaryotic mycelium expands and differentiates to form the large fruit bodies we recognize as mushrooms and toadstools. Diploid formation and meiosis occur within a modified hyphal tip called a **basidium** *(Fig. 6)*.

Four spores are budded from the basidium. Basidia form together to create a hymenium which is highly sensitive to the presence of free water. The hymenium is distributed over sterile, dikaryotic-supporting tissues which

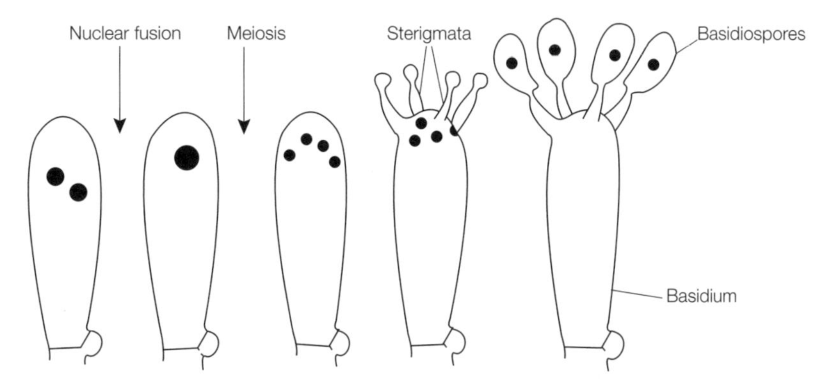

Fig. 6. Basidium formation. Redrawn from from Ingold, C.T. and Hudson, H.J., The Biology of Fungi, *with kind permission from Kluwer Academic Publishers.*

protect it from rain. The hymenium can be exposed on **gills** or **pores** beneath the fruit body, seen in the **toadstools** and **bracket fungi**, or enclosed within chambers as in the **puffballs** and **truffles** (*Fig. 7*).

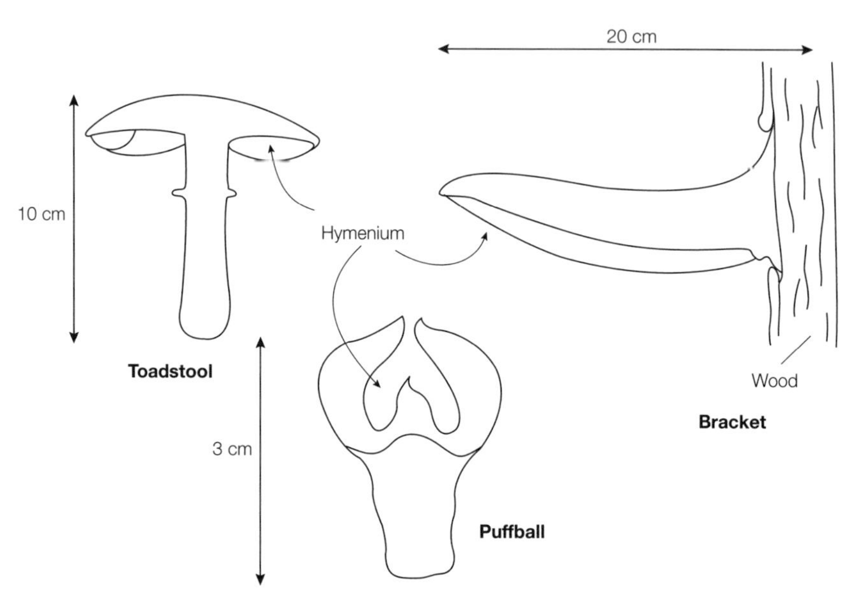

Fig. 7. Structure of sexual sporocarps in the Basidiomycetes. Redrawn from Ingold, C.T. and Hudson, H.J., The Biology of Fungi, *with kind permission from Kluwer Academic Publishers.*

Fungal spores

There are two conflicting requirements fungi have for their spores. Spores must allow fungi to spread, but they must also allow them to survive adverse conditions. These requirements are met by different types of spores. Small, light spores are carried furthest from parent mycelium in air and these are the dispersal spores. They are usually the products of asexual sporulation, the sporangiospores and the conidiospores, and so spread genetically identical individuals as widely as possible. Genetic diversity is maintained by sexual reproduction, and the spore products are often large **resting spores** that withstand adverse conditions but remain close to their site of formation. Spores therefore vary greatly is size, shape and ornamentation, and this variation reflects specialization of purpose.

Spore discharge

Spores that have a dispersal function can be released from their parent mycelium by **active** or **passive** mechanisms (*Fig. 8*). As many spores are wind-dispersed, they are produced in dry friable masses which are passively discharged by wind. Other spores are passively discharged by water droplets splashing spores away from parent mycelium.

Fungal spores can be actively discharged by **explosive** mechanisms. These mechanisms use a combination of an increasing turgor pressure within the spore-bearing hypha, combined with an in-built weak zone of the hyphal wall. This insures that when the hypha bursts the spore discharge is directed for maximum distance. Asci are usually dispersed in this way, and a few sporangia too. Basidiospores are also actively discharged.

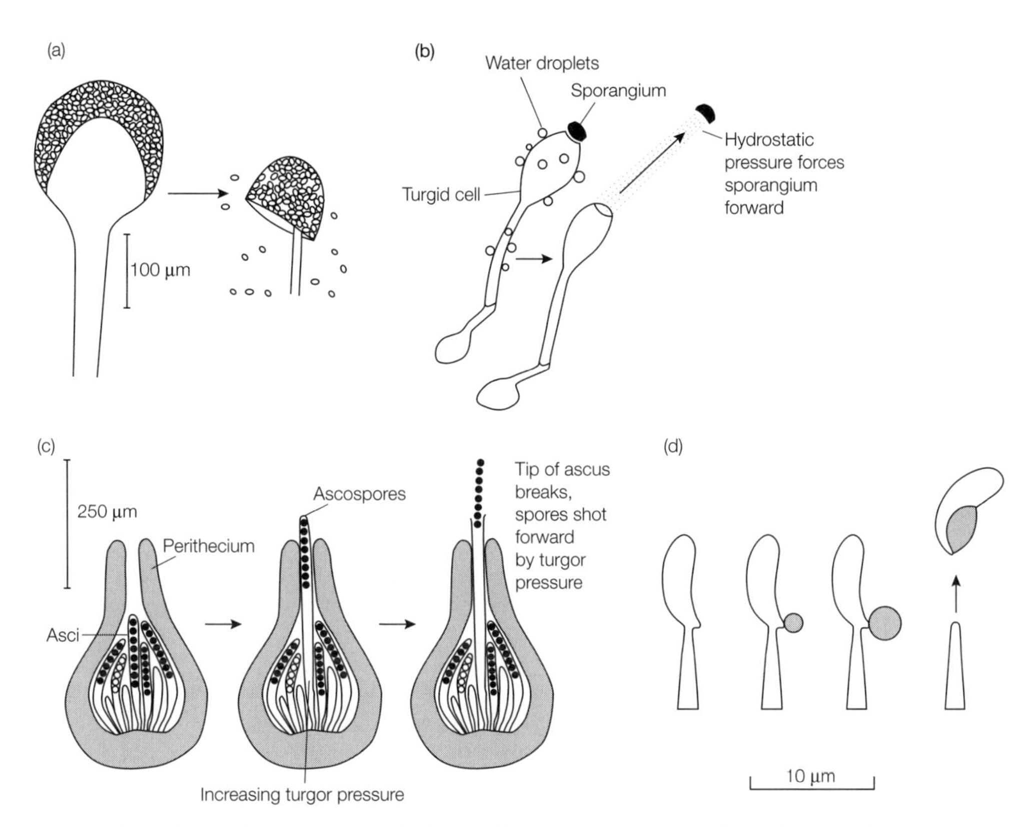

Fig 8. (a) Passive and (b) active spore discharge; (c) Ascospore disharge; (d) basidiospore discharge.

Air spora

Air-borne fungal spores can be carried great distances. Their presence in the air can have an impact on human health as they can cause **allergic rhinitus** (hay fever) and **asthma**. Many plant diseases that cause significant economic losses are air-borne (see Topic H5). Spore clouds can be tracked across continents, and epidemic disease forecasts can be made, depending on weather conditions and air-spora counts.

H4 BENEFICIAL EFFECTS OF FUNGI IN THEIR ENVIRONMENT

Key Notes

Bread and brewing — The products of yeast fermentation (CO_2 and alcohol) are exploited in bread making and alcohol brewing. Both processes enhance the value of the substrate but contribute little to its nutritional value.

Symbioses — Fungi can enter into specialized and intimate, mutually beneficial associations with higher plants, other microbes and animals. The associations can be external to the host cell, as in the ectomycorrhizae and lichens, or inside the cell as in the endomycorrhizae and endophytic fungi.

Decomposers — Fungi are the main agents of decay of plant wastes in the environment, decomposing substrates to CO_2, H_2O and fungal biomass, and releasing other nutrients back to the biosphere.

Biological control — Fungi can be used to control insect pests, weed plants and plant diseases by exploiting their natural antagonistic, competitive and pathological attributes.

Bioremediation — The degradative abilities of fungi can be exploited to decompose man-made pollutants such as hydrocarbons, pesticides and explosives. They may decompose substrates into CO_2 and H_2O by respiratory pathways, or they may reduce toxicity by co-metabolic activity. The toxicity of some compounds can also be increased in this manner.

Industrially important natural products of fungi — Fungi naturally produce antibiotics, immunosuppressants, acids, enzymes and several other classes of useful natural products. They also can be used to produce large quantities of protein, including the popular meat substitute Quorn.

Related topics

Heterotrophic pathways (B1)
Prokaryotes in the environment (F1)
Fungal structure and growth (H1)
Fungal nutrition (H2)
Reproduction in fungi (H3)
Detrimental effects of fungi in their environment (H5)

Bread and brewing

The metabolic products of yeast metabolism are exploited by humans for **bread making** and **alcohol brewing**. Yeast metabolism of flour starch by respiration generates $\mathbf{CO_2}$ which is trapped within the gluten-rich dough and forces bread to rise (see Topic B1). In the brewing of alcohol, yeasts are forced into fermentative metabolism in the sugar-rich, low-oxygen environments of beer wort or crushed grape juice. The fermentative metabolism is inefficient, and only partially metabolizes the available substrates, yielding CO_2 and **ethanol**. Both processes considerably enhance the value of the original substrate while contributing little to its nutritional status!

Symbioses

Fungi can enter into close associations with other microbes and with higher plants and animals. These beneficial associations are termed **symbioses**, and in most cases the symbiotic fungus gains carbohydrates from its associate, while the associating organism gains nutrients and possibly protection from predation and herbivory or plant pathogens (see Topic F1).

Fungal symbioses are common on plant roots, and the symbiotic roots are termed **mycorrhizae** (*Fig. 1*). Their presence enhances plant-root nutrient uptake and plant performance. The fungal association can be predominantly external to the root tissue. These associations are called **ecto**mycorrhizae and can be seen on beech and pine trees, for example. Other associations are predominantly within the plant root, and are termed **endo**mycorrhizae. These associations are seen on the roots of herbaceous species like grasses, but are also found in the roots of tropical species of tree and shrub.

Other associations between fungi and plants can occur within leaves or stems, and these are called **endophytic** fungi. They live almost all their life cycle within

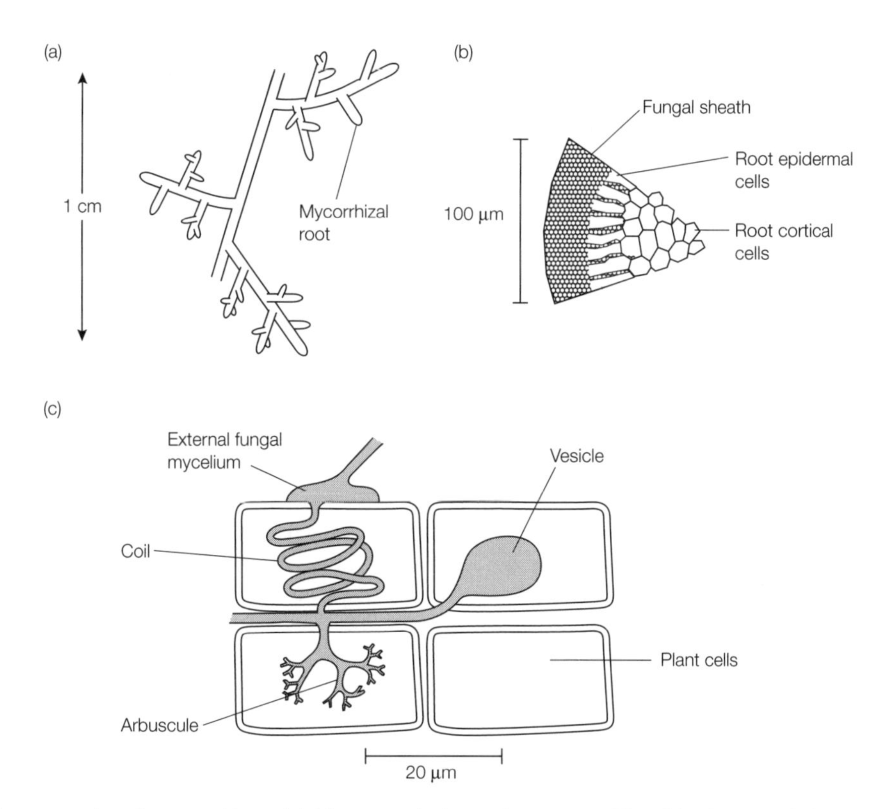

Fig. 1. Structure of ecto- and endo-mycorrhizae. (a) Macromorphology of ectomycorrhiza; (b) micromorphology of ectomycorrhizae; (c) micromorphology of endomycorrhizae. Redrawn from Isaac, S., Fungal-Plant Interactions, *1991, with kind permission from Kluwer Academic Publishers.*

the host, grow very slowly and do not cause any signs of infection. They appear to protect their host from herbivory and fungal infection by the production of metabolites. However, these products can have dramatic effects on herbivorous animals, causing symptoms of fungal toxicosis similar to St Anthony's Fire (see Topic H5).

Some fungi can form very intimate associations with algal species. These associations are termed **lichens**, and they have a form quite distinct from that of either component species, with **crustose**, **foliose** or **fruticose** thalli (*Fig. 2*). They are slow-growing, and are adapted to occupy extreme or marginal environments, like bare rock faces, walls and house roofs. As they are under quite extreme stress they are highly sensitive to pollution, for instance, from acid rain or heavy metals, and their presence or absence in an environment has become a useful indicator of urban and industrial pollution.

Fungi can also associate with insect species in symbioses of varying intimacy. Some species of ants culture specific fungi on cut plant remains within their nests, and then browse on the fungal mycelium that develops. **Termites** have symbiotic fungi within their guts, which in association with a consortium of other microbes, including protista and bacteria, help the termite digest its woody gut contents (see Topic I4)

Decomposers

The degradative processes that fungi perform with their extracellular enzymes are essential to the terrestrial biosphere (see Topics F1 and H2). They are the main agents of decay of cellulosic wastes produced by plants, which in the tropical rain forest can amount to 12 000 kg hectare^{-1} year^{-1}. They decompose this material into CO_2, H_2O and fungal biomass, which in its turn is decomposed by other microbes, returning mineral nutrients like phosphorus, nitrogen and potassium to the biosphere. This process is termed mineralization.

Biological control

The natural attributes of fungi as disease-causing organisms can be exploited by humans to control weed plant populations and insect pests. They are even capable of parasitizing plant disease-causing fungi. This process is termed **biological control** and can be an alternative to the application of chemical pesticides. Applications of fungal propagules to targeted problem populations can cause epidemic disease or overwhelm and monopolize a niche.

Bioremediation

The degradative processes that fungal enzymes catalyze on their natural substrates can be used on other, man-made substrates to provide **biological cleanup** (see Topic F1). Hydrocarbons like oils can be degraded by fungi and other microbes to CO_2 and H_2O by aerobic respiration. These activities are termed bioremediation, and contaminated areas of land can be actively bioremediated by the addition of fungal propagules.

Pesticides, explosives and other recalcitrant molecules can be changed by **co-metabolic** activities of fungi, where enzymes normally used for one metabolic process within the fungus coincidentally catalyze another reaction. The products of co-metabolic reactions are not utilized any further by the fungus, but can sometimes be used by other microbes in the ecosystem. Reactions like these can lead to reduced toxicity of some contaminants, but in other cases can lead to **activation** of the pollutant, leading to an increase in the toxicity of the compound.

Industrially important natural products of fungi

Fungi are able to produce many different types of metabolite that are of commercial importance. These include **antibiotics** (e.g. Penicilliins and Cephalosporins, see Topic F7) and **immunosuppressants** (cyclosporins), important in medicine, **enzymes** that are used in the food industry (e.g. α-amylase, renin), other enzymes (e.g. cellulases, catalase), **acids** (e.g. lactic, citric) and several other products.

Protein extracted from fungi is also an important commercial product. In the 1960s protein from yeasts (single-cell protein) was developed. Currently, fungal protein extracted from *Fusarium graminearum*, known as **Quorn**, is a useful meat alternative.

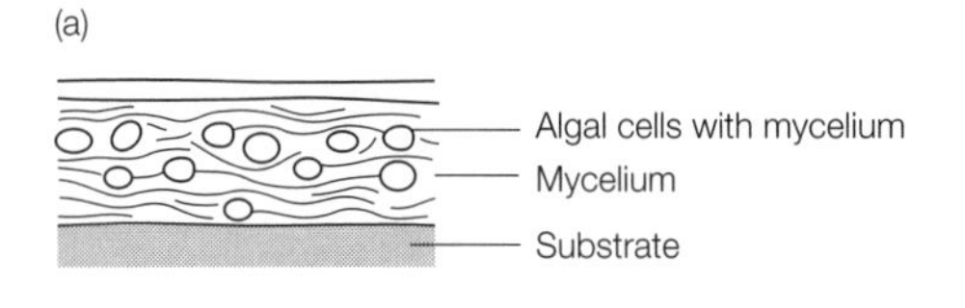

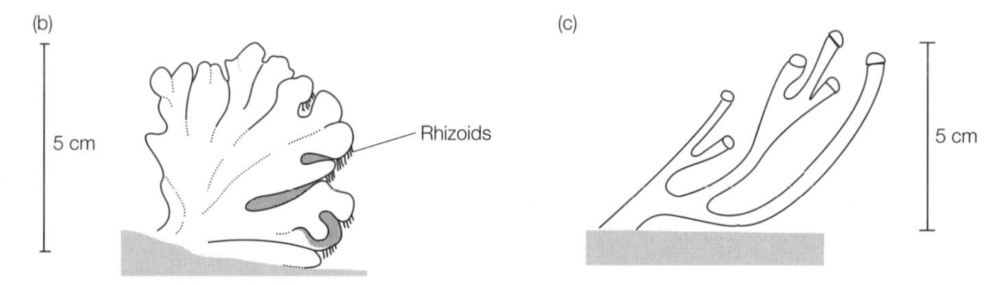

Fig. 2. Lichen structure. (a) Crustose, tightly attached to substrate; (b) foliose, loosely attached to substrate; (c) fruticose, attached only at base.

H5 DETRIMENTAL EFFECTS OF FUNGI IN THEIR ENVIRONMENT

Key Notes

Biodeterioration The degradative activities of fungi in unwanted situations cause significant economic losses. Materials containing large quantities of cellulose, leather and hydrocarbons can be used as substrates by fungi, providing there is an adequate water supply.

Plant disease Fungi are capable of causing significant losses to crops both before and after harvest. However, this can be countered by the use of fungicides, storage conditions that do not favor fungal growth, and the development of resistant plant varieties.

Animal and human disease Fungi can cause both superficial and deep, life-threatening infections of both man and animals. The latter are particularly dangerous in immunocompromised individuals.

Fungal toxicosis Ingestion of fungi or their secondary metabolic products, accidentally or deliberately, can cause intoxication and occasionally death in both humans and animals.

Related topics Fungal structure and growth (H1) Reproduction in fungi (H3)
Fungal nutrition (H2)

Biodeterioration

The same extracellular enzymes that are important in the degradation of leaf litter and the recycling of nutrients in the biosphere can cause massive economic losses when they occur in circumstances where they are not wanted. Fungi can attack and utilize as substrates paper, cloth, leather and hydrocarbons, but also can cause degradative change in other materials, for instance, glass and metal, because of their ability to produce acid as they grow. The supply of water is a key control point in these processes, and keeping substrates dry is an effective way of avoiding these changes.

Plant disease

Fungi are capable of attacking all plant species, causing serious damage and in some circumstances even death. In crop production over half of potential crop yield is lost to plant pathogens, most of it to **fungal disease**. In storage, up to one-third of the harvested product can be lost to post-harvest disease, again mostly as a result of the activities of fungi. Use of **fungicides** can reduce both pre- and post-harvest disease, and **plant-breeding** programs can introduce disease-resistant strains of crop plant. Post-harvest losses can be reduced by storage of products at low temperatures and low moisture levels.

Animal and human disease

Human and animal epidermis can be attacked by fungi, causing superficial damage and discomfort like **ringworm** infections, **athletes foot** and **thrush**. Other deeper, systemic fungal infections of the lung and central nervous and lymphatic systems cause much more serious diseases, for example, **aspergillosis**, **coccidiomycosis**, **blastomycosis**, **histoplasmosis** and **pneumocystis** pneumonia are all caused by fungi. Although most humans experience superficial fungal infections and survive, these deeper diseases are especially dangerous for the immunocompromised patient after transplantation, and the HIV-positive population.

Fungal toxicosis

Accidental or deliberate consumption of wild fungi or fungally contaminated food can lead to poisoning or **toxicosis** of the consumer because some fungi naturally contain toxic metabolites called **mycotoxins**. Deliberate toxicosis can arise from consumption of mushrooms and toadstools that are known to contain naturally hallucinatory drugs like **psilocibins**, which lead to euphoric states followed by extreme gastrointestinal distress. Accidental consumption of misidentified fungal fruit bodies can lead to fatal mushroom poisoning from fungal toxins, causing total liver failure between 8 and 10 h. Consumption of food accidentally contaminated by fungal metabolites also leads to human and animal death. For instance, rye flour contaminated by the ergots of the fungus *Claviceps purpurea* leads to the symptoms of **St Anthony's fire**, where peripheral nerve damage is caused by the presence of **ergometrine** in the fungal tissue. This can be followed by gangrene of the limbs and death. Detection of fungal mycotoxins such as **ochratoxin** in apple juice and **aflatoxin** in peanuts have also caused problems for food producers and consumers and forced improvements in product processing.

I1 Chlorophytan and Protistan taxonomy and structure

Key notes

Chlorophytan and protistan taxonomy

Until fairly recently, the photosynthetic and non-photosynthetic protistan genera were divided into two polyphyletic form groups, the algae and protozoa based on the presence or absence of chloroplasts. Currently, a monophyletic taxonomy of the protista is being developed, based on molecular data.

Chlorophytan and protistan structure

Many members of the Chlorophyta are unicellular, photosynthetic organisms. Some have a filamentous or membranous morphology. They have a cellulose cell wall. Many species are flagellate, and they contain chloroplasts, which vary in structure and pigment content. Members of the protista are heterotrophic or photosynthetic, unicellular eukaryotes. They vary greatly in shape and size, and contain most eukaryotic cell organelles. They also have some unique organelles. There are three major groups within the protista, the Euglenozoa (containing the euglenids and kinetoplastids), the Alveolata (containing the ciliates, dinoflagellates and the apicomplexans) and the Stramenopila (containing the diatoms, Chrysophytes, Oomycetes, opalines and the amebae and slime molds).

Chlorophytan and protistan growth

Growth in most unicellular Chlorophyta and protista is synonymous with longitudinal binary fission. Cenocytic, tubular or filamentous Chlorophyta grow by tip growth like the fungi. Other filamentous or membranous algae grow by intersusception of new cells into the filament. The kinetics of growth are similar to those of bacteria, but in addition to estimations of growth by mass measurement, cell counts and chlorophyll content can be assessed. Rapid cell division can lead to very high cell populations, only limited by nitrogen, phosphate or silicon availability.

Related topics

The microbial world (A1)
Prokaryote cell structure (D2)
Prokaryote growth and cell cycle (D7)
Eukaryotic cell structure (G2)
Cell division and ploidy (G3)
Fungal structure and growth (H1)
Chlorophytan and Protistan nutrition and metabolism (I2)
Beneficial effects of the Chlorophyta and Protista (I4)

Chlorophytan and Protistan taxonomy

In the past the eukaryotic photosynthetic and non-photosynthetic microorganisms were divided into **form groups**, the algae and the protozoa, based on the presence or absence of chloroplasts. The algae were sub-divided into groups based on pigmentation, the number and type of flagellae and other structural characteristics. Protozoa were divided on similar structural characteristics into four **polyphyletic** form groups, the ciliates, flagellates, sporozoans and amoebas. Advances in molecular biology now allow us to begin to create a **monophyletic taxonomy** of the Chlorophyta and Protista and such a scheme includes many former members of the algae, fungi and protista. A suggested taxonomic scheme is shown in Fig. 1 and characteristics of the major monophyletic groups are shown in Table 1. Only those organisms included in the microbial world will be discussed here, the red and brown algae are multicellular organisms and are omitted from consideration.

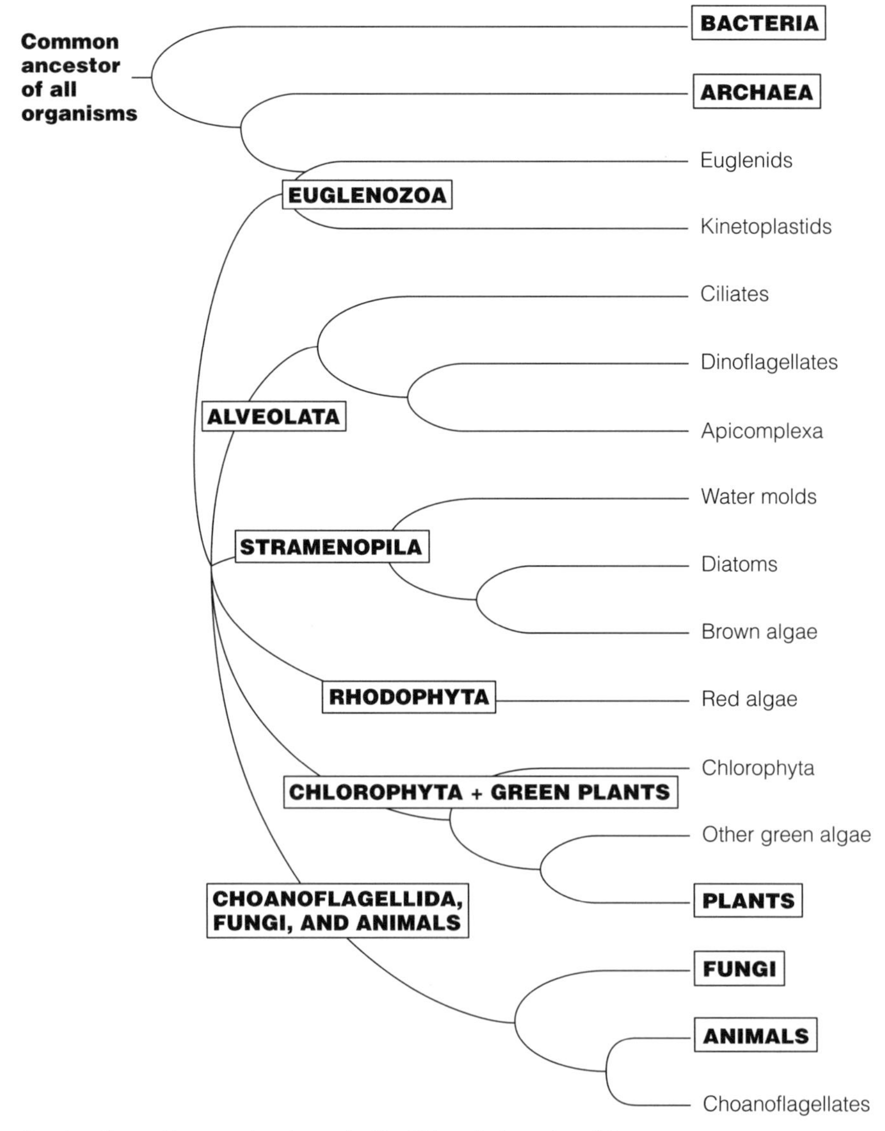

Fig. 1. Current taxonomic scheme for the Chlorophyta and protista.

Table 1. Characteristics of the major monophyletic groups of the Chlorophyta and protista

Group	Common name	Characteristics	Examples
Euglenoza		**Flagellate unicells**	
Euglenoids		Mostly photosynthetic	**Euglena**
Kinetoplastids		Single large mitochondrion	**Trypanosoma**
Alveolata		**Unicellular; sacs (alveolata) below cell surface**	
Pyrrophyta	Dinoflagellates	Golden brown algae	**Peridinium**
Apicomplexa		Apical complex aids host cell penetration	**Plasmodium**
Ciliophora	Ciliates	Cilia; macro- and micro-nucleus	**Paramecium**
Stramenopila		**Motile stages with two unequal flagellae, one with hairs**	
Bacillariophyta	Diatoms	Unicellular, photosynthetic, 2 silicon containing frustules	*Navicula*
Chrysophytes	Golden Brown algae	Multicellular photosynthetic marine species	*Ceratium*
Oomycota	Water molds	Heterotrophic coenocytes	*Saprolegnia*
Chlorophyta	Green algae	Chlorophyll a and b, cellulose cell walls	*Chlamydomonas*

Chlorophytan and protistan structure

The Chlorophyta are a monophyletic group. They range in complexity from unicellular motile or non-motile organisms to **sheets**, **filaments** and **cenocytes** (Fig. 2). They are found in fresh and salt water, in soil and on and in plants and animals. Most Chlorophytan cell walls are formed from cellulose and they may be fibrillar, similar to those of the fungi, and sometimes impregnated with silica or calcium carbonate.

Chlorophytan cells contain nuclei, mitochondria, ribosomes, Golgi and chloroplasts (Topic G2). The internal cell structure is supported by a network of microtubules and endoplasmic reticulum. Chloroplasts in the Chlorophyta are very variable structures and can be large and single, multiple, ribbon-like or stellate chloroplasts with chlorophylls *a* and *b* and carotenoids and they store **starch**. They have a vegetative phase that is haploid, and sexual reproduction occurs when cells are stimulated to produce gametes instead of normal vegetative cells at binary fission.

They often possess flagella that have a 9 + 2 microtubule arrangement within them (see Topic G2) and there may be one or two per cell. They may be inserted **apically**, **laterally** or **posteriorly** and trail or girdle the cell. The flagellum can be a single whiplash or it can have hairs and scales. The presence of **eyespots** near the flagellar insertion point allows the cell to swim towards the light. Movement may be by lateral strokes or by a spiral movement that can push or pull the cell through the water.

Members of the protista are unicellular, heterotrophic or photosynthetic, organisms. They may be motile or non-motile. They contain many of the organelles found in other eukaryotes, including nuclei, mitochondria (absent in some groups), chloroplasts (absent in some groups), ribosomes, endoplasmic reticulum, Golgi vesicles, microtubules and microfibrils. Chloroplasts can be very variable in this group, and their shape and pigment content are useful distinguishing taxonomic features. The numbers of ER membranes that surround the chloroplast, and the numbers of thylakoid stacks within them, are also important indicators of phylogeny. Unique organelles found in the protista include **contractile vacuoles** that control influx of water to the cells due to osmosis. Some of the protista have **hydrogenosomes** where fermentative

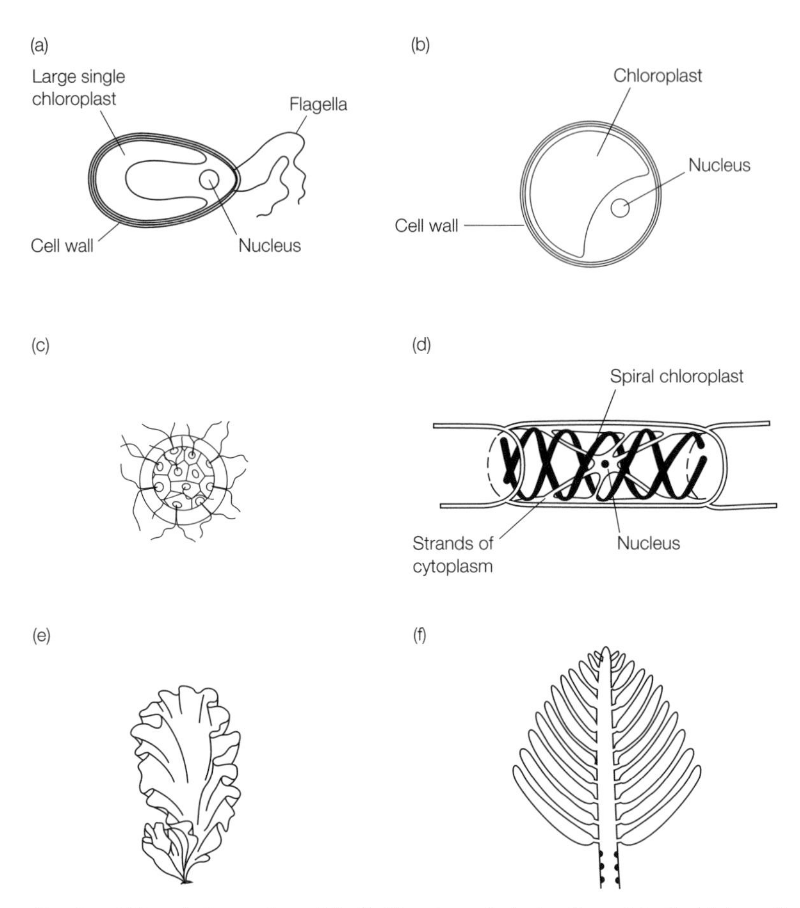

Fig. 2. Chlorophytan cells and thalli. Structure of algal cells and thalli. (a) A motile unicell; (b) non-motile unicell; (c) colonial forms; (d) filamentous forms; (e) membranous (only two cells thick); and (f) tubular. Reproduced from Clegg, C.J., Lower Plants, *1984, with kind permission from John Murray (Publishers) Ltd.*

reactions take place (see Topic G2). Some protista have a cell covering of scales or plates formed of pseudochitin, and this may be impregnated with calcium or silica scales. These scales are secreted from the Golgi bodies within alveoli. Some species have the ability to from thick-walled cysts during a resting phase in the life cycle. These cysts can be very resistant to adverse conditions.

The group is characterized by a considerable variation in morphology (*Fig. 3*). Free-floating species are radially or laterally symmetrical and streamlined into bullet or kidney shapes. At its simplest, the outer surface of protistan cells may be naked plasma membrane. This is seen in the gametes of some species, in amoebae and in some of the intracellular parasites. The simple shape of the amoeba, a simple naked cell with locomotion based on pseudopodia and a phagocytic nutrition, appears to have arisen on a number of occasions in all of the major taxonomic groups. Such a naked cell has advantages in nutrient uptake that outweigh its vulnerability. The naked membrane can be extended with lobes of cytoplasm that are called pseudopodia. These structures are important in locomotion and feeding. This type of movement requires that the cell is in contact with a solid surface. The cell is partitioned into two regions, the viscous ectoplasm and the more liquid endoplasm. Within the ectoplasm are

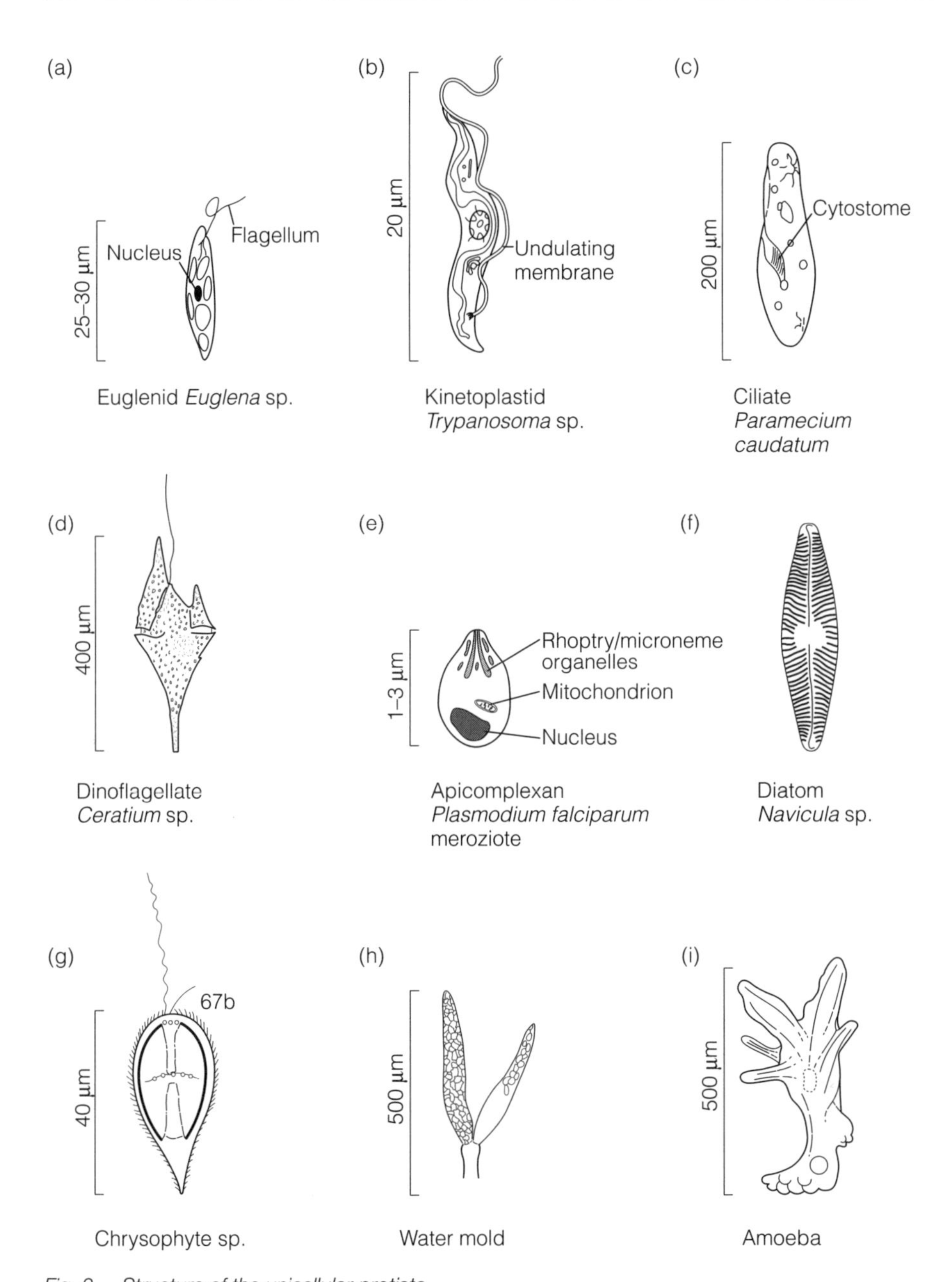

Fig. 3. Structure of the unicellular protista.

myosin and actin filaments whilst in the endoplasm only un-polymerized actin is present. To produce ameboid movement the microfilaments in the ectoplasm produce a pressure on the endoplasm (*Fig. 4*) moving the endoplasm to one part of the cell, and away from another. Motility can also be via flagellae or cilia that propel the cells through water by rhythmic beating. Flagellae typically exhibit a sinusoidal motion in propelling water parallel to their axis (*Fig. 5a*). The undulating action of the flagellum either propels water away from the surface of the cell body or draws water towards and over the cell body. Cilia exhibit an oar-like motion, propelling water parallel to the cell surface (*Fig. 5b*).

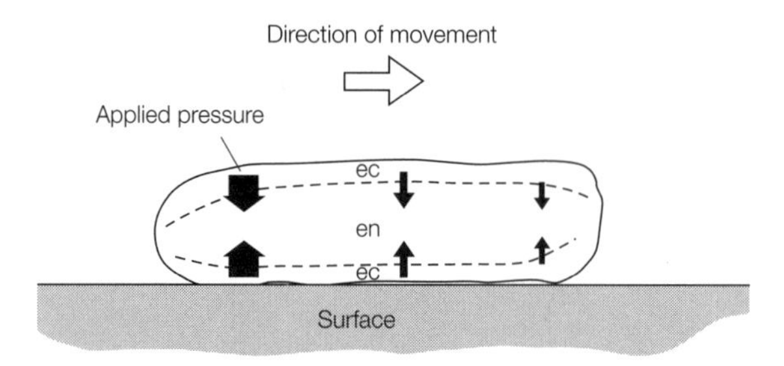

Fig. 4. A proposed mechanism for amoeboid movement. Symbols: ec – ectoplasm; en – endoplasm.

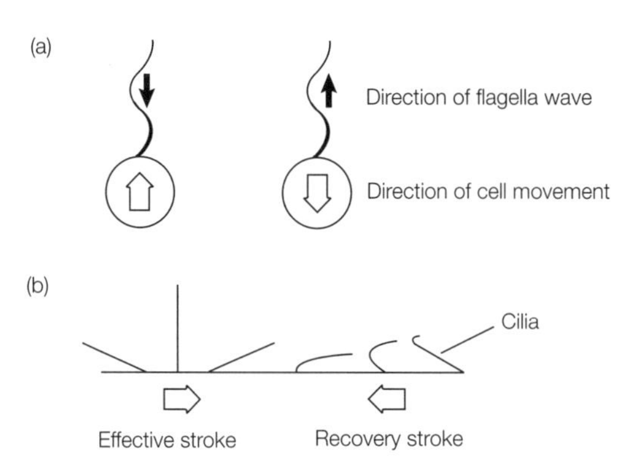

Fig. 5. Motility in (a) flagellates and (b) ciliates.

Within the protista the major groups currently considered to be monophyletic are the **Euglenozoa**, **Alveolata** and **Stramenopila**. However, there are a large number of taxa that have yet to be assigned, and there are new groupings emerging as more molecular data are published.

Euglenozoa

The Euglenozoa contains the **flagellate** forms, the **Euglenids** and **Kinetoplastids**.

Euglenids

Euglenids are unicellular, motile organisms, with two flagellae that originate from a pocket at the anterior end of the cell (see Fig. 3a). They may have a heterotrophic or photosynthetic nutrition. When photosynthetic they contain chloroplasts with chlorophylls *a* and *b* and they store **paramylon** starch. These features suggest that the chloroplasts in this group originated from a Chlorophytan ancestor. They inhabit nutrient-rich (**eutrophicated**) fresh and brackish water. Many species are capable of phagotrophy, and some are wholly saprophytic and devoid of chloroplasts. A few species are parasitic. They have a proteinaceous flexible pellicle that lies underneath the cell membrane very like that of the closely related alveolatan species. The flexibility of this pellicle allows euglenids a characteristic flexibility termed **euglenoid movement** when cells are on solid substrates. The pellicle allows the organism to move through muds and sands. The nuclei appear to contain many chromosomes, and polyploidy appears common.

Kinetoplastids
The sister group to the euglenozoa is the kinetoplastids, parasitic flagellates with a single large mitochondrion and a naked cell membrane (see Fig. 3b). They share many of the characteristics with the euglenoids. The group are named after a large structure within the mitochondria called the **kinetoplast** which contains DNA and associated proteins. The DNA is in the form of **maxicircles** and **minicircles**. Maxicircles encode oxidative metabolism enzymes, minicircles encode unusual RNA-editing enzymes.

Alveolata
The second major group is the alveolata. They have a unique cell surface where the cell plasma membrane is underlain by a layer of vesicles called alveoli, which contain cellulose plates or scales generated from the Golgi. These vesicles fuse with the plasma membrane and provide the cell with an effective cell wall. The plates can be impregnated with silica or calcium carbonate. Within this group are the ciliates, dinoflagellates and the apicomplexa. Some species in the Alveolata have a plasma membrane supported by a proteinaceous pellicle, very similar to that seen in the euglenids.

The ciliates
The ciliates are a largely free-living group, that are characterized by their dense covering of cilia and complex nuclear arrangements (see Fig. 3c). Many species are photosynthetic, having acquired chloroplasts from photosynthetic symbionts.

Dinoflagellates
Dinoflagellates are predominantly unicellular, marine, free-living, motile organisms (see Fig. 3d) Although unicellular, the dinoflagellates are structurally an extremely diverse group of unicellular organisms. Most have two flagellae inserted into the cell at right angles to each other, around the midline of the cell. One is wrapped around the waist of the cell in a groove, the other extends from the posterior of the cell. Their chlorolasts contain chlorophylls *a*, *c*1 and *c*2 and the carotenoid **fucoxanthin**. They store **chrysolaminarin**, a β1–3 linked glucose. Some are capable of phagotrophy; others live within marine invertebrates and are termed **zooxanthallae** (see Topic I4).

Apicomplexans
Apicomplexans are a wholly parasitic group and have a body-form much like an amoeba, but they have an apical complex, a mass of organelles contained within the apical end of the cell (see Fig. 3e). This structure helps the cell invade the host tissue.

The Stramenopila
The Stramenopila contain the **diatoms**, **chrysophytes**, **oomycetes**, and **opalines** plus other large multicellular groups. Also within the stramenopila are several microbial genera with the amoebic body form including the **slime molds.** Many of the Stramenopila have arisen after the symbiotic association between a non-photosynthetic protistan and **red** or **brown** photosynthetic species some time ago in evolutionary history. The pigment content, the numbers of ER layers around the chloroplast, and the numbers of thylakoid lamellae confirm this hypothesis. Those species that contain red pigments from the cryptophytes can

occupy the very deepest layers of the photic zone (see Topic I2). Most species in this group are unicellular, but some are colonial or filamentous. Life cycles are similar to those of the chlorophytes, but the dominant vegetative stage is diploid.

The diatoms

The diatoms (golden brown species of protista) contain chorophylls *a* and *c* and various accessory pigments. Diatoms differ from other members of the golden brown protista because their vegetative stage often lacks flagellae, but they have a **gliding motility** on solid surfaces. They have a silica-containing cell wall composed of a pair of 'nested' shells called **frustules** with a **girdle band** around them. The large half of the shell is termed the **epitheca**, the smaller the **hypotheca** (see Fig. 3f).

Chrysophytes

The chrysophytes contain chlorophylls *a*, *c*1 and *c*2 and have two flagellae inserted into the cell at near right angles to each other. Some species are covered in radially or bilaterally symmetrical scales (see Fig. 3g). They are also characterized by the formation of spores. Spore formation is termed **intrinsic** and it is independent of external conditions. Around 10% of the population will encyst as a zygotic spore in any one generation, allowing populations in optimal growth conditions to maintain genetic diversity and reduce intraspecific competition.

The water molds, Oomycetes and Hyphochytrids

Members of the Oomycota are common water molds, saprophytes or parasites of animals and plants. They share many morphological characters of the chytrids (see Topic H1), but differences include cellulose-containing cell walls, tubular mitochondrial cristae and a life cycle where the dominant somatic phase is diploid rather than haploid or dikaryotic (see Fig. 3h). They also possess biflagellate zoospores instead of the single flagellate zoospores found in the chytrids (see Topic I3). From DNA-sequence analysis their closest relatives appear to be the dinoflagellates. These organisms are capable of causing devastating disease in both plants and fish.

Members of the Hyphochytrids also have cellulose-containing cell walls, and these organisms have only recently been separated from the chytrids. A haploid vegetative stage is dominant in the life cycle. Members of this phyla of organisms are destructive **intracellular parasites** of plants with an absorptive nutrition. They cause enlargement and multiplication of host cells, creating large and unsightly clubbed roots and often death of the plant. The best known disease is **club root** of crucifers. Like the Oomycota, though they share morphological and nutritional similarities with the chytrid fungi, they have some cellulose in their cell walls, and DNA-sequence data also indicate a closer relationship with the dinoflagellate algae than that with the fungi.

Amoebae and slime molds

Amoebae are naked protistan cells that have an absorptive nutrition (see Fig. 3i) They have a wide distribution, living free in soil and water, and as parasites of animals and man. They should not be considered as a monophyletic group as there are amoebic groups aligned closely to the Euglenozoa and the Alveolata. Reproduction in the amoebae is by binary fission of the cell after mitotic division of the nucleus. A single cell splits to form two identical progeny.

Often grouped together, the phyla termed **Dictyosteliomycota** and the **Acrasiomycota** are phyla that contain the cellular slime molds. They exist for most of their life cycle as haploid amoebae that feed within soil by engulfing bacteria. They are uninucleate for most of their life cycle but form a plasmodium at sporulation.

Myxomycota are acellular slime molds. They exist as haploid amoebae for their vegetative stage, but fuse in pairs to form a diploid cell that undergoes repeated mitotic nuclear divisions without cell division, forming a plasmodium. The life cycle is then very similar to the cellular slime molds.

Opalines

A group of specialized ciliate-like organisms, found living saprophytically in the bowels of amphibians. They are covered with closely spaced flagellae.

Chlorophytan and protistan growth

Growth in the unicellular Chlorophyta and protista is synonymous with binary fission. In most unicells, haploid or diploid nuclei undergo mitosis, and the cell then divides longitudinally to form two daughter cells. In some species there are two haploid divisions within the parent cell, followed by the formation of four motile daughter cells (see Topic G3). Some cenocytic filamentous algae grow from the tip of the filament in a way very similar to that of hyphal growth (see Topic H1). Others grow by division of vegetative cells within filaments or sheets.

Accurate estimates of Chlorophytan and protistan growth rates can be made by cell counting or by estimating chlorophyll content of a culture. Kinetics of growth are similar to those seen in the bacteria (see Topic D8), but for photosynthetic species, depletion of nutrients other than carbon leads to culture limitations and the stationary and death phases. Nitrogen, phosphates or silicon are frequently limiting.

12 CHLOROPHYTAN AND PROTISTAN NUTRITION AND METABOLISM

Key notes

Carbon and energy metabolism

The Chlorophyta and pigmented protista are photosynthetic organisms and obtain their carbon and energy requirements by the fixation of CO_2, using photosynthesis. In a terrestrial habitat, light levels are usually adequate to support photosynthesis, but in the aquatic habitat light energy is rapidly absorbed in the top 0.5 m of the water column. Aquatic species have evolved three chlorophylls, *a*, *b* and *c*, and a large number of accessory pigments to allow them to extend the depth to which they can grow. Photosynthetic species use aerobic respiration via glycolysis and the citric acid cycle. Non-photosynthetic protista can obtain their carbon and energy requirements from the environment by diffusion, pinocytosis and phagocytosis. In aerobic species, glycolysis, the citric acid cycle and mitochondrial respiration provide the cell with energy and metabolites. Anaerobic protista may utilize fermentative pathways within the hydrogenosomes.

Oxygen and carbon dioxide

In the terrestrial environment, photosynthetic CO_2 and O_2 requirements are almost always satisfied by atmospheric gases. In the aquatic environment, the solubility of O_2 decreases with temperature and increasing dissolved CO_2 levels, and availability of O_2 therefore becomes limiting in warm waters.

Nitrogen nutrition

No Chlorophytan or protistan species can fix nitrogen and they all must therefore obtain it in a fixed, inorganic or organic form. Most can utilize nitrate or ammonia; some require organic compounds. Nitrogen levels may be limiting to growth in marine environments.

Macronutrients, micronutrients and growth factors

Most nutrients are available to excess in the aquatic environment, but phosphates and silicon are only poorly soluble in water and are often limiting to growth in fresh water. Some species are predominantly autotrophic, but many require an external supply of amino acids, vitamins, nucleic acids and other growth factors; such species are described as auxotrophic.

Water, pH and temperature optima

The members of the Chlorophyta and protista do not survive severe desiccation and are therefore found mostly in damp terrestrial habitats or in water. Phagocytic species have an absolute requirement for liquid water. Most species can tolerate a wide range of pH and temperature. Some are specialized and can inhabit extremely acidic, hot springs while others can complete their entire life cycle below 0°C.

Osmolarity

Fresh-water species have a large difference between internal and osmotic pressure, and they must either have a rigid cell wall or a contractile vacuole to compensate for water uptake. Conversely, marine species of Chlorophyta and protista are roughly isotonic with sea water.

Related topics

Heterotrophic pathways (B1)
Autotrophic reactions (B3)
Growth in the laboratory (D6)
Eukaryotic cell structure (G2)
Beneficial effects of fungi in their environment (H4)
Chlorophytan and protistan taxonomy and structure (I1)
Beneficial effects of the Chlorophyta and protista (I4)

Carbon and energy metabolism

All species of the Chlorophyta and many species of the protista are photosynthetic and therefore gain carbon and energy from the fixation of atmospheric or dissolved carbon dioxide using photosynthesis.

The photosynthetic reactions take place in the chloroplasts (see Topic I1), the light reactions occurring in the chloroplast thylakoids, and the light-independent reactions occurring in the stroma (see Topic B3). The chlorophyll pigments are membrane-bound within the thylakoids, and their properties include the ability to be excited by light. Different chlorophylls accept light energy of different wavelengths, depending on their structure. Accessory pigments, the **carotenoids**, **phycobilins** and **xanthophylls**, also absorb light energy of different wavelengths and pass their excitation to chlorophyll, maximizing the breadth of wavelength over which light energy can be absorbed. The role of accessory pigments is particularly important as they allow photosynthetic organisms to occupy different parts of the **photic zone**, the shallow layer of water where sufficient light penetrates to support photosynthesis. The actual depth to which light will penetrate varies with turbidity and dissolved organic matter content (see Topic I4). Furthermore, light of different wavelengths penetrates water to different degrees (*Fig. 1a*). Red light is absorbed rapidly by water, blue light least. The members of the Chlorophyta occupy only the shallowest of waters.

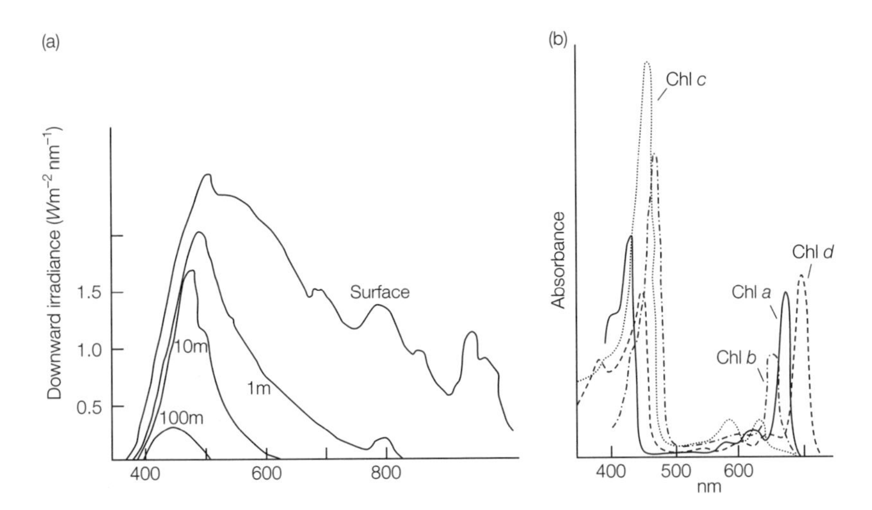

Fig. 1. (a) Downward penetration of different light wavelengths. (b) Absorption spectra of chlorophylls.

Chlorophylls *a*, *b* and *c* are the photosynthetic pigments, with chlorophyll *a* being the primary photosynthetic pigment. The absorption **maxima** of chlorophyll *a* are at 430 and 663 nm (*Fig. 1b*). Other chlorophylls are accessory pigments and have slightly different absorption maxima: chlorophyll *b* at 435 and 645 nm, chlorophyll *c*1 at 440, 583 and 634nm, and *c*2 at 452, 586 and 635 nm. Phycobilins absorb light energy at 565 nm. There are also other accessory pigments found in the different algal groups, the carotenes and xanthophylls. Depending on the different accessory pigments present in individual species, light of different wavelengths can be absorbed and utilized in photosynthesis.

Nutrient uptake in non-photosynthetic protista may be by **diffusion**, **pinocytosis**, or **phagocytosis**. Within nutrient-rich tissues, such as those that intracellular parasites or blood-dwelling parasites live in, nutrient uptake is by diffusion and pinocytosis, the formation and digestion of small **vesicles** from the plasma membrane. These vesicles capture soluble nutrients from the environment.

In most larger, free-living species, particulate organic matter and bacteria are engulfed by phagocytosis. In this process large vacuoles can form anywhere over the cell surface in the amoebae, or from specialized sites such as the cytostome, in ciliates and flagellates. Digestion of the vesicle content occurs by the fusion of lysozyme-containing enzyme vesicles with the phagocytic vesicles. Once digestion is complete enzymes are recycled by the cell and sequestered into small vesicles, whilst cell debris is expelled from the cell by **reverse phagocytosis**, where the food vacuole re-fuses with the plasma membrane.

A large number of symbiotic species can be found in the protista, in association with many different photosynthetic symbionts. Species of Alveolata and Euglenozoa with these symbionts have an autotrophic nutrition, relying on photosynthetic products for carbon and energy sources (see Topic B4). In some cases the cytostome seen in non-photosynthetic protistans is not present. Further evidence of the close taxonomic relationships between the photosynthetic and non-photosynthetic groups is provided by the fact that it is possible to 'cure' some species of their symbionts, and once cured, these protista return to their holozoic nutrition and form food vesicles from a reformed cytostome.

Almost all members of the Chlorophyta are aerobic and use mitochondrial respiration with oxygen as the terminal electron acceptor. In most of the protista, glycolysis and the citric acid cycle provide energy and intermediates for cellular metabolism (see Topic B1). Aerobic protista use mitochondrial respiration, with oxygen as the terminal electron acceptor, generating ATP.

Anaerobic species have a fermentative metabolism (see Topic B1), which results in the incomplete oxidation of substrates. A few species that live in anaerobic lake sediments can use alternative electron acceptors such as nitrate.

There are several modifications to the usual glycolytic pathway (see Topic B1) found in some members of the protista. Some can utilize **inorganic pyrophosphate** rather than ATP, replacing enzymes like pyruvate kinase with pyrophosphate kinase. The advantage of this is that pyrophosphate kinase activity is reversible and can be used to synthesize glucose from other substrates.

In parasitic kinetoplastids glycolysis does not occur in the cytoplasm but in its own organelle, the **glycosome**. Within this organelle a modified form of glycolysis can give rise to glycerol which is the substrate of respiration in this group (see *Fig. 2*). In anaerobic protista fermentative reactions occur within the **hydrogenosome**. Substrates like pyruvate and malate from glycolysis are taken up by this organelle and incompletely oxidized to end products like acetate, in

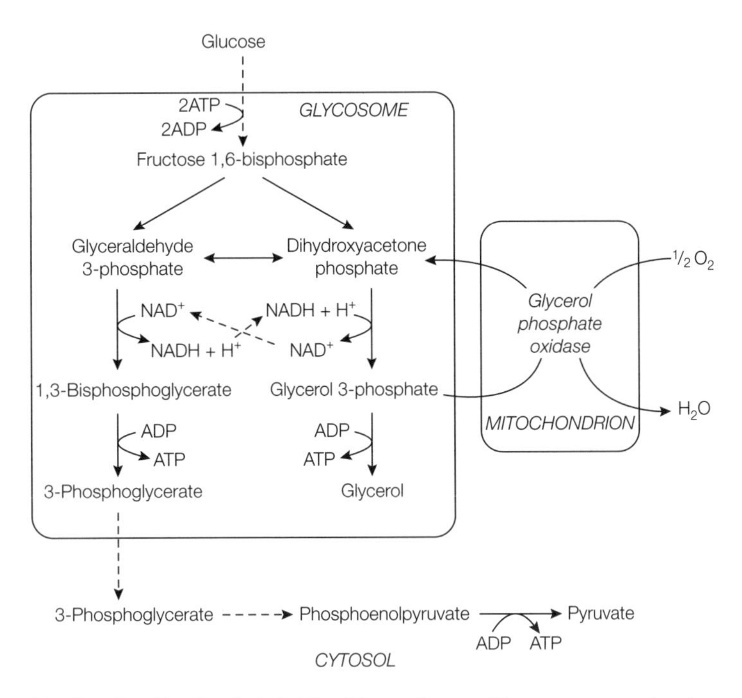

Fig. 2. Aerobic glycolysis in bloodstream forms of trypanosomes, showing compartmentalization of much of the pathway in the glycosome. Dotted lines indicate several reaction steps.

the process forming ATP. Hydrogen is another characteristic end-product of hydrogenosomal metabolism (see *Fig. 3*).

In many protista mitochondrial metabolism resembles that of bacteria respiration, having a branched pathway, part of which is sensitive to cyanide (mammalian-like) and part of which has an alternative terminal oxidase which allows the mitochondrion to operate at low oxygen levels. These modifications are most developed in the specialized parasites, where both cellular and mitochondrial structure change as the parasite passes though different stages of the life cycle.

Oxygen and carbon dioxide

In an aquatic environment temperature influences levels of dissolved oxygen. Water is saturated with oxygen at 14 mg $O_2\,l^{-1}$ at 0°C, but only 9 mg l^{-1} at 20°C. Oxygen utilization by living organisms increases with a rise in temperature; thus, availability of oxygen is likely to limit growth in warmer waters.

Carbon dioxide, the levels of which in water vary inversely with dissolved oxygen, provides carbon to autotrophic species for photosynthesis. Dissolved carbon (as **carbonic acid**, HCO_3^-) is usually present at between 2.2 and 2.5 $nmol^{-1}$, while CO_2 is present at only 10 mmol l^{-1}. Most Chlorophyta and photosynthetic protista utilize carbonic acid. Anaerobic photosynthesis occurs in a few species, using hydrogen sulfide or carbon dioxide as terminal acceptors (see Topic B4).

The absolute requirement for light in photosynthetic species of Chlorophyta and protista means that they must have adaptations, which allow them to remain in the photic zone of their environment. On land this is not problematic, but in the aquatic ecosystem there is a natural tendency for cells to sink, and

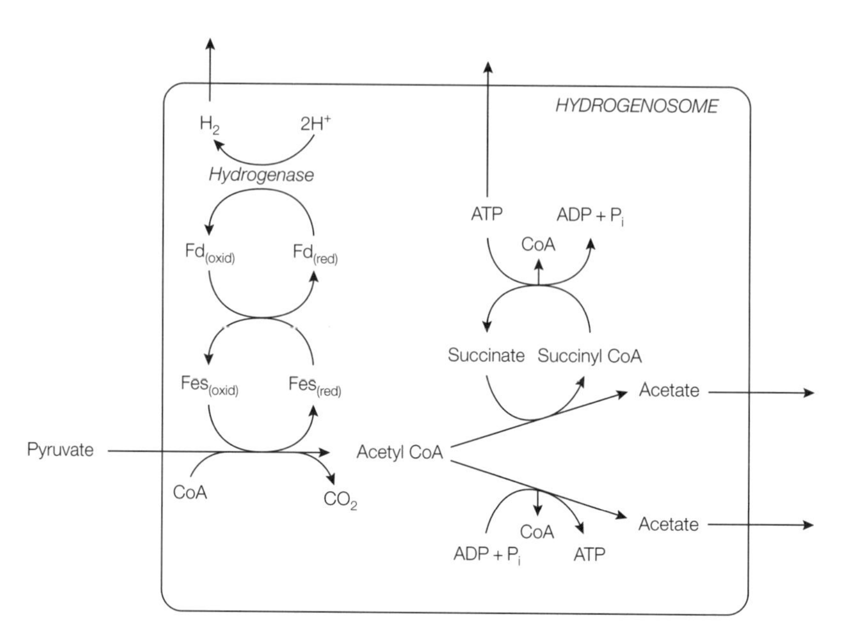

Fig. 3. Hydrogenosomal metabolism in the trichomonads. Fes = Iron-sulfur proteins; Fd = ferrodoxin; oxid = oxidized state; red = reduced state.

thus there are cell modifications that counter this. Many Chlorophytan and protistan cells, zoospores and gametes are flagellate and can swim towards the light (see Topic I1). Other species have structural modifications like spines that increase the resistance of the cell to sinking, and catch water uplift. Some have **gas vacuoles** that increase bouyancy.

Nitrogen nutrition

Members of the Chlorophyta and the protista are heterotrophic for nitrogen and must obtain it in a fixed form such as nitrate, ammonia or amino acids. Amino acid requirement is common amongst the algae (see Growth factors). Poor availability of nitrogen is a common limiting factor to growth in the marine environment. Complex nitrogen sources such as peptides and proteins can be utilized after either extracellular or intracellular enzyme secretion has digested the polymers. Amino acids can also be stored as a nitrogen store.

Macronutrients, micronutrients and growth factors

Carbon is rarely limiting for growth of photosynthetic species, but nitrogen and phosphorus often are. Fresh-water productivity is often limited by phosphate availability. Silicon is often limiting to diatom growth in nutrient poor, **oligotrophic** lakes. Most other nutrients are present to excess in the aquatic environment.

Although some photosynthetic species are completely **autotrophic**, many require an external supply of **vitamins**, frequently thiamin, biotin, B_{12} and riboflavin, purines, pyrimidines and other classes of growth factors. This requirement is termed **auxotrophy**, and it reflects the abundance of dissolved organic matter to be found in their environment, which allows the selection of populations that do not synthesize all their own metabolic requirements (see Topic D6).

Water, pH and temperature optima

Almost all Chlorophyta and protista are limited to damp environments. A few photosynthetic species that are in lichen symbiosis (see Topic H4) are protected from desiccation and can survive extremely dry conditions. Most members of the Chlorophyta and protista are tolerant of a wide range of pH; some are specialized and can inhabit highly acidic environments like those of hot sulfur springs. In highly illuminated, oligotrophic lakes, large changes in pH can be caused by variations in dissolved CO_2 concentrations and these changes can have detrimental effects on the populations of photosynthetic species.

Some dormant stages of protista are capable of withstanding 100°C for several hours. Other species can grow and divide at –2°C in seawater, and specialized 'snow algae' have growth optima between 1 and 5°C. Most species have a temperature optimum between 5 and 50°C.

Osmolarity

Seawater and fresh water have very different **osmolarities**. Fresh water species of Chlorophytes and Protists have an internal osmotic pressure of 50–150 mOs ml^{-1} while fresh water osmolarity is <10 $mOsml^{-1}$. Those species that do not have a rigid cell wall to prevent excessive water uptake have contractile vacuoles that collect and expel excess water from the cell. Marine organisms have cytoplasm that is roughly isotonic with seawater and therefore usually do not require contractile vacuoles.

I3 LIFE CYCLES IN THE CHLOROPHYTA AND PROTISTA

Key notes

Life cycles in the Chlorophyta

Many members of the Chlotrophyta have a haploid vegetative phase, and gametes (motile or non-motile) are formed by the differentiation of a vegetative cell. Zygote formation is followed by meiosis, producing haploid progeny cells. Other species have a diploid vegetative phase and produce haploid gametes by meiosis.

Life cycles in the Protista

Life cycles can be dominated by haploid or diploid phases. Life cycles in the Euglenozoa are characterized by asexual reproduction in flagellate species by longitudinal binary fission. Sexual reproduction is seen in the parasitic flagellates. In the Alveolata, asexual reproduction is by homothetogenic cell division. Sexual reproduction occurs and the dominant phase of the life cycle is diploid. Asexual reproduction in the apicomplexa is characterized by multiple fission. Sexual reproduction occurs in alternative hosts.

Life cycles in the Stramenopila

The diploid vegetative cells of diatoms reproduce asexually by mitosis. For sexual reproduction in round centric diatoms, one or several macrogametes are formed in an oogonium which is then fertilized by motile microgametes produced from the other mating-type diatom to generate a zygote within an auxospore. Pennate diatoms do not display motile microgametes but after meiosis fuse to form a zygote.

Reproduction in the Oomycota and Hyphochytridiomycota

Except for a few terrestrial species, asexual reproduction in Oomycetes is via flagellate motile diploid zoospores released from terminal sporangia. Sexual reproduction involves haploid gametes produced by terminal or sub-terminal antheridia and oogonia. Hypochytrids reproduce via a multinucleate protoplast that is formed in the host cell, then releasing cysts into the soil that germinate into biflagellate zoospores able to invade new host plant roots.

Life cycles in the amoebae and slime molds

Amoebae undergo mitosis and reproduce asexually by binary fission. Sexual reproduction is by fusion of haploid amoebae of different mating types to produce a diploid, followed by meiosis to generate haploid amoebae once more. Amoebae of the cellular slime molds (Dictyosteliomycota and Acrasiomycota) aggregate upon starvation to yield a multicellular pseudoplasmodium that differentiates into a fruiting body and generates haploid spores. Amoebae of the acellular slime molds (Myxomocota) fuse to form a diploid cell (or this is produced via haploid flagellated cells) which grows into a large multinucleate plasmodium. Sporangia arise and meiosis within them generates haploid spores that germinate to release amoebae.

Related topics

Cell division and ploidy (G3) Fungal structure and growth (H1)

Life cycles in the Chlorophyta

The dominant vegetative phase of members of the Chlorophyta can be haploid or diploid. Vegetative growth is associated with mitotic cell division. In chlorophytes with a haploid life history, meiosis occurs at zygote germination and the cells remain haploid for the whole of their vegetative life. Diploid formation occurs only in the zygote. Chlorophytes with a diploid life history are only haploid at gamete formation and after zygote formation continue their life cycles as diploids.

Motile compatible gametes first entangle flagellae, cells then **conjugate** and nuclei fuse to form a zygote. The zygote may remain motile or it may form a thick-walled resting cyst. Meiosis occurs within the zygote and haploid, flagellate cells are released (*Fig. 1*). Similar events are seen in the colonial forms of the Chlorophyta; all cells of a colony may develop into free-swimming gametes after breakdown of the colony structure.

Filamentous Chlorophyta may reproduce sexually by conjugation. In this process, two vegetative cells form a **conjugation tube** between them and fuse. The cellular content from one cell then moves into the other, where nuclear fusion and zygote formation occur. The zygote encysts and meiosis occurs before the emergence of a haploid new filament (*Fig. 2*). Other filamentous species produce motile gametes of two mating types from different vegetative cells. Often one gamete is considerably larger (**macrogamete**) than the other (**microgamete**). In some species, only one motile microgamete is produced and it fuses with a non-motile gamete cell called an **oogonium** to form the zygote.

Life cycles in the Protista

Vegetative growth is associated with mitotic nuclear division followed by cell division. Cell division can be by budding, binary or multiple division. Each of the different groups of protista are characterized by differences in their life cycle. The dominant vegetative phase of protista can be haploid or diploid.

In protista with a haploid life cycle, meiosis occurs at the germination of the diploid zygote. The cells remain haploid for the rest of the vegetative phase, forming haploid gametes that fuse to form the transient diploid. Protista with a diploid life history are haploid only at gamete formation. After formation of the

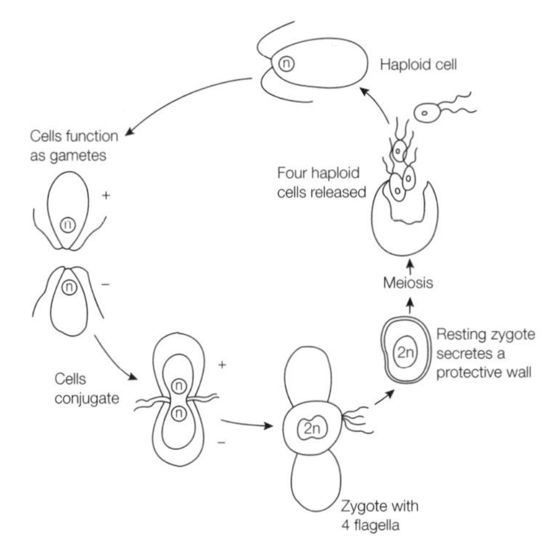

Fig. 1. Sexual reproduction in the unicellular Chlorophyta. Reproduced from Gross, T., Faull, J.L., Kettridge, S. and Springham, D., Introductory Microbiology, *1995, with kind permission from Kluwer Academic Publishers.*

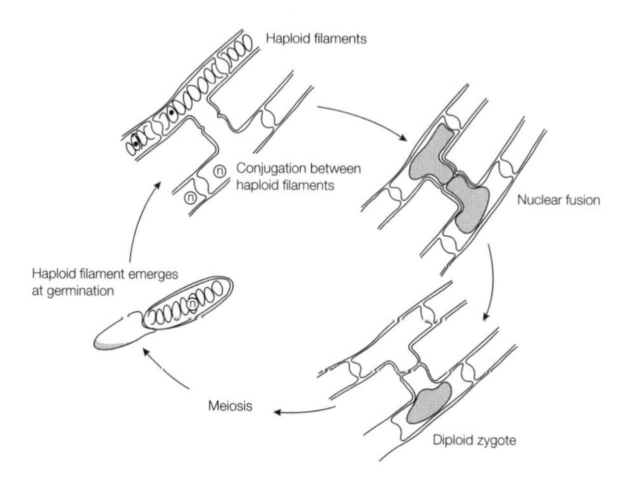

Fig. 2. Conjugation between two filamentous Chlorophytan cells. Reproduced from Gross, T., Faull, J.L., Kettridge, S. and Springham, D., Introductory Microbiology, *1995, with kind permission from Kluwer Academic Publishers.*

diploid zygote they continue their life cycle as diploids until gamete formation necessitates meiosis.

Euglenozoa

Reproduction in the flagellates is characterized by mitosis of the nucleus followed by cell division, which occurs along the longitudinal axis of the cell. This produces mirror image progeny (see Topic G3). Sexual reproduction is seen in some species of this group. The group includes many of the parasitic species including trypanosomes (see *Fig. 3*).

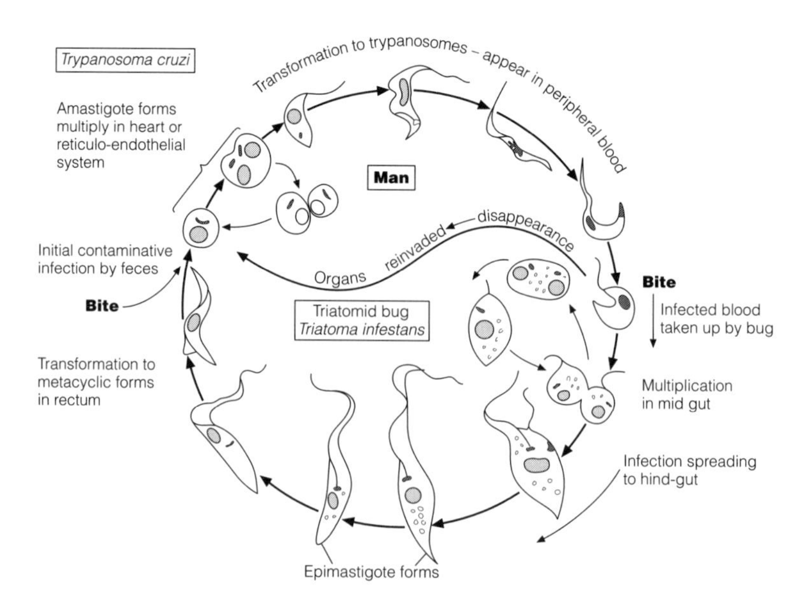

Fig. 3. Life cycle of Trypanosoma cruzi *in man and in the triatomid bug* Triatoma infestans*; other arthropods, bed bugs, ticks and keds can also act as vectors. (After Brumpt, 1949.)*

Alveolata

Dinoflagellates are haploid and have unusual chromosomes, which are condensed throughout their life cycle. The chromosomes contain very little histone. Sexual reproduction commences when motile cells differentiate to become **macro-** or **microgametes** (*Fig. 4*). The gametes fuse to from a zygote. Meiosis occurs followed by the degeneration of three of the four nuclear products. A haploid, motile cell then emerges from the zygote.

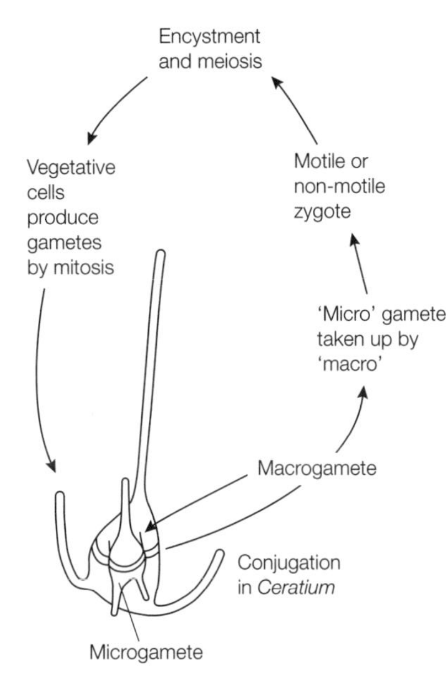

Fig. 4. Reproduction in the dinoflagellates. Reproduced from Gross, T., Faull, J.L., Kettridge, S. and Springham, D., Introductory Microbiology, *1995, with kind permission from Kluwer Academic Publishers.*

Apicomplexa

Apicomplexan protists are parasites with a complex life cycle. There are both diploid and haploid phases and often two host species are infected in a life cycle. The group is characterized by a type of cell division that is called multiple division or **shizogony**. During this process, multiple division of haploid nuclei occurs, producing many progeny. These progeny are then released into body fluids like blood where they rapidly enter new host cells and establish themselves as intracellular parasites. The malarial parasite *Plasmodium* spp. is an example of an apicomplexan parasite (see *Fig. 5*).

Ciliates

Ciliate asexual reproduction is also by binary fission after mitosis of the nucleus. Cell division is described as **homothetogenic**, across the narrow part of the cell. Ciliates have two types of nuclei, a **micronucleus** that is diploid, contains little RNA and a lot of histone, and a **macronucleus** which is polyploid and controls day-to-day cellular activities. There may be hundreds of macronuclei in one ciliate, and up to 80 micronuclei. During asexual mitotic cell division, both of these nuclei divide to provide progeny cells with at least one copy of both nuclei. During sexual reproduction the macronucleus degenerates and the micronucleus undergoes meiosis to form gametes. Gametes fuse to form a new diploid nucleus. This new nucleus divides, one copy of the nucleus remaining as the new micronucleus, and the others differentiating to provide a new polyploid macronucleus (see *Fig. 6*).

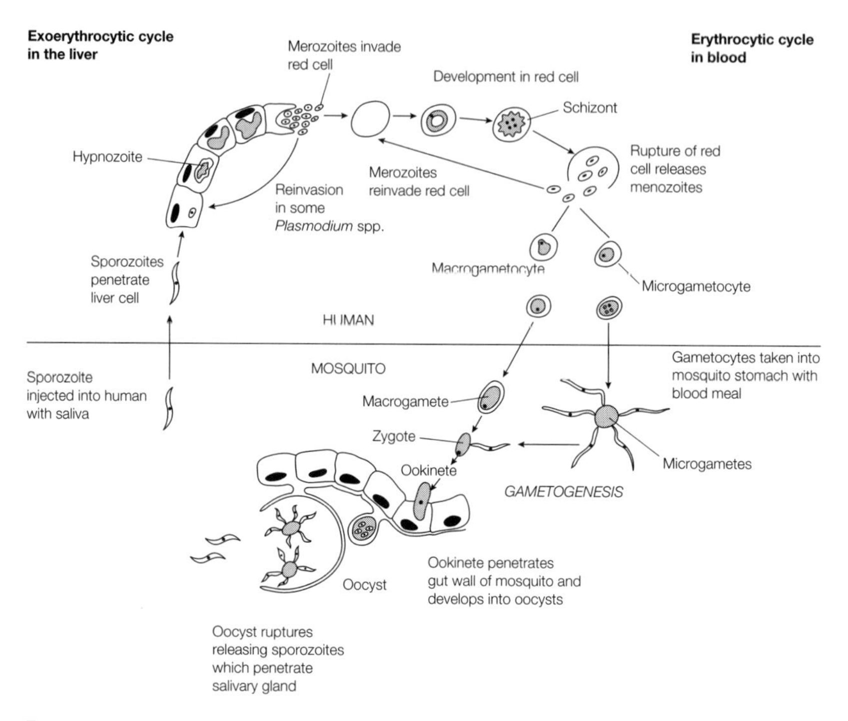

Fig. 5. Life cycle of Plasmodium *sp.*

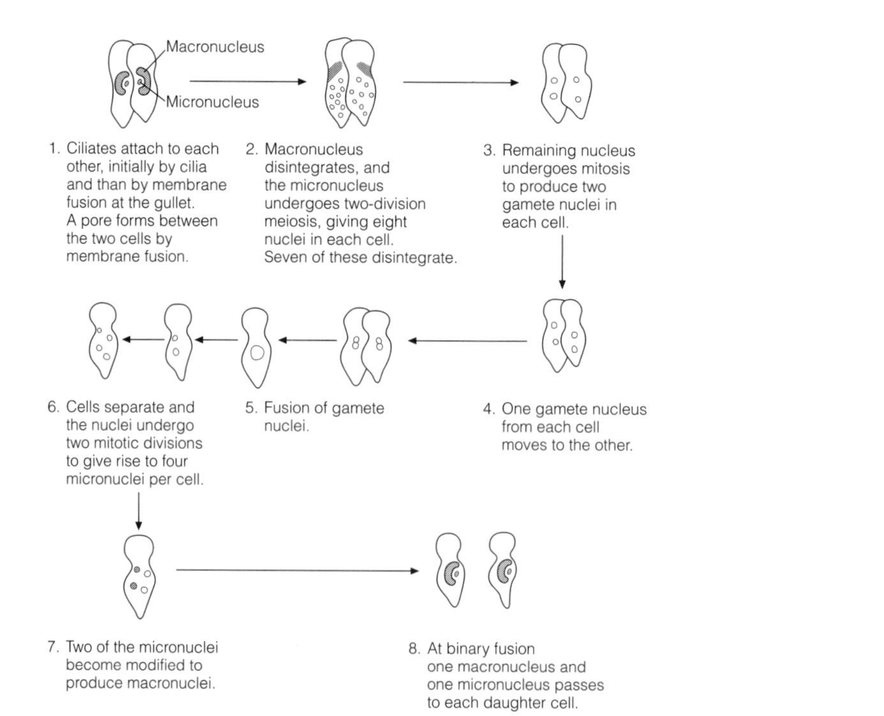

Fig. 6. Sexual reproduction in the ciliates. Redrawn from Gross et al. Introductory Microbiology *1993.*

This unusual state of nuclear dualism appears to confer an advantage to the ciliates by having a separate genetic store in the micronucleus, and enhances RNA synthesis by the macronucleus. This allows them to be very adaptable to changing environmental conditions.

Life cycles in the Stramenopila

Within the Stramenopila there are a number of life cycle patterns.

The diatoms

Vegetative cells are diploid, and repeated mitotic divisions lead to a reduction in cell volume as daughter cells synthesize new frustules that fit within the inherited parent frustule. Once a 30% reduction in volume has been reached diatoms either produce a resting spore (or **auxospore**) to regain cell size or they reproduce sexually (see *Fig. 7*). Sexual reproduction begins by the formation of gametes after meiosis. In round, centric diatoms a single or multiple macrogamete forms within an **oogonium**. Motile microgametes are formed within the other mating-type diatom. These microgametes are released and fertilization of the oogonium leads to the formation of a zygote within an auxospore. The auxospore enlarges and secretes a new pair of full-size frustules. The diploid, vegetative life cycle then continues. The long, thin diatoms, termed pennate, do not form motile gametes, but after a meiotic division fuse somatically to form a zygote.

Reproduction in the Oomycota and Hyphochytridiomycota

Oomycetes are filamentous, coenocytic organisms. Asexual reproduction is characterized by the production of flagellate, motile, diploid zoospores from terminal sporangia. Sexual reproduction occurs after meiosis and gamete production in terminal or sub-terminal antheridia and oogonia. A few species are terrestrial and produce non-motile gametes from sporangia and Oogonia.

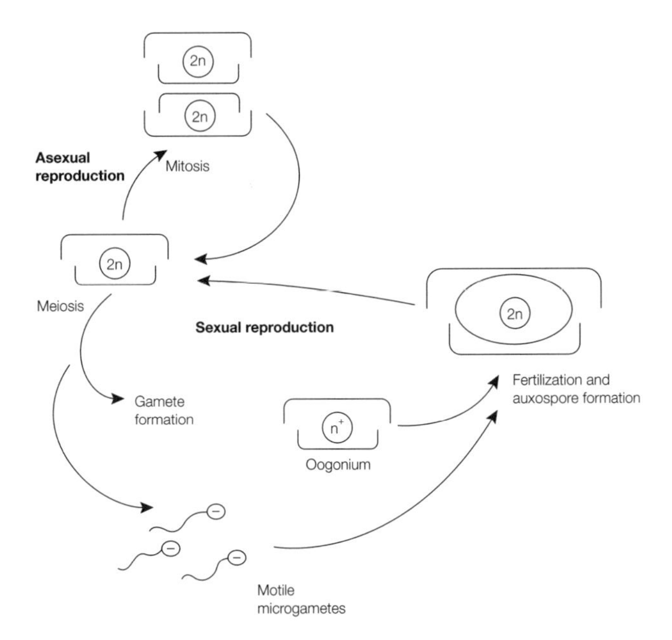

Fig. 7. Life cycle of the diatoms.

The life cycle of the plant pathogen *Phytophthora infestans* is typical of this terrestrial group (*Fig. 8*).

Hyphochytrids reproduce by the formation of multinucleate, unwalled protoplasts within an enlarged host cell. Cyst formation within the host cell occurs, and these are released into the soil on breakdown of the plant root where they can persist for years. The presence of a suitable host plant root breaks dormancy, and the cyst germinates by producing anteriorly biflagellate zoospores that swim to the host root and actively penetrate it.

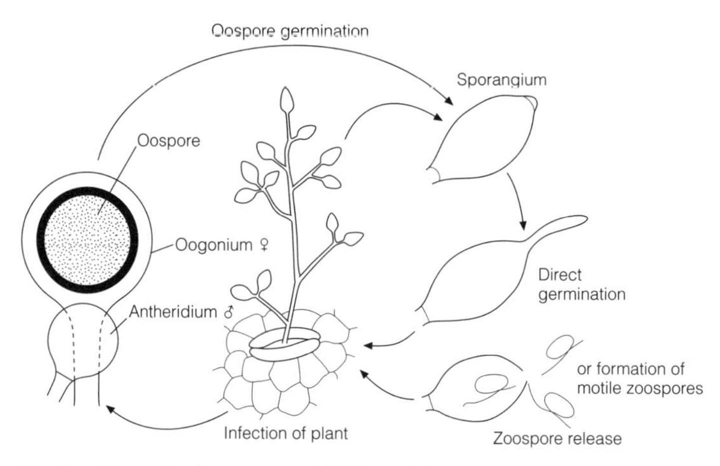

Fig. 8. Life cycle of a water mold, Phytophthora infestans. *Redrawn from Deacon,* Modern Mycology *with permission of Blackwell Science.*

Life cycles in the amoebae and slime molds

Amoebae

Amoebae have a very wide distribution, living free in soil and water, and as parasites of man and animals. They should not be considered as a monophyletic group. Reproduction in amoebae is by binary fission of the cell after mitotic division of the nucleus. A single cell splits to form two identical progeny. Sexual reproduction is seen in some species, the haploid amoebae differentiating into gametes of different mating types which can then fuse to form the diploid zygote. Meiosis then occurs and the haploid phase of the life cycle continues.

Cellular slime molds

Starvation of amoebae of the Dictyosteliomycota and Acrasiomycota cellular slime molds leads to their aggregation into multicellular **pseudoplasmodium**, which eventually differentiates into a fruiting body consisting of a foot, stalk and sporangium. Haploid spores are formed which are disseminated and give rise to haploid amoebae (*Fig. 9*).

Acellular slime molds

In the Myxomycota, the acellular slime molds, under certain conditions, reproduction will commence by the fusion of amoebae, or amoebae can produce flagellate cells which fuse, to form a diploid cell (*Fig. 10*). This diploid cell grows, feeds and undergoes repeated mitotic nuclear divisions without cell division, forming a large plasmodium. Sporangia are produced within which meiosis occurs to generate haploid spores. Each spore germinates to release an amoeba to complete the life cycle.

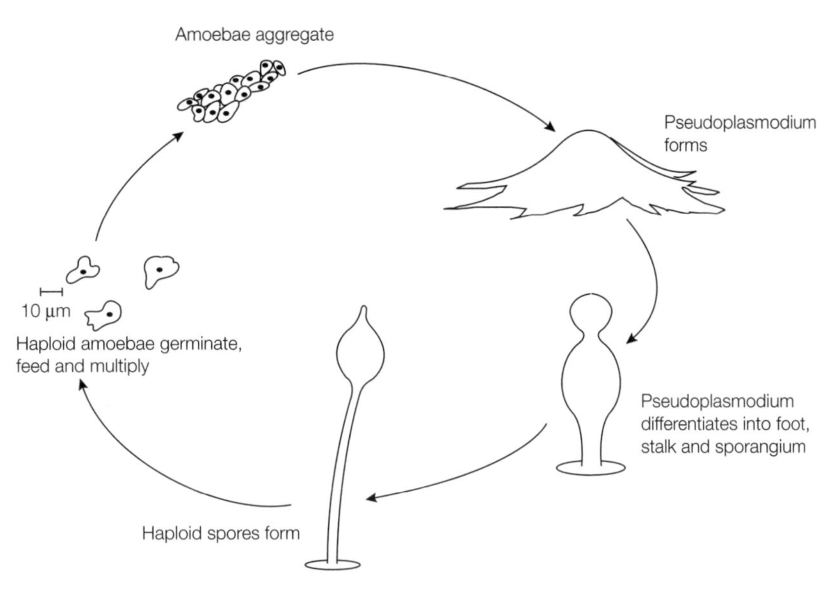

Fig. 9. Life cycle of a cellular slime mold. Redrawn from Sleigh, M., Protozoa and Other Protists, *1991, with kind permission from Cambridge University Press.*

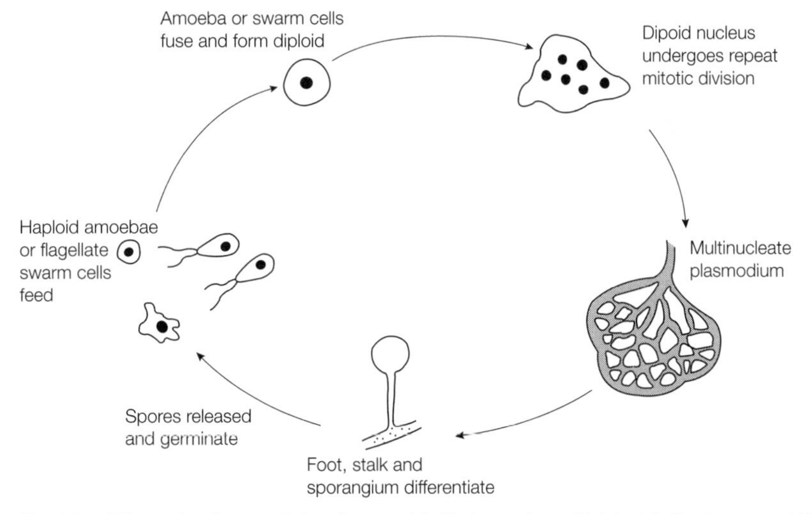

Fig. 10. Life cycle of an acellular slime mold. Redrawn from Sleigh, M., Protozoa and Other Protists, *1991, with kind permission from Cambridge University Press.*

I4 Beneficial effects of the Chlorophyta and Protista

Key notes

Primary productivity

Primary productivity in the oceans due to Chlorophytan and photosynthetic protistan photosynthesis is estimated to be ~ 50 × 109 tonnes per annum. Carbon can be released as dissolved organic matter (80–90%) or as particulates (10–20%). The photosynthetic cells, associated saprophytic bacteria and dissolved organic matter together form the basis of the aquatic food chain.

Symbiosis

Symbiotic associations can occur between species of Chlorophyta, protista, fungi and animals. In these associations, carbohydrates from photosynthesis are exchanged for nutrients from the fungus or animal partner. Protista can have internal symbionts that can be bacteria or other protists. Photosynthetic symbionts provide photosynthate or other factors for the heterotrophic protistan partner. Bacterial endosymbionts can provide vital cell metabolites. Ectosymbiotic bacteria may provide protista with motility or alternative electron acceptors. Some protista are important within gut and rumen microflora of large animals, as part of a complex cellulolytic community.

Diatomaceous earth

Diatomaceous earth is formed from the silica-containing shells of diatoms. It has several commercial uses that take advantage of its chemically inert but physically abrasive qualities.

Bioluminescence

Bioluminescence is found in several of the protistan phyla and is associated with luciferin–luciferase reactions that occur within scintillons in the protistan cell. The precise function of this reaction is unknown, although it may provide defense against predators.

Related topics

Beneficial effects of fungi in their environment (H4)
Chlorophytan and protistan taxonomy and structure (I1)
Chlorophytan and protistan nutrition and metabolism (I2)
Life cycles in the Chlorophyta and protista (I3)

Primary productivity

In the ocean the total plant biomass (as **phytoplankton**) is estimated to be 4×10^9 tonnes (dry weight). Annual net **primary production** is around 50×10^9 tonnes. Productivity ranges from 25 g carbon m^{-3} in oligotrophic waters to 350 g carbon m^{-3} in eutrophic waters. This is released as **dissolved organic matter** (**DOM**), which accounts for 80–90% of organic matter present in the sea. The DOM is used as a nutrient source by heterotrophic bacteria; populations between 10^4 and 10^6 ml^{-1} are commonly found in nutrient-poor (**oligotrophic**) waters, and much higher populations are found in nutrient-rich (**eutrophic**) waters (see Topic I5).

The cells (both alive and as dead particulates), DOM and bacterial populations form the base of the aquatic food chain.

Symbiosis

Protista can form beneficial associations with bacteria, fungi, Chlorophyta, other protista and insects and higher animals. These symbioses can be specific and physically and physiologically intimate or they can be relatively non-specific and merely loose ecological associations.

Endosymbionts are symbionts that live inside other organisms. Photosynthetic species can occur within Alveolata and Euglenozoa (and others) and their presence allows the protistan dual organism to adopt a phototrophic habit. The photosynthetic partner is confined to a membrane-bound vacuole, but it is capable of cell division. There is a two-way exchange of materials where the products of nitrogen metabolism of the heterotrophic protistan and utilized by the photosynthetic partner and the products of photosynthesis are utilized by the heterotrophic partner.

The **zooxanthellae** are symbiotic dinoflagellates that are found as coccoid cells within animal cells. They are enclosed in intracellular double membrane-bound vacuoles, which remain undigested. They are found in protista, hydroids, sea anemones, corals and clams where they provide glycerol, glucose and organic acids for the animal, and the symbiotic alga gains CO_2, inorganic nitrogen, phosphates and some vitamins from the animal. **Reef-building corals** are only able to build reefs if they have their symbiotic photosynthetic partner, and many of the animal hosts are at least partly dependent on the algal partner for carbohydrates. **Radiolaria**, responsible for massive primary productivity in the oceans, are wholly dependent on their photosynthetic symbiont for carbohydrate.

Many aerobic protista contain bacteria as endosymbionts. *Amoeba proteus* has a symbiotic Gram-negative bacterium which is essential to the survival of the amoeba, and *Paramecium aurelia* has specific symbiotic algae that are responsible for the secretion of killer factors that are important in competition within environments and mating.

There are a large number of bacterial ectosymbionts. Spirochetes and many other species of bacteria are often attached to the euglenozoal pellicle, arranged in very specific patterns. For example *Myxotricha paradoxa* is an inhabitant of termite guts. Its motility within this environment depends on the co-ordinated movement of adherent spirochetes.

Ecto- and endosymbiotic bacteria are also found associated with anaerobic alveolata from sulfur-rich environments. *Kentrophorous lanceolata* has a dense mat of sulfur bacteria on its dorsal surface. These symbionts provide alternative electron acceptors to the protistan.

Protista can be symbiotic with insects and higher animals, and are typically found within fermentative guts as mentioned above. They are part of a complex ecosystem where cellulose is broken down by fungi and bacteria, and their metabolic products are fermented by the protista and bacteria to provide fatty acids for their hosts. Some of the protistan population also predate the bacterial population to provide control of numbers.

Diatomaceous earth

At death, diatom cells fall through the water column to the sea-bed. The inert nature of the frustule silica, silicon dioxide (SiO_2), means that it does not decompose but accumulates, eventually forming a layer of diatomaceous earth. This material has many commercial uses to humans, including filtration, insulation and fire-proofing and as an active ingredient in abrasive polishes and reflective paints. Recently, it has been used as an insecticide, where the abrasive qualities of diatomaceous earth are used to disrupt insect cuticle waxes, causing desiccation and death.

Bioluminescence

Many of the marine dinoflagellates are capable of **bioluminescence**, where chemical energy is used to generate light. The light is in the blue–green range, 474 nm, and can be emitted as high-intensity short flashes (0.1 s) either spontaneously, after stimulation, or continuously as a soft glow.

Bioluminescence is created by the reaction between a tetrapyrrole (**luciferin**), and an oxygenase enzyme (**luciferase**). The entire reaction is held within membrane-bound vesicles called **scintillons**, where the luciferin is sequestered by **luciferin-binding protein** (**LBP**) and held at pH 8. Release of light is stimulated by either a cyclical or mechanical stimulation of the scintillon, which leads to pH changes within it. Luciferin is released from LBP, and luciferase is able to activate it. Activated luciferin exists for a very brief time before it returns to its inactivated form, releasing a photon of light energy (*Fig. 1*).

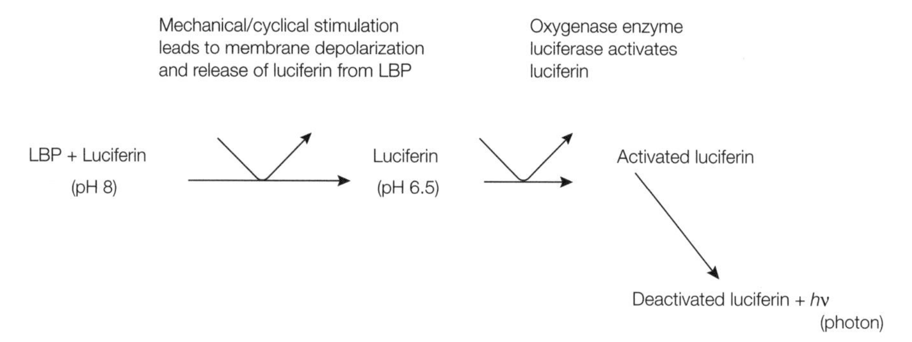

Fig. 1. Luciferin-luciferase reaction within scintillon.

The distribution of scintillons within the photosynthetic cell varies over 24 h. During the night they are distributed throughout the cytoplasm, but in daylight they are tightly packed around the nucleus. The function of bioluminescence for the organism is uncertain. It appears to be useful in defense against predators. It has become important in medicine and science because the coupled system of luciferin/luciferase can be used to mark cells. Once tagged with the bioluminescent marker, marked cells can be mechanically sorted from other cells, visualized by microscopy or targeted for therapy.

I5 Detrimental effects of Chlorophyta and Protista

Key Notes

Parasitic relationships

A wide range of organisms can be parasitized by members of the protista. They may enter via the mouth and colonize the gut where they can cause asymptomatic, mild disease. More specialized parasites cause greater effects on the gut, and cause disease symptoms. Some highly specialized parasites can enter via the gut but move into other tissues to complete their life cycles. Blood parasites are transferred via vectors from previously infected hosts. They have a highly specialized life cycle. A colorless Chlorophytan, Prototheca, can also parasitize damaged skin tissue.

Toxic blooms

Toxic blooms occur when nutrient-rich water supports population explosions of some photosynthetic species of protista. Some of these species can produce toxins that affect marine animals and humans that come into contact with them.

Eutrophication

The presence of excess nutrients in an aquatic ecosystem will support high levels of bacterial and eukaryotic microbial growth. The oxygen demand for this growth is great and will exceed supply, creating anoxic conditions. This is called eutrophication.

Related topics

Chlorophytan and protistan taxonomy and structure (I1)
Chlorophytan and protistan nutrition and metabolism (I2)
Lifecycles in the Chlorophyta and protista (I3)
Beneficial effects of the Chlorophyta and protista (I4)

Parasitic relationships

Parasitism of other organisms by protista is a very common nutritional strategy. Hosts can range from other members of the protista, including aquatic and higher plants, to many species of multicellular animals.

The lowest level of specialization is seen in those that inhabit the intestines of animals as commensals. They gain entry via food or water and exit via the feces. Many species live on gut bacteria. They are transient visitors, able to encyst before they are shed to ensure they survive adverse conditions outside the host. Many of these species have a fermentative metabolism to enable them to survive the low oxygen levels of the gut. A more specialized group attaches itself to the host gut wall to remain within the gut for longer periods. They again gain their nutrition from the passing gut contents and bacteria. These may cause some irritation to the gut epithelium.

The third group derive their nutrition from the gut epithelium and cause considerable damage in doing so. They have a number of pathogenicity factors including toxins and proteolytic enzymes called proteolysins. A more specialized form of parasitism is found in species that infect orally but then migrate into other tissues for part of their life cycle.

Parasites that have a life cycle that includes a blood-borne phase are highly specialized. They require vectors to reach their chosen niche, usually the blood-sucking insects. Part of the life cycle is completed in the vector. Trypanosomes and the malarial parasites (see Topic I3) are two of the most significant human protistan parasites. During the blood-borne phase of disease, parasite metabolism can be anaerobic and mitochondrial metabolism is switched off (see Topic I2).

Prototheca is a common soil-dwelling alga that has lost its chlorophyll and lives saprophytically. It is an opportunist pathogen that can enter wounds on the feet, where it can cause a subcutaneous infection. The initially small lesion can spread through the skin, producing a crusty, warty lesion. It may spread to lymph nodes and the disease can become debilitating. In rare cases it can grow rapidly in the bloodstream, causing rapid death in ailing or immunosuppressed animals and humans.

Toxic blooms

When environmental conditions for growth are optimal, photosynthetic protista can grow exponentially, leading to high local populations (see Topic I1). This can occur in several species of marine dinoflagellates that contain poisons which are toxic to fish, invertebrates or mammals, depending on the species of alga and the class of compound they produce. The toxins are accumulated in the digestive glands of shellfish and when consumed by man cause **paralytic shellfish poisoning**. Many of these compounds are neurotoxic. **Saxitoxin**, accumulated in shellfish and accidentally consumed, causes numbness of the mouth, lips and face which reverses after a few hours.

Toxins can also accumulate in higher animals like fish. For example **ciguatoxin** from *Gambierdiscus toxicus* accumulates in muscle tissue of grouper and snapper and when eaten causes gastric problems, central nervous system damage and respiratory failure.

Toxins can also be formed by other groups of algae. Some species of diatoms can produce **domoic acid**, which can accumulate in mussels and, when consumed by humans, causes **amnesic shellfish poisoning**, a short-term loss of memory but which can occasionally cause death. Some of the golden-brown protista that give rise to these types of poisoning are highly pigmented members of the Chrysophycae, and high cell densities can be seen as so-called **red tides** in the sea. Other toxins from species of the Prymnesiophycae can affect gill function of fish and molluscs.

Eutrophication

The presence of high populations of photosynthetic protista, their DOM products and the high numbers of bacteria that can be supported by large amounts of organic matter lead to a condition known as **eutrophication**. Water is in a nutrient-rich state, biomass is high and oxygen demand exceeds supply. Anoxic conditions rapidly develop, leading to the death of aerobic organisms.

The decomposition of dead material by bacteria leads to further demands for oxygen, and the entire environment becomes anaerobic, allowing for the growth of anaerobic bacteria, and the production of methane, hydrogen sulfide, hydrogen and many other products of anaerobic metabolism (see Topic I4). Eutrophication is commonly seen around untreated sewage outfalls and dairy farm waste run-off, and in fresh-water streams polluted by nitrate-rich agricultural field run-off.

J1 VIRUS STRUCTURE

Key Notes

Definitions

Viruses are obligate intracellular parasites and vary from 20–200 nm in size. They have varied shape and chemical composition, but contain only RNA or DNA. The intact particle is termed a 'virion' which consists of a capsid that may be enveloped further by a glycoprotein/lipid membrane. Viruses are resistant to antibiotics.

Methods of study

Virus morphology has been determined by electron microscopy (EM) (using negative staining), thin-section EM (using negative staining), immunoelectron microscopy (using negative staining), electron cryo-microscopy and X-ray crystallography.

Virus symmetry

Virus capsids have helical or icosahedral symmetry. In many cases the capsid is engulfed by a membrane structure (the virus envelope). Helical symmetry is seen as protein sub-units arranged around the virus nucleic acid in an ordered helical fashion. The icosahedron is a regular-shaped cuboid which consists of repetitions of many protein sub-units assembled so as to resemble a sphere.

Virus envelopes

Virus envelopes are acquired by the capsid as it buds through nuclear or plasma membranes of the infected cell. Envelopes may contain a few glycoproteins, for example, human immunodeficiency virus (HIV), or many glycoproteins, for example, herpes simplex virus (HSV). The virus envelope contains the receptors, which allow the particle to attach to and infect the host cell.

Related topic

Bacteriophage (E7)

Definitions

Viruses are obligate intracellular parasites which can only be viewed with the aid of an electron microscope. They vary in size from approximately **20–200 nm**. In order to persist in the environment they must be capable of being passed from host to host and of infecting and replicating in susceptible host cells. A virus particle has thus been defined as a structure which has evolved to transfer nucleic acid from one cell to another. The nucleic acid found in the particle is either **DNA** or **RNA**, is single- or double-stranded and linear or segmented. In some cases the nucleic acid may be circular. The simplest of virus particles consists of a protein coat (sometimes made up of only one type of protein which is repeated hundreds of times) which surrounds a strand of nucleic acid. More complicated viruses have their nucleic acid surrounded by a protein coat which is further engulfed in a membrane structure, an envelope consisting of virally coded glycoproteins derived from one of several regions within the infected cell during the maturation of the virus particle. The genetic material of these complex viruses encodes for many dozens of virus specific proteins.

The complete fully assembled virus is termed the **virion**. It may have a glycoprotein envelope which has **peplomers** (projections) which form a 'fringe' around the particle. The protein coat surrounding the nucleic acid is referred to as the **capsid** (see *Figs 1* and *2*). The capsid is composed of morphological units or capsomers. The type of capsomer depends on the overall shape of the capsid, but in the case of **icosahedral** capsids the capsomers are either **pentamers** or **hexamers**. Capsomers themselves consist of assembly units that comprise a set of structure units or **protomers**. **Structure units** are a collection of one or more non-identical protein subunits that together form the building block of a larger assembly complex (e.g. virus proteins VP1, VP2, VP3 and VP4 of picornaviruses). The combined nucleic acid–protein complex which comprises the genome is termed the **nucleocapsid**, which is often enclosed in a core within the virion.

Methods of study

While the **electron microscope** (EM) had been known for many years, the invention of the **negative-staining technique** in 1959 revolutionized studies on virus structure. In negative-contrast EM, virus particles are mixed with a heavy metal solution (e.g. sodium phosphotungstate) and dried onto a support film. The

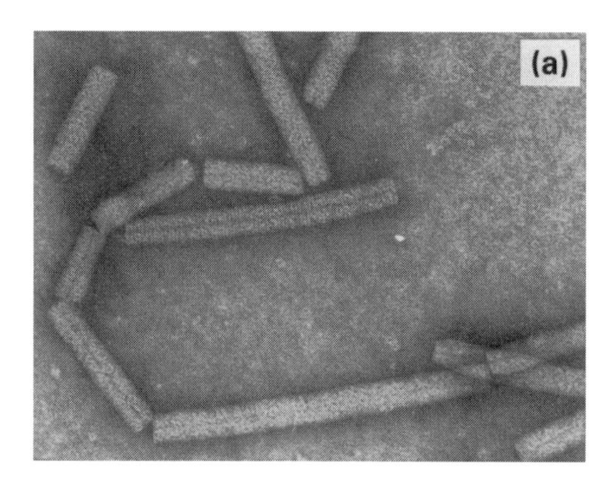

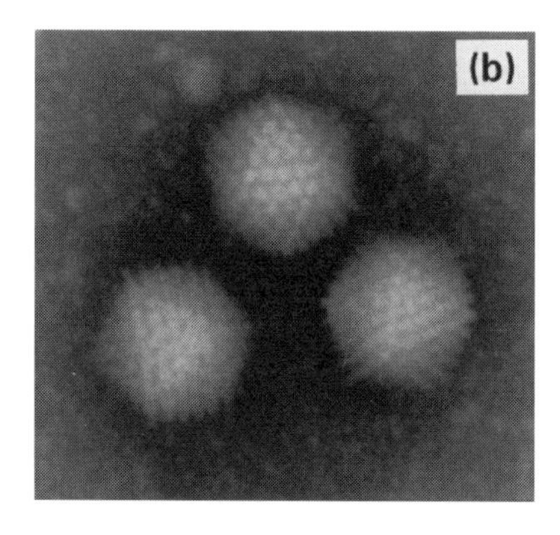

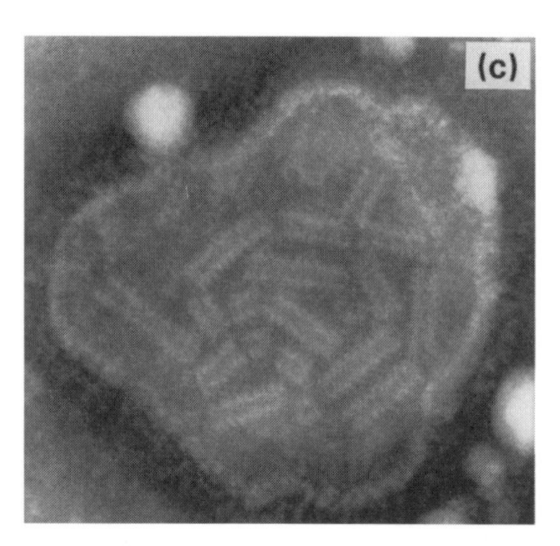

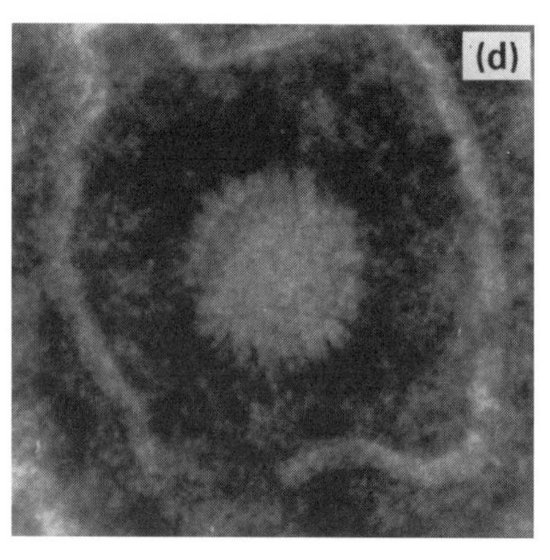

Fig. 1. Examples of viruses from main groups according to 'standard' morphology: (a) unenveloped/helical (tobacco mosaic virus); (b) unenveloped/icosahedral (adenovirus); (c) enveloped/helical (paramyxovirus); (d) enveloped/icosahedral (herpesvirus). From Harper, D., Molecular Virology, *2nd edn, © BIOS Scientific Publishers Limited, 1998. Photographs courtesy of Dr Ian Chrystie, Department of Virology, St Thomas' Hospital, London, and Professor C. R. Madeley, Department of Virology, Royal Victoria Infirmary, Newcastle-upon-Tyne.*

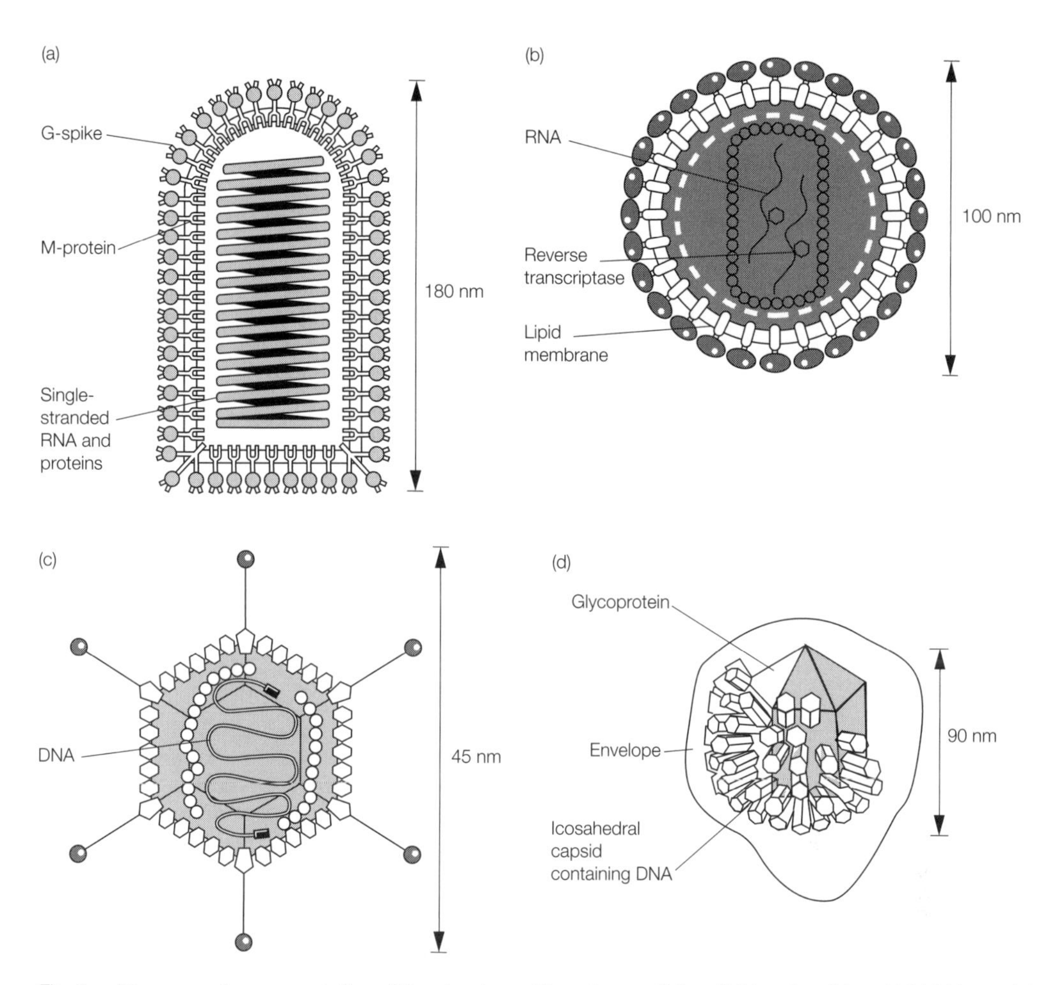

Fig. 2. Diagrammatic representation of the structure of four virus particles. Rabies virus (a) and HIV (b) have tight-fitting envelopes, herpes simplex (d) has a loose-fitting envelope, whereas adenovirus (c) is non-enveloped. The diagrams are based on electron microscopic observation and molecular configuration exercises. Redrawn from Phillips and Murray, Biology of Disease, *1995, with permission from Blackwell Science Ltd.*

stain provides an electron-opaque background against which the virus can be visualized. Observation under the electron microscope has thus allowed the definition of **virus morphology** at the 50–77 Å resolution level. In addition, negative staining of thin sections of infected cells has allowed definition of structures which appear during virus maturation and their interactions with cellular proteins. **Immunoelectron microscopy** is used to study those viruses which may be present in low concentrations or grow poorly in tissue culture (e.g. Norwalk virus). These clumps of virus are more readily observed. **Electron cryomicroscopy** reduces the risk of seeing artifacts as may be unavoidable by negative staining. High concentrations of virus are rapidly frozen in liquid ethane while on carbon grids. Electron micrographs can be digitized and three-dimensional reconstruction performed. Resolutions of 9 Å have been achieved using this method. **X-ray diffraction** of virus crystals is the ultimate in determining the ultrastructure of virion morphology. At present, only simple viruses can be crystallized. More complex viruses are analyzed by attempting to form crystals of sub-particular molecules. The X-ray diffraction pattern of the virion particle allows mathematical processing, which can predict the molecular configuration of the virus particle (*Fig. 3*).

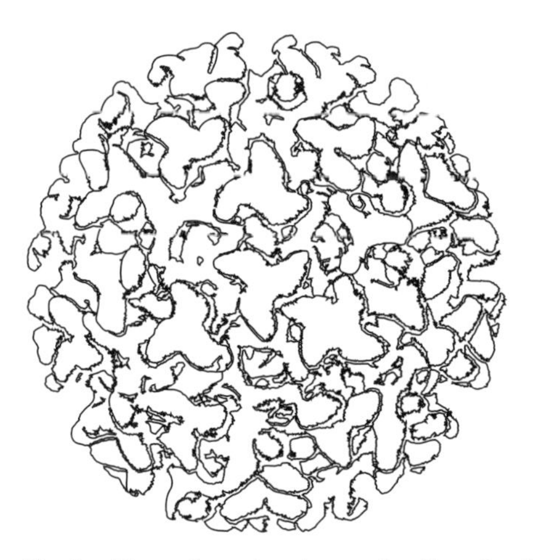

Fig. 3. Three-dimensional reconstruction of an icosahedrally symmetric virus particle.

Virus symmetry

The capsids of virions tend to have one of two symmetries – **helical** or **cuboid.** Helical symmetry can be loosely described as having a 'spiral staircase' structure. The structure has an obvious axis down the center of the helix. The subunits are placed between the turns of the nucleic acid. A diagram of such a structure (tobacco mosaic virus) is shown in *Figure 4*, and an electron micrograph in *Figure 1a*. Animal viruses with a similar capsid structure include measles, rabies and influenza. Most animal viruses have spherical or cuboid symmetry. Obtaining a true sphere is not possible for such structures and hence subunits come together to produce a cuboid structure which is very close to being spherical. The 'closed shell' capsid is usually based on the structure referred to as an icosahedron. A regular icosahedron, formed from assembly of identical subunits, consists of **20 equilateral triangular faces, 30 edges and 12 vertices and exhibits 2-, 3- and 5-fold symmetry** (see *Fig. 4*). The minimum number of capsomers required to construct an icosahedron is 12, each composed of five identical subunits. Many viruses have more than 12. A model of such a structure is shown in *Figure 4,* although in adenoviruses projecting fibers are also present, which distinguishes this capsid from that of other viruses. The maturation and assembly of these structures is very complex; indeed, much of how it happens is unknown. Icosahedral structures are, however, usually formed via a complex but structured array of molecular-assembly procedures which eventually give rise to the mature capsid. These may be self-assembly processes or may involve virus non-structural proteins acting as **scaffolding** proteins which do not finish up in the mature capsid.

Virus envelopes

Many viruses in addition to having a capsid also contain a virus-encoded **envelope**. Most enveloped viruses bud from a cellular membrane (plasma membrane, e.g. influenza virus, or nuclear membrane, e.g. herpes simplex virus). Within this virus lipid/protein bilayer are a number of inserted virus-

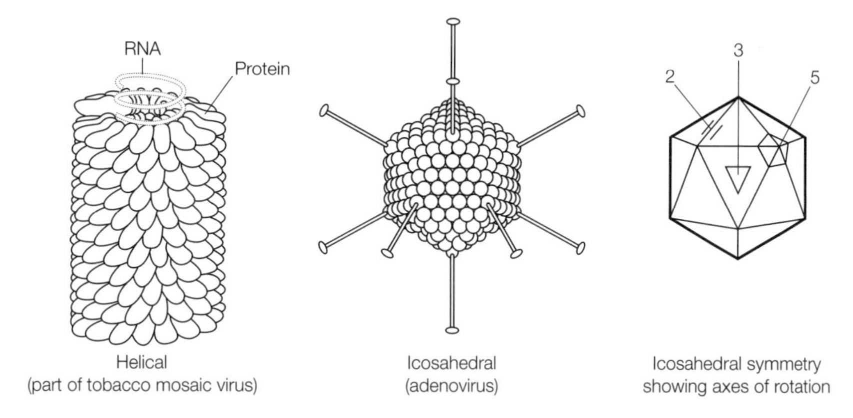

Fig. 4. Diagrammatical representation of helical and icosahedral symmetry. From Harper, D. Molecular Virology, *2nd edn, © BIOS Scientific Publishers Limited, 1998.*

encoded **glycoproteins**. The envelopes can be amorphous (e.g. the herpes virion, *Figs 1* and *2*) or tightly bound to the capsid (e.g. HIV). Thus, the lipid of the envelope is derived from the cell, the glycoprotein being encoded by the virus. Quite how the process of budding occurs is largely unknown.

J2 Virus taxonomy

Key Notes

Virus taxonomy The classification of viruses involves the use of a wide range of characteristics (morphology, genome, physicochemical and physical properties, proteins, antigenic and biological properties) to place viruses in orders, families, genera and species.

Virus orders These are groupings of families of viruses that share common characteristics. There are three orders, Mononegavirates, Nidovirates and Caudovirates.

Virus families These are groupings of genera of viruses that share common characteristics and are distinct from the members of other families. They are designated by the suffix -viridae.

Virus genera These are groupings of species of viruses which share common characteristics. These are designated by the suffix -virus.

Virus species These represent a polythetic class of viruses that constitutes a replicating lineage and occupies a particular ecological niche.

Related topics Prokaryote taxonomy (D1) Reproduction in fungi (H3)
Taxonomy (G1)

Virus taxonomy

Initial attempts to classify viruses were based on their pathogenic properties, and the only common feature to many of the viruses placed together in such groupings was that of **organ tropism** (e.g. viruses causing hepatitis or respiratory disease). There were more important aspects (e.g. **virus structure** and **composition**) which led virologists to believe that these initial attempts of classification were far from adequate. In the late 1950s and early 1960s, hundreds of new viruses were beginning to be isolated and the need for a different classification method became essential. In 1966 the **International Committee on Nomenclature of Viruses (ICNV)** was established at the International Congress of Microbiology held in Moscow. As the present classification scheme has evolved, an acceptance of the characteristics to be considered and their respective weighting has become universal.

The seventh report of the **ICTV (International Committee on Taxonomy of Viruses)** recorded a universal taxonomy scheme consisting of three orders, 56 families and 233 genera (with more than 4000 members). The system still contains hundreds of unassigned viruses, largely because of lack of data. Also, new viruses are still being discovered.

As techniques in molecular biology, serology and EM have advanced, taxonomy has been based on more and more characteristics. Thus, factors now taken into consideration (see *Table 1*) include virion morphology, physicochemical and physical properties, the genome, virus proteins, lipids, carbohydrates,

Table 1. Some properties of viruses used in taxonomy

Virion properties
- Morphology
  - Virion size
  - Virion shape
  - Presence or absence and nature of peplomers
  - Presence or absence of an envelope
  - Capsid symmetry and structure
- Physicochemical and physical properties
  - Virion molecular mass (M_r)
  - Virion buoyant density (in CsCl, sucrose, etc.)
  - Virion sedimentation coefficient
  - pH stability
  - Thermal stability
  - Cation stability (Mg^{2+}, Mn^{2+})
  - Solvent stability
  - Detergent stability
  - Irradiation stability
- Genome
  - Type of nucleic acid (DNA or RNA)
  - Size of genome in kb/kbp
  - Strandedness: ss or ds
  - Linear or circular
  - Sense (positive-sense, negative-sense, ambisense)
  - Number and size of segments
  - Nucleotide sequence
  - Presence of repetitive sequence elements
  - Presence of isomerization
  - G + C content ratio
  - Presence or absence and type of 5′ terminal cap
  - Presence or absence of 5′ terminal covalently linked protein
  - Presence or absence of 3′ terminal poly (A) tract
- Proteins
  - Number, size and functional activities of structural proteins
  - Number, size and functional activities of nonstructural proteins
  - Details of special functional activities of proteins, especially transcriptase, reverse transcriptase, hemagglutinin, neuraminidase and fusion activities
  - Amino acid sequence or partial sequence
  - Glycosylation, phosphorylation, myristylation property of proteins
  - Epitope mapping
- Lipids
  - Content, character, etc.
- Carbohydrates
  - Content, character, etc.
- Genome organization and replication
  - Genome organization
  - Strategy of replication
  - Number and position of open reading frames
  - Transcriptional characteristics
  - Translational characteristics
  - Site of accumulation of virion proteins
  - Site of virion assembly
  - Site and nature of virion maturation and release

Antigenic properties
- Serologic relationships, especially as obtained in reference centers

Biologic properties
- Natural host range
- Mode of transmission in nature
- Vector relationships
- Geographic distribution
- Pathogenicity, association with disease
- Tissue tropisms, pathology, histopathology

Table 2. Taxonomic chart of selected virus families

Family	Characteristics	Typical members	Diseases caused
Poxviridae	dsDNA, 'brick'-shaped particles; largest virus	Vaccinia	Laboratory virus
		Variola	Smallpox (now eradicated)
Herpesviridae	dsDNA, icosahedron capsid enclosed in an envelope, latency in host common	Herpes simplex	'Cold' sores, genital infections
		Varicella-zoster	Chicken pox, shingles
		Cytomegalovirus	Febrile illness or disseminated disease in immunosuppression
		Epstein–Barr virus	Glandular fever. Virus is also associated with certain malignancies, e.g. Burkitt's lymphoma
Adenoviridae	dsDNA, icosahedron with fiber structures, non-enveloped	Adenoviruses (many types)	Respiratory and eye infections, tumors in experimental animals
Papillomaviridae	ds circular DNA, 72 capsomeres in capsid, non-enveloped	Human papilloma viruses	Warts, association with some cancers (e.g. cervical cancer)
Hepadnaviridae	One complete DNA minus strand with 5′ terminal protein, DNA circularized by an incomplete plus strand, 42 nm enveloped particle	Hepatitis-B virus	Serum hepatitis, association with hepatocellular carcinoma
Paramyxoviridae	ssRNA, enveloped particles with 'spikes'	Parainfluenza virus	Respiratory tract infection ('croup')
		Measles virus	Measles
		Respiratory syncytial virus	Bronchiolitis

Orthomyxoviridae	Eight segments of ssRNA, enveloped particles with 'spikes', helical nucleocapsid	Influenza virus	Influenza
Reoviridae	10–12 segments of dsRNA, icosahedron, non-enveloped	Rotavirus	Infantile diarrhea
Picornaviridae	ssRNA, 22–30 nm particle of cubic symmetry, non-enveloped	Poliovirus Coxsackie virus Rhinovirus Heptatitis-A virus	Poliomyelitis Myocarditis Common cold Infectious hepatitis
Togaviridae	ssRNA, enveloped particles, icosahedron nucleocapsid	Rubella virus	German measles
Flaviviridae	ssRNA, enveloped particles, icosahedron	Yellow fever virus, Hepatitis C virus	Yellow fever, Hepatitis C
Rhabdoviridae	ssRNA, bullet-shaped, enveloped particle	Rabies virus	Rabies
Filoviridae	ssRNA, pleomorphic enveloped (mainly bacilliform or filamentors)	Marburg	Hemorrhagic fever
Bunyaviridae	ssRNA, enveloped spherical or pleomorphic, helical capsids	Hantavirus	Hemorrhagic fever
Retroviridae	ssRNA, enveloped particles with icosahedral nucleocapsid, employ reverse transcriptase enzyme to make DNA copy of genome on infection	Human T lymphotropic virus-1 Human immunodeficiency virus (HIV)	Adult T cell leukemia and lymphoma Acquired immune deficiency syndrome (AIDS)

and antigenic and biological properties. It has, however, been estimated that to define a virus appropriately some 500–700 characters must be determined. Over the next few years the ICTV plan is to create a readily accessible database which will allow cataloging of viruses down to strains. *Table 2* is a brief summary of some well known animal viruses and the diseases they cause.

Virus orders

These represent groupings of families of viruses that share common characteristics which make them distinct from other orders and families. Orders are designated by the suffix **-virales**. Three orders have been approved by the ICTV: **Mononegavirales**, which consists of the families Bornaviridae, Paramyxoviridae, Rhabdoviridae and Filoviridae (these are single-stranded negative-sense non-segmented enveloped RNA viruses): **Nidovirales**, which consist of the Coronaviridae and Arteriviridae (these are ss positive sense enveloped icosahedral viruses): **Caudovirales** (which includes two phage families).

Virus families

These represent groupings of genera of viruses that share common characteristics and are distinct from the members of other families. Families are designated by the suffix **-viridae**. Families as groupings have proved to be excellent models for classification. Most of the families have **distinct virion morphology**, **genome structure** and **strategies of replication**. Examples include Picornaviridae, Togaviridae, Poxviridae, Herpesviridae and Paramyxoviridae. In some families (e.g. Herpesviridae) the complex relationships between individual members has led to the formation of sub-families, which are designated with the suffix **-virinae**. Thus, the Herpesviridae are further classified into the Alphaherpesvirinae (e.g. herpes simplex virus) the Betaherpesvirinae (e.g. cytomegalovirus) and the Gammaherpesvirinae (e.g. Epstein–Barr virus).

Virus genera

Virus genera are groupings of species of viruses which share common characteristics and are distinct from the members of other genera. They are designated by the suffix -**virus** (e.g. genus Simplexvirus and genus Varicellovirus of the Alphaherpesvirinae). The criteria for designating genera vary from family to family, but include genetic, structural and other differences.

Virus species

A virus species is defined as 'a polythetic class of viruses that constitutes a replicating lineage and occupies a particular ecological niche'. Members of a polythetic class are defined by more than one property. At present the ICTV is examining carefully the properties which can be included in determining species. The division between species and strains is a difficult one.

J3 Virus proteins

Key Notes

Overview

Viral proteins, encoded by the viral genome, are either structural (capsid, envelope) or non-structural (e.g. enzymes, oncoproteins, inhibitors of cell macromolecular synthesis and interaction with MHC presentation). They may be essential or non-essential in tissue culture replication.

Methodology

Structural proteins are studied, following virus purification, using a range of techniques including sodium dodecyl sulfate (SDS)–polyacrylamide gel electrophoresis, Western blotting and immunoprecipitation. Non-structural proteins are examined by, for example, pulse–chase experiments, or the use of protease and glycosylation inhibitors.

Protein synthesis and complexity

Viral proteins are synthesized by the translation of viral mRNAs on cellular ribosomes. Proteins are often processed following synthesis (e.g. proteolytic cleavage, glycosylation, myristoylation, acylation and palmitoylation). In many viruses, protein synthesis is controlled at the levels of transcription and translation, which makes virus replication quite an efficient process. This control is usually directed by the virus genome.

Structural proteins

Structural proteins are either nucleocapsid, matrix or envelope proteins. They have a role in protecting the viral genome and in delivering the genome from one host to another via receptors on host cells. They also have a major role in the assembly of the virion. In most viruses, structural proteins are produced in abundance late in the replicative cycle.

Non-structural proteins

These may be carried in the virion (but are not part of the virion architecture) where they have enzymic activity which is necessary for initiating infection. Others are not destined for the virion but have roles in the infected cell. These roles include switching off host-cell nucleic acid and protein synthesis, polymerase, protease and kinase activities, DNA-binding activity and gene regulation.

Related topics

Virus replication (J7)
Viruses and the immune system (J9)
Virus vaccines (J10)
Antiviral chemotherapy (J11)

Overview

The **coding capacity** of viral genomes varies from <5 to >100 genes. This in turn is reflected in the different complexities of virus particles and their respective replicative cycles. Viral proteins are either **structural** (part of the virion architecture), **non-structural** but in the virion (usually enzymes) or **non-structural** and present only in infected cells, never the virion (these have a range of functions). The limited number of virus proteins synthesized means that many of them are multifunctional.

Virus proteins are often referred to as being **essential** or **non-essential**. The former are an absolute requirement for the virus in order for it to complete a replicative cycle and to produce infectious virions. The latter can be deleted from the virus genome without seriously affecting virus growth in tissue culture. However, this may not reflect the *in vivo* situation.

Methodology

The study of virus structural proteins as assembled virions has required the **purification of viruses** from infected cells or infected-cell supernatants, ensuring that they are free from surrounding contaminating proteins. This is usually achieved by a series of **differential centrifugation** steps, followed by **sucrose gradient-density centrifugation**. Centrifugation through a sucrose gradient usually results in a sharp band of virus at a specific location on the gradient. This is harvested for further studies. It is normal to study **radiolabeled virions** by a variety of techniques including **SDS–polyacylamide gel electrophoresis** (separation based on size), **Western blotting** (reaction with antibodies) and **immunoprecipitation** (precipitation of proteins with antibodies).

The location of virus proteins within cells can be determined by **differential staining, immunofluorescence and confocal microscopy** techniques, with monoclonal antibodies to specific virus protein epitopes being the key reagents. Use of, for example, **pulse–chase** experiments, **protease** or **glycosylation inhibitors** have allowed studies on protein processing.

Gene sequencing and **amino acid prediction** of proteins has allowed a number of predictions to be made concerning virus protein structure and function, including cleavage sites, membrane spanning, glycosylation. Sequencing allows identification of conserved areas in genes.

Protein synthesis and complexity

As was outlined earlier, all proteins are synthesized from a mRNA template which is translated on the host-cell ribosomes. The mRNA may be (a) transcribed from the viral genome DNA (e.g. HSV), (b) complementary to the viral genome RNA (**negative-strand RNA viruses,** e.g. influenza) or (c) **viral** genomic RNA (**positive-strand RNA viruses,** e.g. polio).

During the growth cycle, in more complex viruses, virus-specific proteins are synthesized at various times post-infection, for example, structural proteins may be produced late on in infection. Simple viruses (e.g. polio) do not have such levels of **temporal control**.

The amino acid sequence of the protein contains a number of **motifs** which determine its **post-translational modification**, its location in the cell (by **signaling**) and its **secondary** and **tertiary** structure.

Proteins may be the result of **proteolytic cleavage** of a larger **precursor molecule** (e.g. polio polyprotein, hepatitis C polyprotein, HIV gag-polymerase complex). They may be **phosphorylated**, the degree of phosphorylation often determining the functional activity of the protein (e.g. N, NS proteins of rhabdoviruses). The attachment of sugar residues to proteins (**glycosylation**) is either N- or O-linked and for many virus proteins (usually envelope glycoproteins) the sugar residues may account for 75% of the protein weight. **Myristylation, acylation** and **palmitoylation** are other post-translational modifications.

Structural proteins

Whatever the complexity, size or shape of a virus particle it is the role of the structural protein to provide **protection** for the viral genome and to allow delivery of virus particles from one host to another. Structural proteins are in either the **nucleocapsid**, **matrix** or **envelope** of the particle (*Fig. 1*).

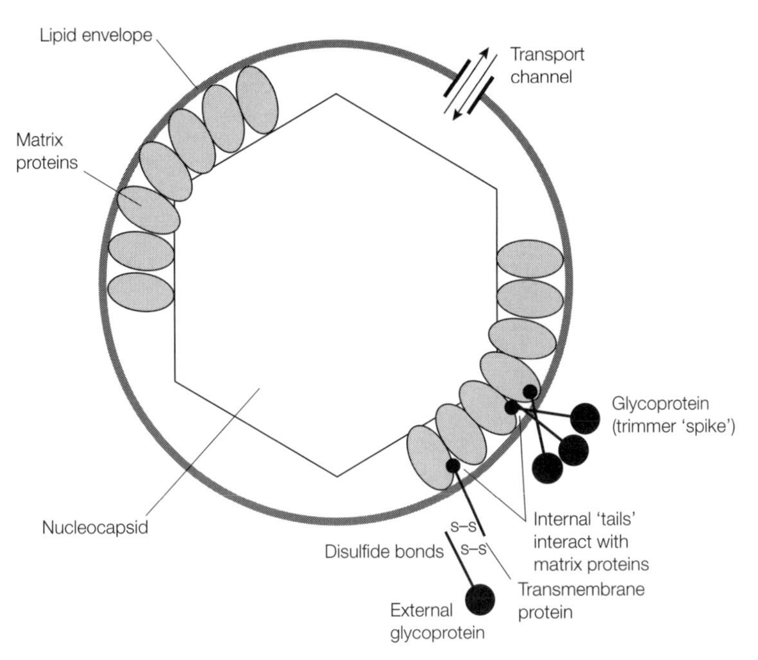

Fig. 1. Several classes of proteins are associated with virus envelopes. Matrix proteins link the envelope to the core of the particle. Virus-encoded glycoproteins inserted into the envelope serve several functions. External glycoproteins are responsible for receptor recognition and binding, while transmembrane proteins act as transport channels across the envelope. Host-cell derived proteins are also sometimes found to be associated with the envelope, usually in small amounts. Redrawn from Cann, A., Principles of Molecular Virology, *1993, with permission from Academic Press.*

Nucleocapsid proteins either self-assemble (e.g. TMV, polio) or assemble with the help of scaffolding proteins (e.g. HSV) to form the helical and icosahedral structures described earlier (Topic J1). The polio capsid is relatively simple with four proteins, VP1, VP2, VP3 and VP4. These assemble via a precursor structure (the procapsid) which contains VP0 (a precursor protein) and VP1 and VP3. VP0 cleaves to give VP2 and VP4 as the capsid acquires its nucleic acid. The reovirus capsid is much more complex and is composed of a double-shelled icosahedron structure (*Fig. 2*). Capsid proteins, in addition to protecting the genome and 'assembling' to form structural virions, must also interact with the nucleic acid genome during the assembly process. **Packaging** of the large piece of nucleic acid into a confined area requires complex folding of the genome and intimate association through chemical bonding with selected capsid proteins. Examples are rhabdovirus N protein and influenza virus NP protein whose positively charged amino acids interact with the negatively charged nucleic acid to encourage packaging. Nucleic acids also contain **packaging sequences** which result not only from nucleotide sequence but also from the **secondary** and **tertiary structures** of the genome. Packaging becomes even more complex when one considers a multigenome virus (e.g. influenza) where eight different segments of RNA have to be packaged into the virion nucleocapsid.

The **envelope** which surrounds many virus particles is acquired by nuclear membrane, plasma membrane or endoplasmic reticulum budding. Within the envelope are glycolipid proteins.

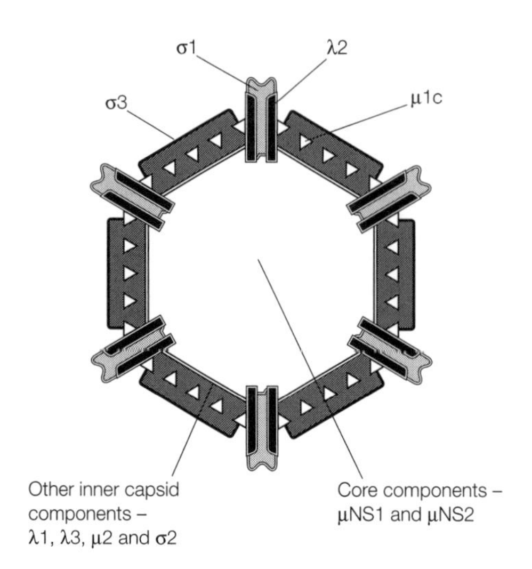

Fig. 2. Reovirus particles consist of an icosahedral, double-shell arrangement of proteins surrounding the core. Redrawn from Cann, A., Principles of Molecular Virology, *1993, with permission from Academic Press.*

Matrix proteins are internal proteins whose function, when present, is to link the internal nucleocapsid proteins to the envelope. They are usually not glycosylated and may contain **transmembrane anchor domains** or are associated with the membrane by hydrophopic patches on their surface or by protein–protein interactions with envelope glycoproteins. In HSV the space between the envelope and the capsid is referred to as the **tegument**.

External glycoproteins are anchored in the envelope by **transmembrane domains**. Most of the structure of the protein is on the outside of the membrane, with a relatively short internal tail. Many of these glycoproteins are **monomers** which group together to form the spikes visible in the electron microscope. They are the major **antigens** of enveloped viruses and include the G spikes of rabies virus, the gp120 spike of HIV and the HA spike of influenza.

The detailed structure of many of these molecules has been determined by **X-ray crystallography** and **cryoelectron microscopy**. One of the first molecules to be analyzed in this detail was the **hemagglutinin** of influenza. This excellent work resulted in the detailed molecule shown in *Fig. 3*.

Certain structural proteins in either naked capsids (e.g. polio) or enveloped viruses (e.g. HSV) initiate viral infection by attaching to **receptor sites** on host cells. Immune responses to these virus proteins often protect against viral infection and our understanding of this interaction forms the basis of modern vaccine design (see Topic J10). *Table 1* lists examples of specific receptor sites available on host cells for attachment by viruses.

Non-structural proteins

Non-structural proteins may be carried within the virus particle (but not as part of the 'architecture') or appear only in infected cells. The number and function of these proteins varies greatly from one virus family to another, depending on the complexity of the viral genome and the replicative cycle. Many have enzymic activity (e.g. the **reverse transcriptase, protease** and **integrase** of retroviruses, the **thymidine kinase** and **DNA polymerase** of HSV). These enzymes

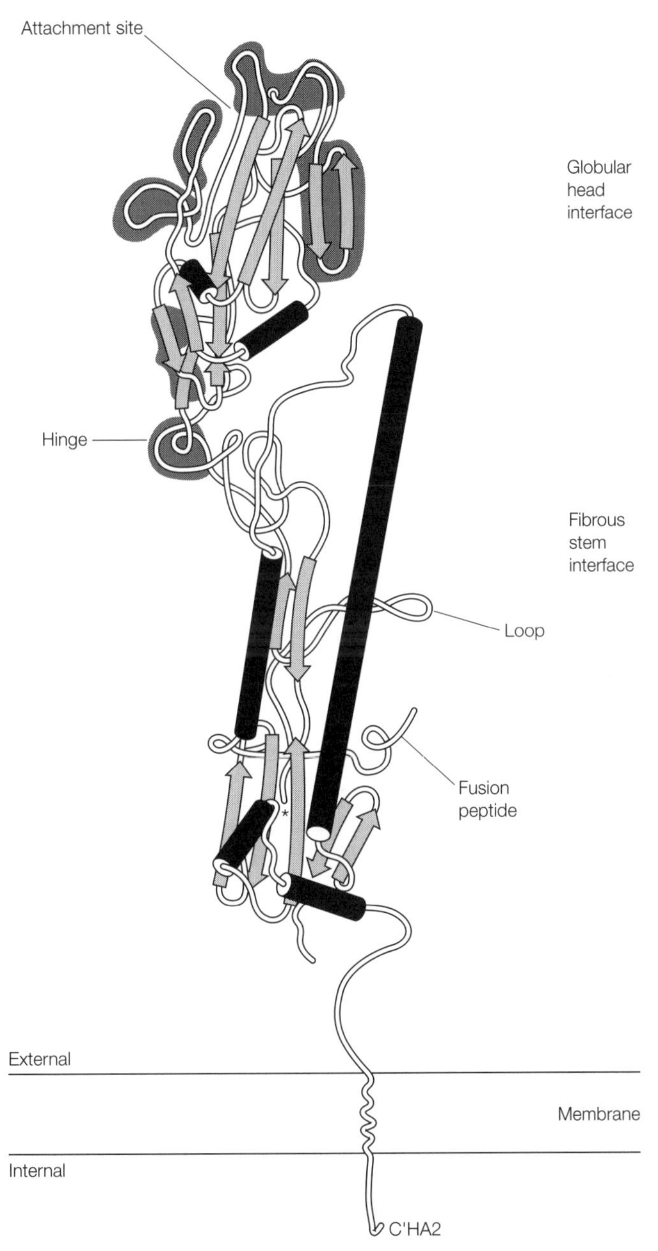

Fig. 3. The structure of the hemagglutinin of influenza. Redrawn from Wiley et al., Nature, *vol. 289, pp. 373–378, 1981, with permission from Macmillan Magazines Limited.*

are key targets for antiviral drugs (see Topic J11). Others have a role in virion assembly, acting as scaffolding proteins upon which capsids are assembled. Many (e.g. HIV **Tat** HSV **tegument proteins**) have regulatory roles in transcription and others form part of nucleic acid synthesis complexes (e.g. **DNA helicase**, DNA **binding proteins**). The shut-off of host cell macromolecular synthesis is usually the function of viral non-structural proteins. Viruses

Table 1. *Examples of receptors for viruses that infect humans*

Family	Virus	Cellular Receptor
Adenoviridae	Adenovirus type 2	Integrins $\alpha_v\beta_3$ and $\alpha_v\beta_5$
Coronaviridae	Human coronavirus 229E	Aminopeptidase N
Coronaviridae	Human coronavirus OC43	*N*-Acetyl-9-*O*-acetylneuraminic acid
Hepadnaviridae	Hepatitis B virus	IgA receptor
Herpesviridae	Herpes simplex virus	Heparan sulfate proteoglycan plus mannose-6-phosphate receptor
Herpesviridae	Varicella zoster virus	Heparan sulfate proteoglycan?
Herpesviridae	Cytomegalovirus	Heparan sulfate proteoglycan plus second receptor
Herpesviridae	Epstein–Barr virus	CD21 (CR2) complement receptor
Herpesviridae	Human herpesvirus 7	CD4 (T4) T-cell marker glycoprotein
Orthomyxoviridae	Influenza A virus	Neu-5-Ac (neuraminic acid) on glycosyl group
Orthomyxoviridae	Influenza B virus	Neu-5-Ac (neuraminic acid) on glycosyl group
Orthomyxoviridae	Influenza C virus	*N*-Acetyl-9-*O*-acetylneuraminic acid
Paramyxoviridae	Measles virus	CD46 (MCP) complement regulator
Picornaviridae	Echovirus 1	Integrin VLA-2 ($\alpha_2\beta_1$)
Picornaviridae	Poliovirus	IgG superfamily protein
Picornaviridae	Rhinoviruses	ICAM-1 adhesion molecule
Poxviridae	Vaccinia	Epidermal growth factor receptor
Reoviridae	Reovirus serotype 3	β-Adrenergic receptor
Retroviridae	Human immunodeficiency virus	CD4 (T4) T-cell marker glycoprotein and Chemokine co-receptor
Rhabdoviridae	Rabies	Acetylcholine receptor

which induce tumors in their host by **transforming** normal cells do so by producing **viral oncogenes** (non-structural virus proteins) or by activating cellular oncogenes. Other proteins associated with cell transformation include, for example, the large **T antigen** of SV40 and the **EBNA** protein of Epstein–Barr virus. More recently, non-structural proteins have been associated with **anti-apoptosis**, **anti-cytokine** activity and interference with MHC antigen presentation (i.e. immune evasion – see Topic J9): *Table 2* lists the functions of HIV non-structural proteins as an example of selected activities.

Table 2. *Proteins of the human immunodeficiency virus (HIV) type 1*

gag	Polyprotein, cleaved by the viral aspartyl proteinase, produces the core proteins
pol	Polymerase/reverse transcriptase, aspartyl proteinase and integrase
env	Polyprotein (gp160), cleaved by cellular enzymes to produce gp41 and gp120
Auxiliary genes	
vif	Required for production of infectious virus
vpr	Function unknown
tat	Up-regulates viral mRNA synthesis by binding to the TAR (*trans*-activator response element) in the transcripts from the long terminal repeat. Made from a spliced mRNA
rev	Required by mRNA transport. Made from a spliced mRNA
vpu	Virion release, receptor degradation
nef	Down-regulates CD4 & MHC, essential for virus pathogenicity

J4 VIRUS NUCLEIC ACIDS

Key Notes

Types of viral genome

Viral genomes are diverse in size, structure and nucleotide make up. They can be linear, circular, dsDNA, ssDNA, dsRNA, ssRNA, segmented or non-segmented.

Techniques of study

Many viral genomes have been sequenced and the repertoire of their coding potential determined. Genomes are studied by a range of techniques including restriction enzyme analysis, buoyant density, thermal denaturation, nuclease sensitivity and EM.

Large DNA viruses

Herpesviruses and adenoviruses have large DNA genomes which vary in their structures. DNA viruses often have terminal inverted repeats as part of their genome structure. They may have coding capacity for up to 80 virus proteins.

Small DNA viruses

Parvovirus (ssDNA) and polyomaviruses (dsDNA) have small DNA genomes. The coding capacity may be increased by use of overlapping genes and both strands of DNA.

RNA viruses (non-segmented)

These are either positive sense (e.g. picornaviruses, coronaviruses, flaviviruses and togaviruses) or negative sense (e.g. paramyxoviruses and rhabdoviruses). Positive-sense genomes act as mRNAs, negative sense genomes need a cRNA to act as mRNA.

RNA viruses (segmented)

Segmented viral genomes (e.g. orthomyxoviruses) are those which are divided into two or more physically distinct molecules of nucleic acid packaged into a single virion.

Related topics

Structure and organization of DNA (C1)
DNA replication (C2)

Types of viral genome

The viral genome carries the nucleic acid sequences which are responsible for the **genetic code** of the virus. In infected cells the genome is **transcribed** and **translated** into those amino acid sequences which make up the viral proteins, be they structural or non-structural products. The genome, or in some cases a transcript of it, forms the basis of the template upon which new genomes are synthesized before assembly of progeny virions. Viral genomes are either **linear** or **circular** and composed of **RNA** or **DNA**. They vary in size from 3500 nucleotides (e.g. small phage) to 280 kb pairs, that is, 560 000 nucleotides (e.g. some herpesviruses). These sequences must be capable of being decoded by the host cell and thus the control signals must be recognized by host factors usually in association with viral proteins. Because of their small size, viral genomes have evolved to make **maximal** use of their nucleotide coding potential. Thus, **overlapping** genes and **spliced mRNAs** are common.

Characterization of the viral genome is based on a number of parameters: (1) **composition of the nucleic acid (i.e. DNA or RNA);** (2) **size and number of strands;** (3) **terminal structures;** (4) **nucleotide sequence;** (5) **coding capacity;** (6) **regulatory signal elements, transcriptional enhancers, promoters and terminators.**

Some points of note are:

- dsDNA genomes (e.g. Poxviridae, Herpesviridae and Adenoviridae) are usually the largest genomes, ssDNA genomes (e.g. Parvoviridae) being smaller.
- dsRNA genomes (e.g. Reoviridae) are all segmented.
- ssRNA genomes are classified as being positive (+) sense or negative (–) sense. Viral genomes that are + sense can act as mRNAs and are usually infectious without the nucleocapsid proteins (e.g. Picornaviridae, Caliciviridae, Coronaviridae, Flaviviridae and Togaviridae). Negative-sense RNA genomes are usually not infectious unless accompanied by nucleocapsid proteins which have enzymic activity (transcriptases). These enzymes transcribe the negative-sense RNA into a complementary strand (cRNA) which acts as mRNA (e.g. Orthomyxoviridae, Paramyxoviridae, Rhabdoviridae and Filoviridae).
- Most ssRNA genomes are single molecules with the exception of, for example, Orthomyxoviridae (influenza) which have segmented genomes. Retroviridae have a ssRNA viral genome which replicates via a dsDNA intermediate, this being facilitated by a reverse transcriptase enzyme carried within the virion.
- The DNA genome of Hepadnaviridae is synthesized within hepatocytes via an RNA intermediate. These different modes of viral replication are summarized in *Fig. 1*.

Techniques of study

The advances in molecular-biological techniques over the last decade have made analysis of viral nucleic acids simpler, more efficient and quick. Viral genomes from most of the families have been totally **sequenced** and the open reading frames, and in many cases the gene products, characterized. This allows comparison with known sequences of genes stored on computer databases and has revealed fascinating similarities between viral and various eukaryotic genes which have been conserved through the process of evolution. Genes can be cloned into a variety of vectors and analyzed by a number of techniques (e.g. site-directed mutagenesis) to study the role of individual amino acids in determining the structural and functional integrity of the coded protein (see *Instant Notes in Biochemistry* in this series for an excellent review of these techniques).

DNA viruses are often shown diagramatically as linear molecules with **restriction enzyme** sites scattered throughout the genome. There are dozens of restriction enzymes, which are used to digest DNA into small segments at very specific nucleotide sequences. These segments are separated by size on agar gels following electrophoresis. Each DNA genome has a restriction enzyme **map** which characterizes it. This was not possible for RNA genomes until a reverse transcriptase enzyme was used to make a copy (c) DNA molecule from the RNA template, the cDNA being 'cut' with restriction enzymes.

Viral nucleic acids can be characterized by their melting temperature (T_m), their buoyant density in cesium chloride gradients, their 'S' value in sucrose gradients, their infectivity (or lack of), their nuclease sensitivity and their electron microscopic appearance.

(a) dsDNA genome: pox, herpes, adeno, papova, irido

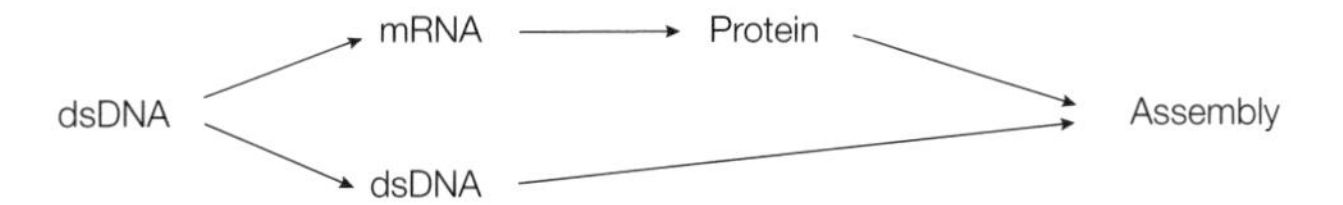

(b) ssDNA genome: parvo

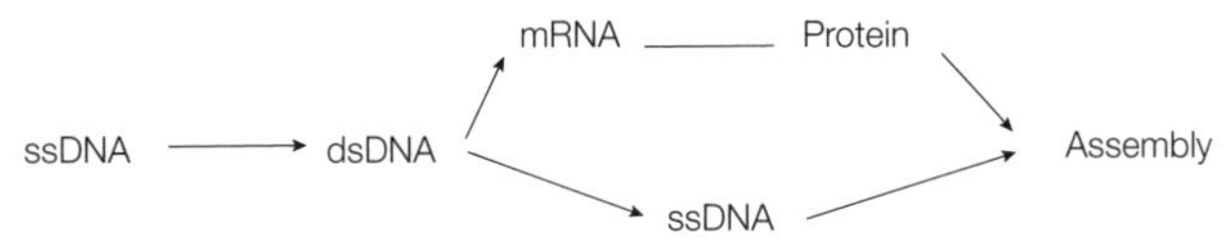

(c) ss/dsDNA genome using ssRNA intermediate: hepadna

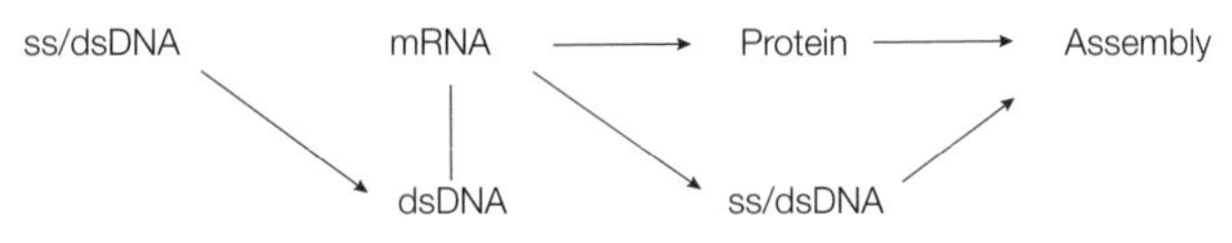

(d) dsRNA genome: reo, birna

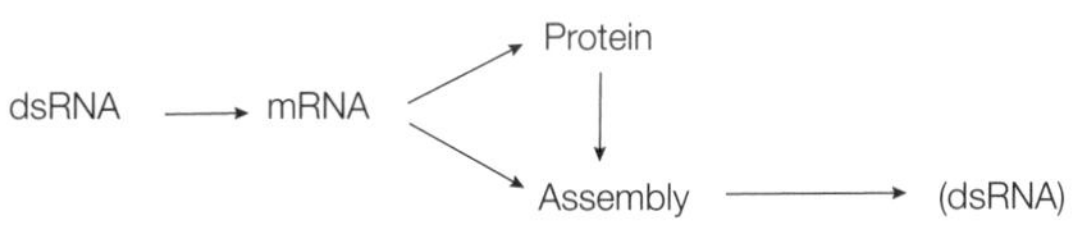

(e) +sense ssRNA genome: picorna, calici, corona, toro, toga

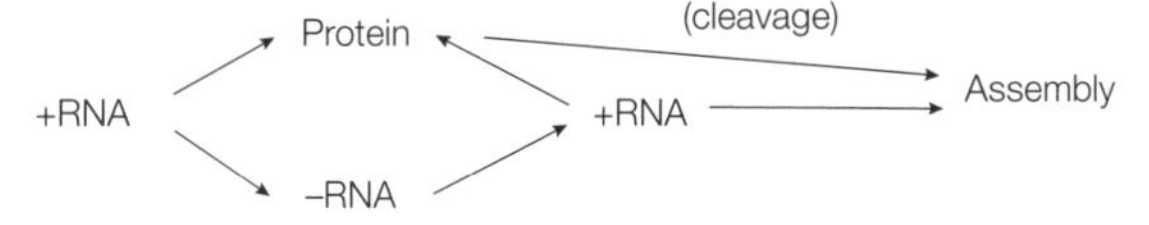

(f) –sense ssRNA genome: orthomyxo, paramyxo, rhabdo, filo (bunya, arena)

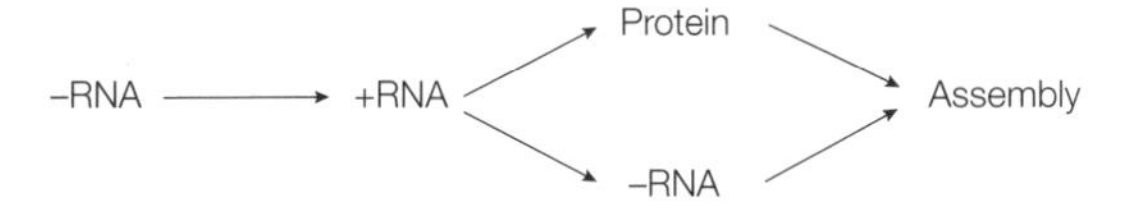

(g) ssRNA genome using ds DNA intermediate: retro

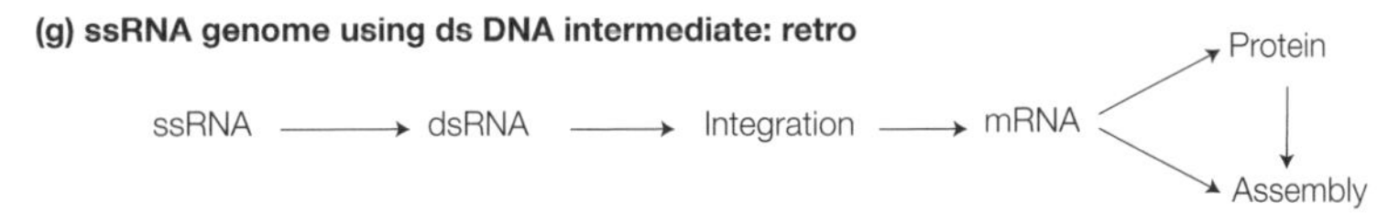

Fig. 1. General methods of viral replication. From Harper, D., Molecular Virology, *2nd edn, © BIOS Scientific Publishers Limited, 1998.*

Large DNA viruses

Herpesviridae vary in size from herpes simplex virus and varicella zoster virus (120–180 kbp) to cytomegalovirus and HHV-6 (180–230 kbp). The DNA codes for >30 virion proteins and >40 non-structural proteins (found only in infected cells). The structure of the DNA, while showing minor variations between members of the group is unique in that several **isomers** of the same molecule can exist. The genome of herpes simplex virus consists of two covalently-joined sections, a **unique long** (U_L) and a **unique short** (U_S) region, each bounded by **inverted repeats**. These repeats allow structural rearrangements of the unique regions, thus creating four isomers all of which are functionally equivalent (*Fig. 2*). Herpesvirus genomes also contain **multiple repeated sequences** which vary greatly between viruses, thus giving them a larger or smaller molecular weight than average.

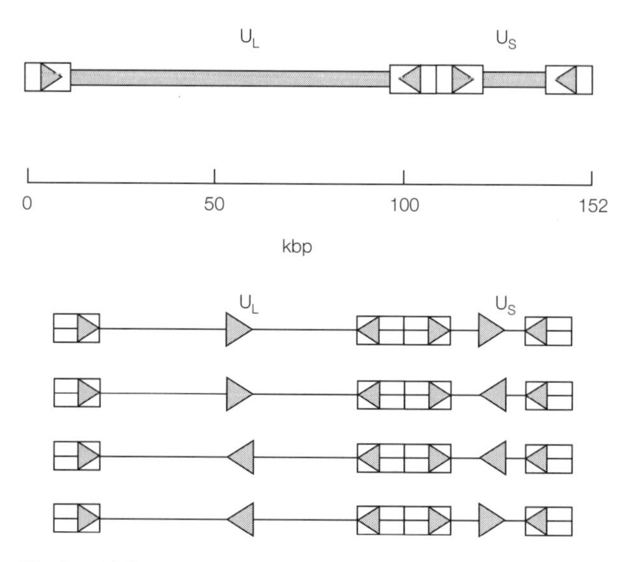

Fig. 2. (a) Some herpes virus genomes (e.g. herpes simplex virus) consist of two covalently joined sections, U_L and U_S, each bounded by inverted repeats. (b) This organization permits the formation of four different forms of the genome. Reproduced from Cann, A., Principles of Molecular Virology, *1993, with permission from Academic Press.*

The genome of adenoviruses is smaller, with a size of 30–38 kbp. Each virus contains 30–40 genes. The terminal sequence of each strand is a 100–140 bp **inverted repeat** and thus the denatured single strands can combine to form a **pan-handle** structure, which has an important role in replication. The genome has a 55 kb terminal protein at the 5′ end which acts as a primer for the synthesis of new DNA strands (*see Fig. 3*).

Whereas in herpesviruses each gene has its own promoter the adenovirus genome has **clusters** of genes which are served by a common promoter.

Small DNA viruses

Parvovirus genomes are linear non-segmented ssDNA of about 5 kb. Most of the strands that appear in virions are of negative (–) sense. These very small genomes contain only two genes, **rep**, which encodes proteins involved in transcription, and **cap,** which encodes the coat proteins. The ends of the genomes have **palindromic sequences** of about 115 nucleotides which form hairpin structures essential for the initiation of genome replication.

Polyoma viruses have dsDNA, 5 kbp in size and circular. The DNA in the virion is **super-coiled** and is associated with four cellular histones (H2A, H2B, H3 and H4). The virus has six genes, this having been achieved by using **both strands** of the DNA for coding and making use of **overlapping genes** (see *Fig. 4*).

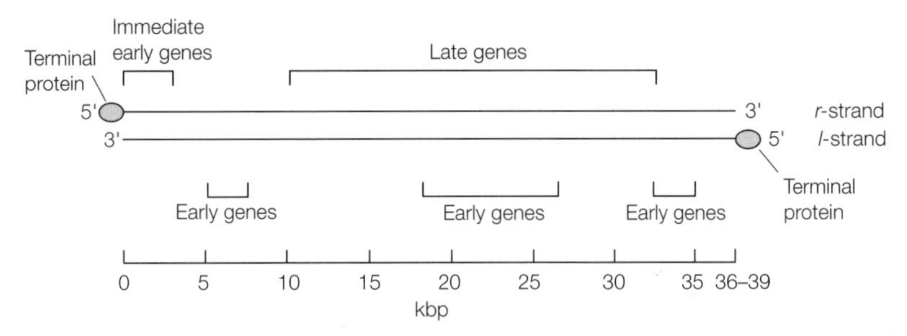

Fig. 3. Organization of the adenovirus genome. Reproduced from Cann, A., Principles of Molecular Virology, *1993, with permission from Academic Press.*

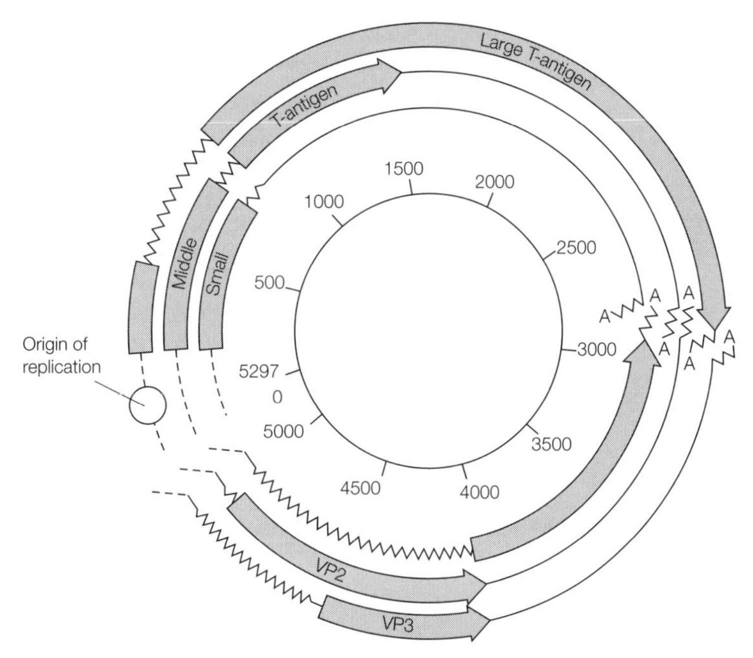

Fig. 4. The complex organization of the polyomavirus genome results in the compression of much genetic information into a relatively short sequence. Reproduced from Cann, A., Principles of Molecular Virology, *1993, with permission from Academic Press.*

RNA viruses (non-segmented)

RNA viruses tend to have smaller genomes than most DNA viruses. They vary in size from 30 kb (for example, coronaviruses) to 3.5 kb (small phage, for example, Qβ). They are either **positive sense** or **negative sense** as described earlier. Most (not all) positive-sense ssRNA genomes have a **cap** that protects the 5′ terminus of the RNA from attack by phosphatases and other nucleases and promotes mRNA function at the level of translation. In the 5′ cap the terminal 7-methylguanine and the penultimate nucleotide are joined by their 5′-hydroxyl groups through a triphosphate bridge. This 5′–5′ linkage is inverted relative to the normal 3′–5′ phosphodiester bonds in the remainder of the polynucleotide chain. In most picornaviruses this **cap** is replaced by a small protein, VPg. Some viruses, e.g. picornaviruses, flaviviruses, have a complex hairpin-like RNA structure – an **internal ribosome entry site** (IRES) where translation is initiated. The 3′ end of most positive-sense viral genomes, like most eukaryotic mRNAs, is **polyadenylated.**

Viruses with negative-sense RNA genomes are usually larger and encode more genetic information than positive genomes. The genomic organization of selected RNA viruses is shown in *Figure 5.*

RNA viruses (segmented)

Segmented virus genomes are those which are divided into two or more **physically distinct molecules** of nucleic acid, all of which are packaged into a single virion. They have the advantage of carrying genetic information in smaller strands, which makes them less likely to break due to shearing. This breakage is a hazard for large RNA molecules. On the downside, of course, is the fact that these viruses must have an elaborate packaging mechanism which allows a representative of each strand to be packaged into the virion. Influenza virus, a negative ssRNA virus, has eight segments each coding for one or two viral proteins (*Fig. 6* and *Table 1*).

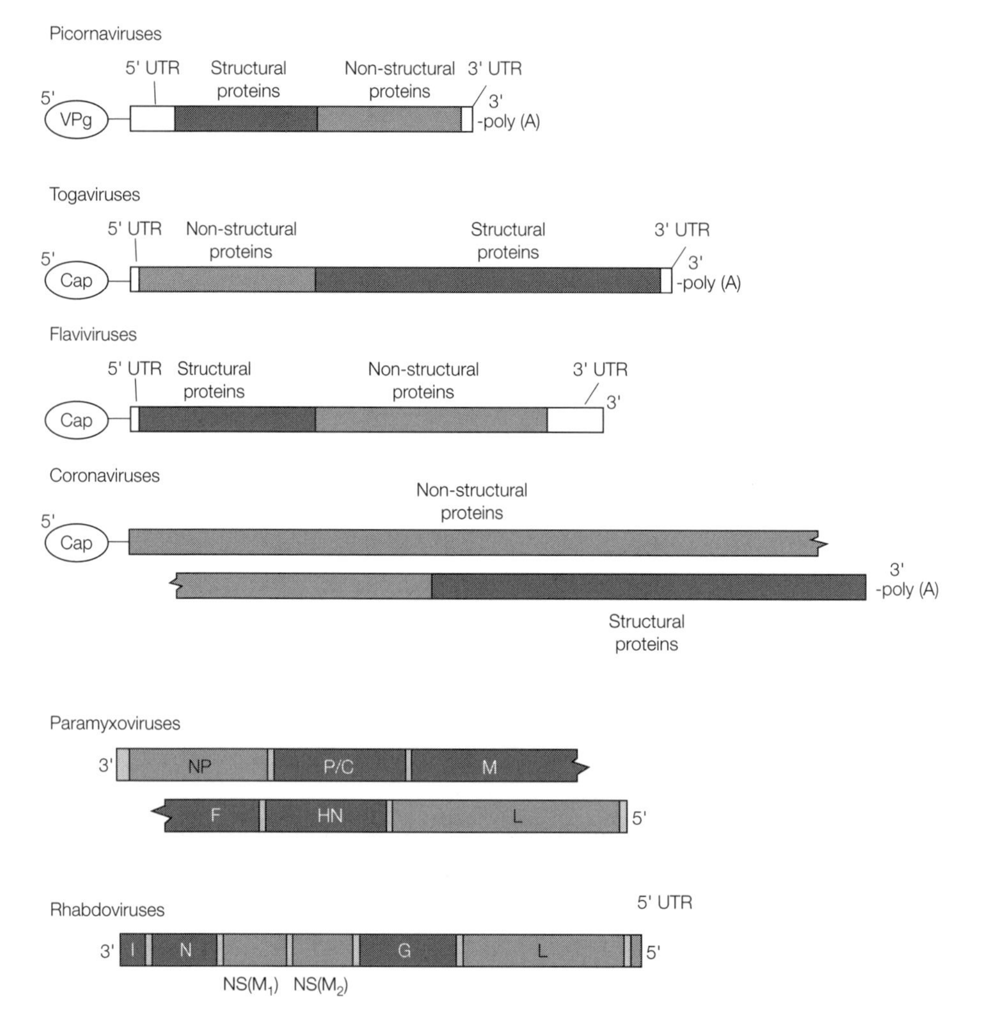

Fig. 5. Diagrammatic representation of selected RNA virus genomes showing structural and non-structural genes. 5′ UTR are regions of the genes that are not translated into proteins. Reproduced from Cann, A., Principles of Molecular Virology, *1993, with permission from Academic Press.*

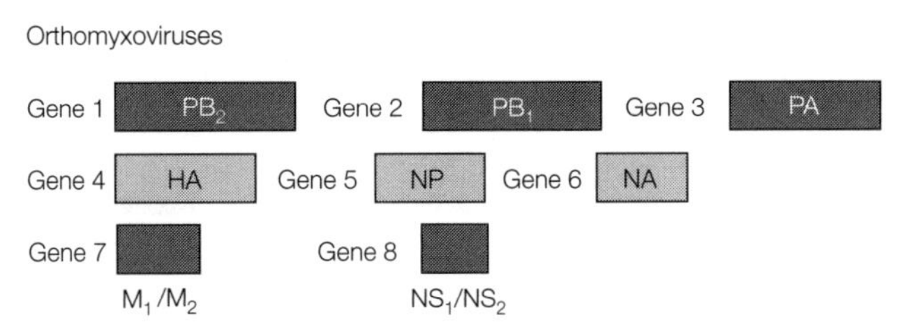

Fig. 6. Relative size and coding of orthomyxovirus RNA segments. Reproduced from Cann, A., Principles of Molecular Virology, *1993, with permission from Academic Press.*

Table 1. Segments of the influenza virus genome

Segment	Size (nt)	Polypeptide(s)	Function
1	2341	PB_2	Transcriptase: cap binding
2	2341	PB_1	Transcriptase: elongation
3	2233	PA	Transcriptase: (?)
4	1778	HA	Hemagglutinin
5	1565	NP	Nucleoprotein: RNA binding; part of transcriptase complex
6	1413	NA	Neuraminidase
7	1027	M_1	Matrix protein: major component of virion
		M_2	Integral membrane protein – ion channel
8	890	NS_1	Non-structural (nucleus) function unknown
		NS_2	Non-structural (nucleus + cytoplasm) function unknown

J5 Cell culture and virus growth

Key Notes

Historical perspective

The major thrust in virology research was only possible after the derivation of defined synthetic growth medium and the *in vitro* culture of cells. This allowed viruses to be grown, purified and studied outside of the host, leading to the development of the first virus vaccines.

Method of cell culture

Single-cell suspensions are seeded into tissue culture quality glass or plastic vessels where they attach to the surface and divide to form a monolayer of cells. Cells can be grown in a full range of vessels but must be carefully nurtured and repassaged, ensuring they are free from contamination. Most cells grow at 37°C.

Media and buffer

Defined growth media contain a balanced salts solution supplemented with amino acids, vitamins, glucose, serum, antibiotics, a buffer and pH indicator. Such media are used routinely to support the growth of tissue-culture cells and can be either manufactured from the individual constituents or bought ready-made commercially.

Cell culture types

Cells are either primary (very limited cell passage), diploid cell strains (up to 50–60 cell passages) or continuous cell lines (unlimited cell passage).

Virus growth in culture

Small aliquots of virus are added to cell monolayers at low or high multiplicity of infection and, following replication, progeny virus is harvested and assayed. The virus is either cell-associated or released into the culture medium. Following titration, the virus is stored at either –20°C or –70°C until required.

Use of embryonated eggs

For some viruses, growth in embryonated hens eggs is the preferred culture method. Influenza virus is grown in the allantoic cavity of the embryonated egg.

Related topics

Virus assay (J6)
Virus replication (J7)

Historical perspective

In addition to helping solve the mysteries of virus diseases, knowledge of virus structure, replication and host interactions, virus research has extended our understanding of the fundamentals of eukaryotic biology (e.g. protein:protein interactions, nucleic acid transcription and translation, evolution). Few of these studies would have been possible without the means of growing viruses outside their normal hosts. For most viruses this means growing them in **cell culture**. Cell culture is the art of growing cells *in vitro*. In addition to deriving cells for culture, this technique relied heavily on the development of **media** and **buffers** that supported cell growth *in vitro*. Most of these techniques did not become available until the early 1950s, which consequently saw an upsurge in the study of viruses.

There are now over 3200 characterized cell lines derived from over 75 species, which are kept in, for example, the American Type Culture Collection and the European Collection of Animal Cell Cultures.

Method of cell culture

Cells are propagated on **glass** or **plastic flasks** of various sizes (as required) or in vast vessels or vats. The technique relies on good **aseptic technique**, thus keeping the cultures free of fungal and bacterial contamination.

Single-cell suspensions of cells of known concentration are **seeded** into a sterile flask along with appropriate **growth medium**. The flask, usually a plastic or glass bottle, is incubated at the appropriate temperature (usually 37°C) in the flattened position. Cells adhere to the surface and begin to replicate, forming a **monolayer** of cells which adhere to each other, in addition to being anchored to the surface of the flask. After a few days the metabolic activity of the cells means that the growth medium will be 'spent' and the cells, unless **reseeded**, will deteriorate and die. Thus, the cell monolayer is treated with trypsin and/or versene solution to create single cells again. These cells are used for **seeding** into fresh flasks. The continued seeding of cells is referred to as cell passage. Cell monolayers are used for virus growth and assay and to examine many aspects of virus–host interactions (see *Fig. 1*).

In addition to growing as a monolayer, some cell types may also be able to grow in **suspension**, where they do not anchor themselves to the surface of the flask or adhere to each other (e.g. hybridoma cells which secrete monoclonal antibodies).

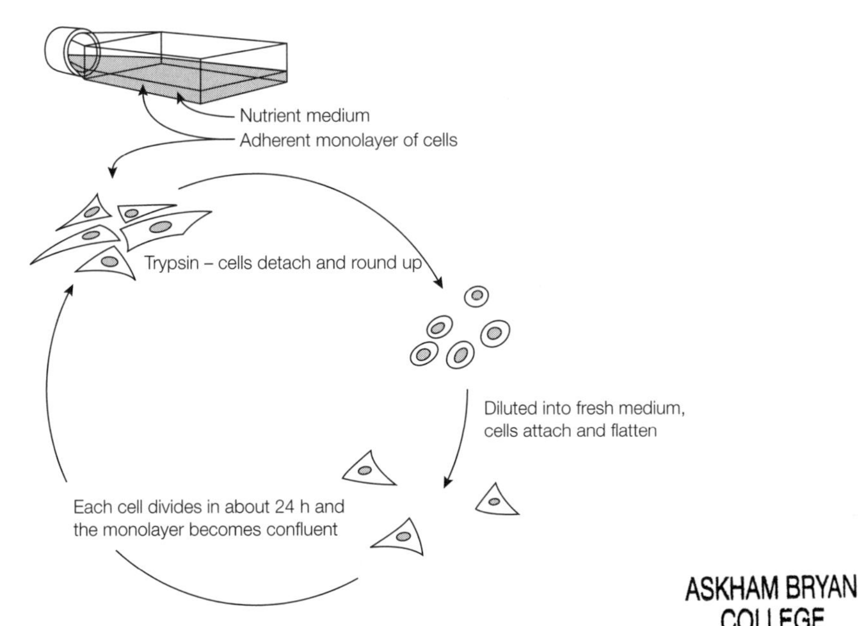

Fig. 1. Cell culture. Redrawn from Dimmock and Primrose, 1994, Introduction to Modern Virology, 4th edn, with permission from Blackwell Science Ltd.

Media and buffer

Most growth media in use today are chemically defined, but are usually supplemented with 5–20% serum (which contains the stimulants necessary for cell division). Serum-free medium with added stimulants is used for some purposes. Media contain an isotonic **balanced salts solution**, supplemented with amino acids, vitamins and glucose, for example, Eagle's minimal essential medium (MEM), formulated by Eagle in the 1950s. In addition to serum, MEM is also supplemented with antibiotics, usually penicillin and streptomycin, to help prevent bacterial contamination. Generally, cells grow well at pH 7.0–7.4 and hence phenol red is added to act as a pH indicator. It is red at pH 7.4, orange at pH 7.0 and yellow at pH 6.5. In the alkaline direction it becomes bluish at pH 7.6 and purple at pH 7.8.

Culture media also require buffering under two sets of conditions: (a) use of open flasks, in which exposure to oxygen causes the pH to rise; and (b) high cell concentrations when CO_2 and lactic acid are produced, causing the pH to fall. A buffer is thus incorporated into the medium, and in 'open flask' conditions the incubator is purged with an exogenous source of CO_2. The buffer used by most laboratories relies on the bicarbonate–CO_2 system, and hence bicarbonate solution is added to the growth medium.

Reagents for use in media preparation and cell culture must be sterilized. This is achieved by autoclaving (moist heat), hot-air ovens (dry heat), membrane filtration or, in the case of plastic ware, by irradiation.

Cell culture types

Primary cells are freshly isolated cells that are derived directly from the tissue of origin. The tissue source for the majority of primary cell cultures is either laboratory animals (e.g. monkey kidney cells) or human pathology specimens (e.g. human amnion cells). Tissue samples are incubated with a proteolytic enzyme (usually trypsin) overnight with a number of washes, etc., in order to produce a single-cell suspension. The harvested cells are then seeded into appropriate flasks with growth medium. Cell cultures are usually either epithelial (cuboid-shaped) or fibroblastic (spindle-shaped), but it is usual for primary cultures to contain both morphological types. Primary cultures are sensitive to a wide range of viruses and are routinely used in diagnostic laboratories for growth of fresh virus isolates (from patients). Unfortunately, primary cells usually die after only a few passages. Some cells, when grown *in vitro* change so as to lose their ability to support virus growth. Hepatocytes are an example, where *in vivo* they do, but *in vitro* do not, support hepatitis C replication.

Many **cell lines** will continue to grow for more than four to five passages but subsequently die after 50–60 passages. These cell lines, usually referred to as **diploid cell strains**, are often derived from fetal lung tissue. They have the normal chromosome number and are usually fibroblastic (e.g. MRC-5 cells).

Continuous cell lines are capable of continued passage in tissue culture. They may be epithelial (e.g. Hela cells) or fibroblastic (e.g. baby hamster kidney cells, BHKs) and are derived from tumorous tissue (e.g. Hela cells) or by the sudden **transformation** of a primary cell (e.g. BHK cells). They are **heteroploid** (i.e. have aberrant chromosome numbers).

Virus growth in culture

Most experiments in virology have required virus growth in culture although, nowadays, more experiments rely entirely on cloned genes and expressed proteins outside of cell cultures. Historically, however, it has been those viruses which grow well in cell culture that have been studied in most detail. Lack of *in vitro* growth has seriously curtailed progress in research, vaccine production and development of antiviral drugs, for example hepatitis B and C viruses.

Viruses are grown in cultures to create **virus stocks**. The passaged virus is stored at –70°C and referred to as a **master-stock, sub-master stock**, etc., depending on its **passage number**. It is important in virology to record the passage history in tissue culture of the virus being used.

Virus stocks are grown by infecting cells at a low **multiplicity of infection (m.o.i.)** that is, approximately 0.1–0.01 infectious units per cell. Virus attaches to cells and goes through several replicative cycles in the cell culture. After a few days, virus is harvested from the extracellular medium surrounding the cultured cells or from the cells themselves, which are lysed by freezing and thawing or by using an ultrasonic bath. The virus so derived is quantitated by an infectivity assay.

When large numbers of virus particles are required (e.g., for virus purification), cell cultures are infected at a high m.o.i. (e.g. 10 infectious units per cell). This ensures that all cells are synchronously infected and that only one round of replicative cycle will ensue. Virus is harvested as above at the end of the cycle. Infected cells yield various numbers of new (progeny) virus particles ranging from 10–10 000 particles per cell.

Use of embryonated eggs

For some viruses (e.g. influenza virus) cell culture is not the chosen procedure for virus growth and instead a fertilized chick embryo is used. The fertilized embryo has a complex array of membranes and cavities which will support the growth of viruses (see *Fig. 2*). Small aliquots of influenza virus are inoculated into the allantoic cavity of the egg. The virus then attaches to and replicates in the epithelial cells lining the cavity. Virus is released into the allantoic fluid and harvested after two days growth at 37°C. Influenza vaccines are propagated in this way.

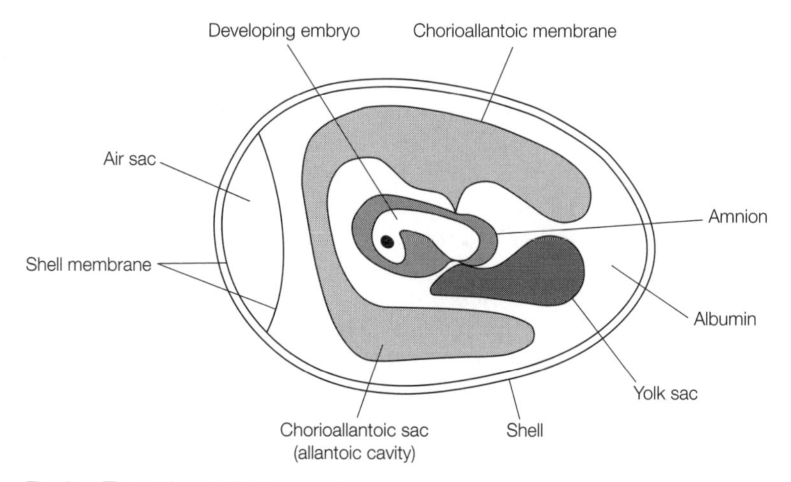

Fig. 2. The embryonated egg.

J6 VIRUS ASSAY

Key Notes

Virus infectivity — This is defined as the ability of a virus particle to attach, penetrate and undergo an infectious cycle in a susceptible host cell, usually resulting in cell damage.

Virus dilution — Virus is diluted using factors of two, five or 10 in an appropriate solution (e.g. buffer or growth medium) prior to an infectivity assay.

Plaque assay — This is a focal assay used to detect zones of cytopathic effect (CPE), also known as plaques, in a monolayer of healthy cells. One infectious virus produces one plaque.

$TCID_{50}$ — The tissue culture infective $dose_{50}$ ($TCID_{50}$) is defined as that dilution of virus which will cause CPE in 50% of a given batch of cell cultures.

Particle counting — The total number of virus particles (infectious and non-infectious) is determined by counting with the aid of an electron microscope. This allows determination of the particle/infectivity ratio.

Hemagglutination — The agglutination of red blood cells (RBCs) by some viruses is referred to as hemagglutination (HA). This agglutination forms the basis of an assay which measures the number of HA units per unit volume in a given suspension.

Related topics Cell culture and virus growth (J5) Virus replication (J7)

Virus infectivity

In order to 'reproduce', viruses need to be capable of replicating in a susceptible host cell. This **replicative cycle** is accompanied by a number of biochemical and morphological changes within the cell which usually results in the death of the cell. The accompanying morphological changes (e.g. cell rounding or fusion) are referred to as the cytopathic effect (CPE). A particular type of CPE is often a characteristic of specific virus growth and can be used when attempting to identify an unknown virus. The appearance and detection of CPE regularly forms the basis of **infectivity assays**, designed to determine the number of **infectious units of virus per unit volume**, and is the infectivity **titer** (e.g. **plaque forming units** (pfus) per milliliter). An infectious unit is thought of as being the smallest amount of virus that will produce a detectable biological effect in the assay (e.g. a pfu). Infectivity assays are either **quantal,** an 'all or none' approach (e.g. tissue culture infective dose 50 ($TCID_{50}$)) or **focal,** detection of a focus of infection (e.g. a plaque assay).

Virus dilution

Virus titers are determined by making accurate serial dilutions of virus suspensions. Such dilutions are usually done using factors of two, five or 10. For routine use, **10-fold dilutions** are usually carried out. It is important to use a new sterile pipette for the transfer of volumes between each dilution and to mix thoroughly the dilution before further transfer. Once diluted, virus should be assayed as soon as possible as most viruses rapidly lose infectivity at room temperature.

Plaque assay

The plaque assay quantifies the number of **infectious units** in a given suspension of virus. **Plaques** are localized discrete foci of infection denoted by zones of cell lysis or cytopathic effect (CPE) within a monolayer of otherwise healthy tissue culture cells. Each plaque originates from a **single infectious virion,** thus allowing a very precise calculation of the virus titer. The most common plaque assay is the **monolayer** assay. Here, a small volume of virus diluent (0.1 ml) is added to a previously seeded confluent tissue culture cell monolayer. Following adsorption of virus to the cells, an **overlay medium** is added to prevent the formation of **secondary** plaques. Following incubation, the cell sheets are 'fixed' in **formol saline** and stained, and the plaques counted. For statistical reasons, 20–100 plaques per monolayer are ideal to count, although the actual number that can be easily counted is often dependent on the size of the plaque and the size of the vessel used for the assay. Typical plaques are shown in *Fig. 1*, and the assay procedure is summarized in *Fig 2*. The **infectivity titer** is expressed as the number of plaque forming units per ml (**pfu ml^{-1}**) and is obtained in the following way:

$$\text{pfu ml}^{-1} = \frac{\text{plaque number}}{\text{dilution} \times \text{volume (ml)}}$$

For example, if there is a mean number of 100 plaques from monolayers infected with 0.1 ml of a 10^{-6} dilution then the calculation is:

$$\frac{100}{10^{-6} \times 0.1} = 1 \times 10^{9} \text{ pfu ml}^{-1}$$

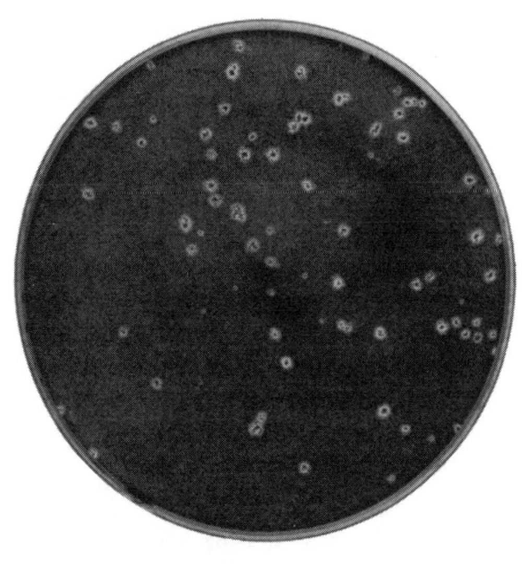

Fig. 1. Herpes virus plaques on a tissue culture monolayer.

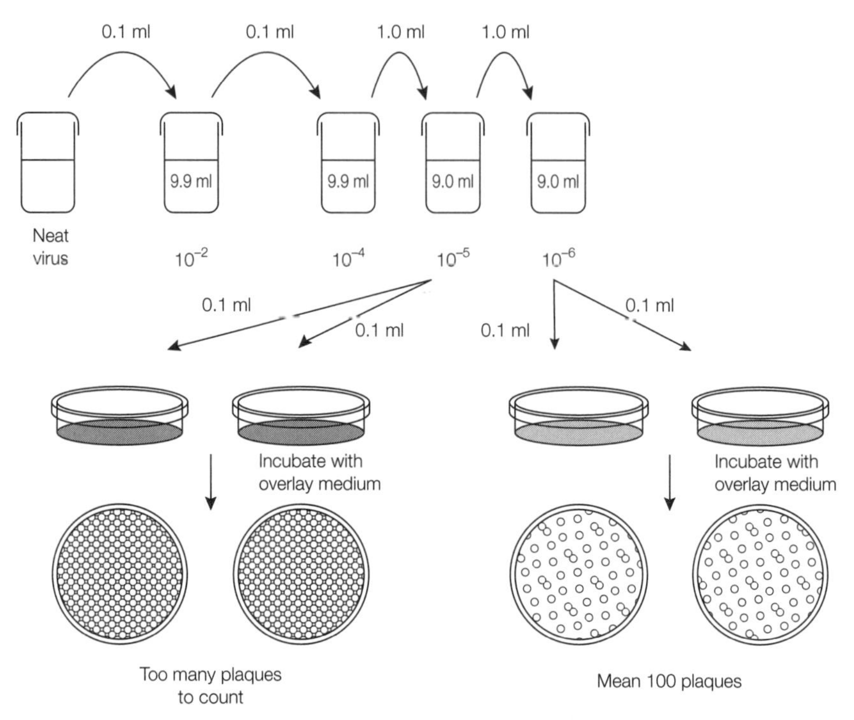

Fig. 2. Diagrammatic representation of virus plaque assay (see text for calculation).

$TCID_{50}$

The $TCID_{50}$ is defined as **that dilution of virus required to infect 50% of a given batch of inoculated cell cultures**. The assay relies on the presence and detection of CPE. Host cells are grown in confluent healthy monolayers, usually in tubes, to which aliquots of virus dilutions are made. It is usual to use either five or 10 repetitions per dilution. During incubation the virus replicates and releases progeny virus particles into the supernatant which in turn infect other healthy cells in the monolayer. The CPE is allowed to develop over a period of days, at which time the cell monolayers are observed microscopically. Tubes (the 'test units') are scored for the presence or absence of CPE. In this quantal assay the data are used to calculate the $TCID_{50}$ i.e. the dilution of virus which will give CPE in 50% of the cells inoculated. *Table 1* shows some typical data.

By using the data in *Table 1* the following calculation can be made:

$$TCID_{50} = \log_{10} \text{ of highest dilution giving 100\% CPE} + \tfrac{1}{2} - \frac{\text{total number of test units showing CPE}}{\text{number of test units per dilution}}$$

$$= -6 + \tfrac{1}{2} - \tfrac{9}{5} = -7.3 \; TCID_{50}$$

or $10^{-7.3}$ $TCID_{50}$ unit vol.$^{-1}$

The titer is therefore $10^{7.3}$ $TCID_{50}$ per unit vol.$^{-1}$

Particle counting

Not all virus particles are infectious. Indeed in many cases for every one infectious particle up to 100 or more **non-infectious particles** may be produced from an infected cell. The total number of particles can only be determined by

Table 1. Data used to calculate $TCID_{50}$ (See text for calculation)

Log_{10} of virus dilution	Infected test units (e.g. infected tubes)
–6	5/5
–7	3/5
–8	1/5
–9	0/5

counting them with the aid of an **electron microscope**. The counting procedure relies on the use of **reference particles** which are usually latex beads of uniform diameter. The principle is that if viruses can be mixed with reference particles of known concentration (i.e. a number per unit volume), a simple determination of the ratio of virus to reference particles will yield the virus count. Latex and virus particles are distinguished after **negative staining** with phosphotungstate. The ratio of total particles to infectious particles is termed the **particle/infectivity ratio,** which is important to know when, for example, monitoring virus purification, or determining the state or age of a virus suspension.

Hemagglutination

Many viruses have the ability to agglutinate RBCs, this being referred to as **hemagglutination**. In order for the reaction to occur, the virus should be in sufficient concentration to form cross-bridges between RBCs, causing their agglutination. Non-agglutinated RBCs will form a **pellet** in a hemispherical well, whereas agglutinated RBCs form a **lattice-work** structure which coats the sides of the well. This phenomenon forms the basis of an assay which determines the number of **hemagglutinating particles** in a given suspension of virus. It is not a measure of infectivity, but is one of the most commonly used **indirect methods** for the determination of virus titer. The assay is done by **end-point titration**. Serial two-fold dilutions of virus are mixed with an equal volume of RBCs and the wells are observed for agglutination. The end point of the titration is the **last dilution showing complete agglutination,** which by definition is said to contain one **HA unit**. The HA titer of a virus suspension is therefore defined as being the reciprocal of the highest dilution which causes complete agglutination and is expressed as the number of HA units per unit volume. An example upon which a calculation of the HA titer can be made is shown in *Fig. 3*. The end point in this figure is 1/512. If 0.2 ml virus dilution was added per well the HA titer would be 512 HA units per 0.2 ml or 2560 HA units ml^{-1}.

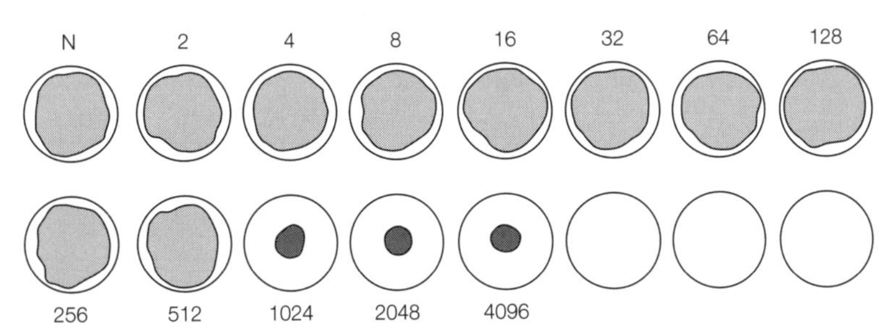

Fig. 3. Diagram of a sample hemagglutination assay. Serial doubling dilutions of virus shows agglutination end-point at 1:512.

J7 VIRUS REPLICATION

Key Notes

Replicative cycle

As obligate intracellular parasites, viruses must enter and replicate in living cells in order to 'reproduce' themselves. This 'growth cycle' involves specific attachment of virus, penetration and uncoating, nucleic acid transcription, protein synthesis, maturation and assembly of the virions and their subsequent release from the cell by budding or lysis.

Attachment, penetration and uncoating

Attachment is a very specific interaction between the virus capsid or envelope and a receptor on the plasma membrane of the cell. Virions are either engulfed into vacuoles by 'endocytosis' or the virus envelope fuses with the plasma membrane to facilitate entry. Uncoating is usually achieved by cellular proteases 'opening up' the capsid.

Transcription and translation

Using cellular and virus-encoded enzymes and 'helper' proteins, nucleic acid is usually transcribed in a controlled fashion. Control is also exercised at the level of mRNA concentration (apart from some simple viruses, e.g. polio). Nucleic acid is synthesized by virus-encoded enzymes. Translated proteins may undergo post-transitional modification (e.g. cleavage, glycosylation, phosphorylation).

Maturation, assembly and release

Subunits of capsids assemble via 'sub-assembly' structures, with or without the help of scaffolding proteins. Envelopes, when present, are acquired by capsids budding through the nuclear or plasma membrane.

Related topics

Cell culture and virus growth (J5)
Virus assay (J6)
Antiviral chemotherapy (J11)

Replicative cycle

The complexity and range of virus types is echoed in the various strategies they adopt in their replicative cycles. Viruses, as **obligate intracellular parasites,** must **attach** to or **enter** host cells in order to undergo a 'reproductive' cycle. This cycle is highly dependent on the metabolic machinery of the cell, which in most cases the virus takes over and orchestrates towards its own replication, usually **inhibiting host-cell protein and nucleic acid synthesis**. The outcome is the production of hundreds of progeny virions which leave the infected cell (by **lysis** or **budding**), killing the cell and spreading to infect more host cells and tissues. This replicative or **growth cycle** can be analyzed in tissue culture cells and is often referred to as the **one-step** growth cycle. The cycle has a number of stages – attachment and penetration, nucleic acid synthesis and transcription, protein synthesis, maturation, assembly and release. A typical pattern for a growth curve is shown in *Figure 1*. Following attachment and penetration by virus, cells are lysed and titrated for infectious virus particles (pfus) at various times post-infection. Plotting $\log_{10}$ pfu versus time gives the characteristic curve which has an **eclipse period** (where no new virions have been formed) followed by a **logarithmic expansion phase** until **peak** virus titers are reached when the

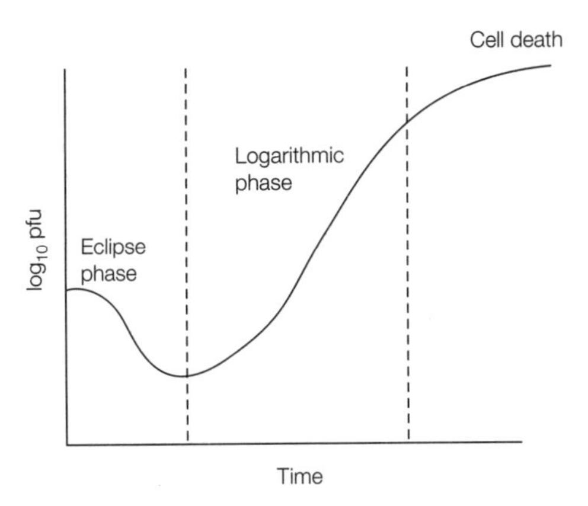

Fig. 1. A typical 'one-step' growth or replicative cycle of a virus.

cell usually dies and virions are released. The shape of this curve varies greatly between viruses – for most bacteriophage it takes less than 60 min and for many animal viruses it can exceed 24 h before maximum titers are reached. The replication cycles of three animal viruses have been selected below for further study and to compare their strategies – herpes simplex (a DNA virus), poliovirus (RNA virus) and HIV (a single-stranded RNA virus which replicates via a DNA intermediate). Their strategies are typical of many viruses, although divergent strategies also exist!

Attachment, penetration and uncoating

Attachment is mediated by a **specific interaction** between the virus and a **receptor** on the plasma membrane of the cell (*Fig. 2*). Indeed it is the presence of such a receptor that determines the **cell tropism** and **species tropism** of the virus. These receptors have cellular functions other than providing a binding

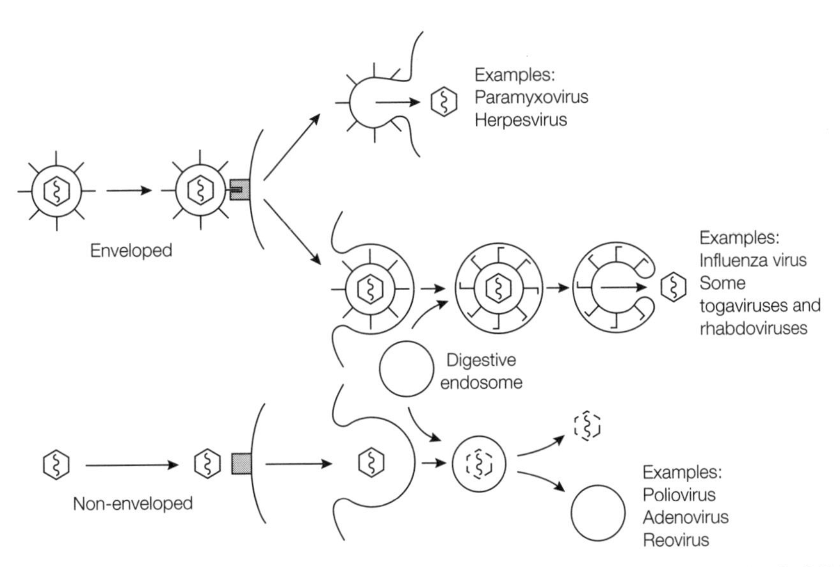

Fig. 2. Methods of virus entry. From Harper, D., Molecular Virology, *2nd edn, © BIOS Scientific Publishers Limited, 1998.*

site for viruses, but have been paramount in affecting virus evolution. Thus, the herpes simplex virus (HSV) binds to a heparin sulfate proteoglycan plus mannose-6-phosphate receptor via at least two virus-coded envelope glycoproteins. One of the four proteins in the poliovirus capsid attaches to a receptor of the **Ig protein superfamily** (only on primate cells). HIV, via its major envelope glycoprotein (**gp120**), attaches to the **CD4 receptor** found predominantly on human T4 lymphocytes. Recently, two further ligand receptors were shown to have a role in HIV attachment.

For HSV and HIV, penetration of the virus across the plasma membrane is achieved by **fusion** of the viral envelope with the membrane, releasing the nucleocapsid into the cytoplasm of the cell (*Fig. 2*). The naked capsid of poliovirus, however, is taken up by the process of **endocytosis**, the membrane invaginating to engulf the capsid, resulting in the formation of a **vacuole** which transports the capsid into the cytoplasm. The virion is later released from this vacuole (*Fig. 2*).

Transcription and translation

(i) **HSV**. HSV, a large DNA virus, replicates mainly in the **nucleus** of the cell, although of course protein synthesis and post-translational modification take place in the cytoplasm. The genome encodes for dozens of virus-specific proteins, many with **enzymic** activity (e.g. **thymidine kinase**, **DNA polymerase**) – such proteins are usually **non-structural** (i.e. will not finish up in the virion). Others are structural proteins and will form the capsid, envelope and tegument (the structure between the capsid and the envelope).

Viruses need proteins at different times and in different concentrations throughout their growth cycles and hence the virus shows transcriptional control. Depending on the timing of their expression in the virus replication cycle, HSV genes are classified as either **immediate early**, **early** or **late** (α, β or γ). The mRNA produced encodes proteins that have control functions, switching on subsequent genes. All proteins, of course, are formed on cytoplasmic host-cell ribosomes and remain in the cytosol or are directed to the endoplasmic reticulum where they undergo the **post-translational** events (e.g. **glycosylation**, **phosphorylation**) that give them their final identity. Eventually, these proteins find their way (specifically directed, i.e. **chaperoned**) back to the nucleus for assembly.

The double-stranded viral DNA is synthesized by the viral **DNA polymerase** in association with a number of **DNA-binding proteins**. This enzyme has formed the target for a number of **antiviral drugs** as it is significantly different to the host-cell DNA polymerase (see *Fig. 3*).

(ii) **Poliovirus**. Poliovirus is much simpler in its replicative procedures and control, indeed it lacks fine control! The poliovirus genome RNA strand also acts as mRNA (termed a **positive**-sense RNA virus) and is immediately translated into one long **polyprotein** which is subsequently cleaved into a number of structural and non-structural poliovirus proteins. Included are the structural proteins VP1 and VP3 and the **precursor** protein VP0. Non-structural proteins include a **protease** and **RNA polymerase**. Replication takes place in the **cytoplasm**, indeed enucleate cells will support poliovirus replication. The ss RNA, under the direction of the viral **RNA polymerase** and cellular factors, replicates via a series of ds **replicative intermediate** molecules which act as template for the synthesis of new positive strands. These are then destined to act as mRNA for further rounds of protein synthesis, or become genomes in newly formed progeny virions (see *Fig. 4*).

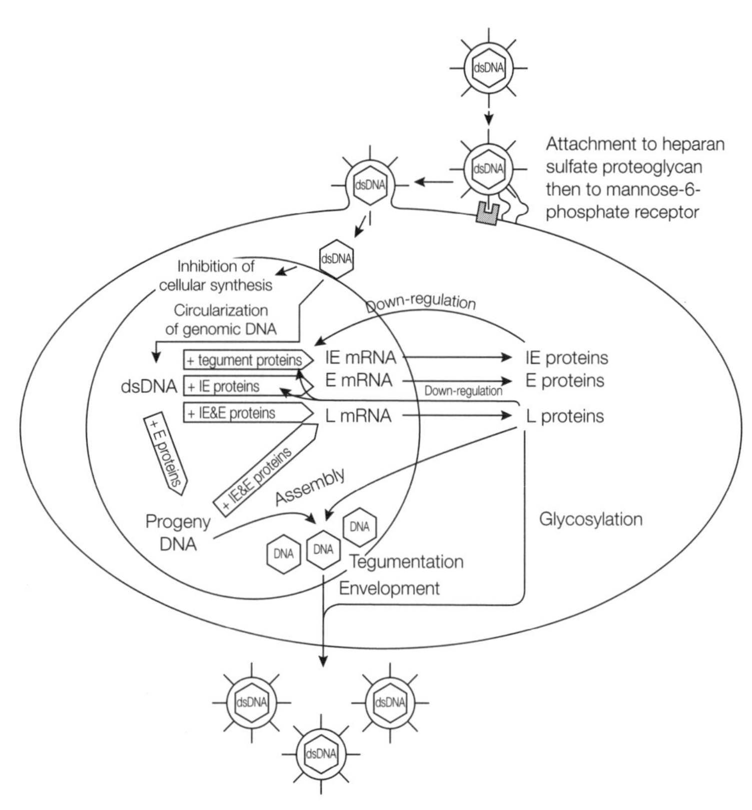

Fig. 3. Herpes simplex virus replication (a dsDNA virus) showing regulation of by immediate early (IE), early (E) and late (L) proteins. From Harper, D., Molecular Virology, *2nd edn, © BIOS Scientific Publishers Limited, 1998.*

(iii) **HIV**. HIV is a member of the unique family of viruses (**Retroviridae**) that carry a **reverse transcriptase enzyme** in their particles, the enzyme catalyzing the formation of DNA from a RNA template. Thus, HIV goes through a series of events (see *Fig. 5*), which terminates in the formation of a ds DNA circularized molecule, formed from an initial input of HIV ss RNA. This molecule, under the direction of a virally encoded integrase, is inserted into the host DNA as a **provirus**. DNA synthesis is now under the control of the cell – when a daughter cell is produced the provirus is reproduced at the same time. The transcription of viral mRNA is under viral control (from the long terminal repeat (LTR) region of its genome) and a series of **mRNA molecules of various sizes** are transcribed. Translated proteins are in some cases **proteolytically cleaved** by virus **protease** into smaller functional proteins. HIV envelope proteins (gp120 and gp45) are further processed in the endoplasmic reticulum before being laid down in the plasma membrane of the cell. **Full-length copy RNA** molecules are also transcribed from the provirus DNA, these forming **progeny RNA strands** destined to be encapsidated.

Maturation, assembly and release

As proteins and nucleic acid are synthesized in the infected cell they are channeled to various locations for virion assembly. Capsids assemble in the **nucleus** or the **cytoplasm**. The steps of capsid assembly vary, depending on the

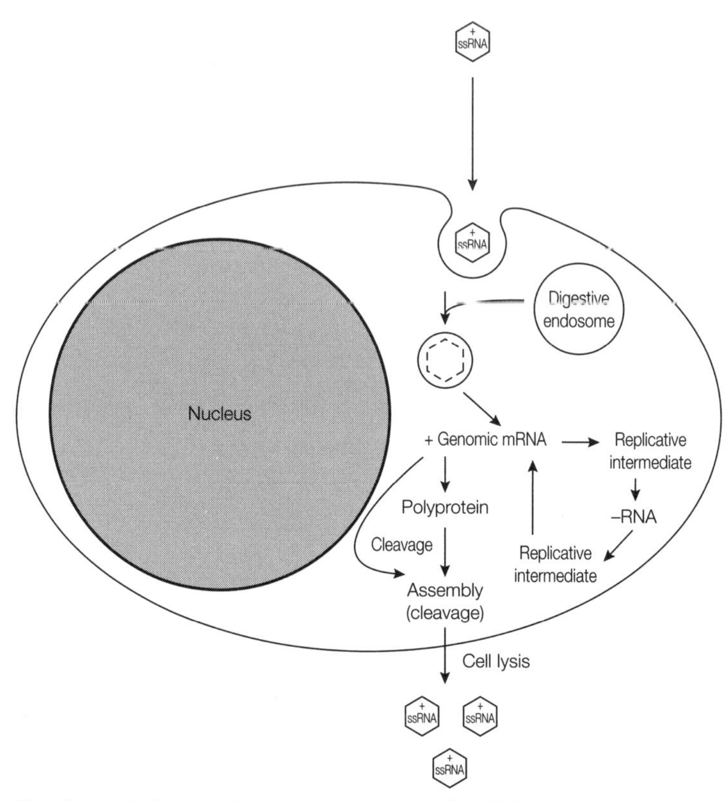

Fig. 4. Poliovirus replication. From Harper, D., Molecular Virology, *© BIOS Scientific Publishers Limited, 1994.*

complexity of the mature capsid, and for enveloped viruses the site of envelope acquisition is either the **nuclear membrane, plasma membrane** and rarely the **endoplasmic reticulum**. Prior to capsids 'budding' through these membranes, virus-specific proteins would have been laid down in the membrane.

(i) **HSV**. Mature HSV capsids are assembled in the nucleus via a number of **precursor forms**. Assembly is assisted by **scaffolding proteins,** which do not finish up in the mature capsid but facilitate protein–protein and protein–nucleic acid bonding as the subunits of the capsid come together. The mature capsid (containing the DNA genome) buds through the nuclear membrane to acquire its envelope. Virions are channeled through the ER to the plasma membrane where they are released. Many herpesviruses invade adjacent cells by the process of **cell–cell fusion**. Thus, the plasma membrane of an infected cell fuses with an adjacent normal cell, facilitating the entry of progeny virions which undergo a further replication cycle. In tissue culture this phenomenon can be seen as large areas of **multinucleate fused cells (syncytia)**.

(ii) **Poliovirus**. Poliovirus is a relatively simple icosahedral capsid with four proteins making up the capsid (VP1, VP2, VP3 and VP4). A further virus-coded protein, VPg, is attached to the ssRNA, serving as a recognition protein. This capsid self-assembles without the need for scaffolding proteins, but does produce an 'immature' capsid form prior to the RNA being inserted into the virion. Thus, the subunits of the capsid assemble via 'pentamers' to form a nucleic acid-free capsid shell containing VP0, VP1 and VP3. As the RNA is

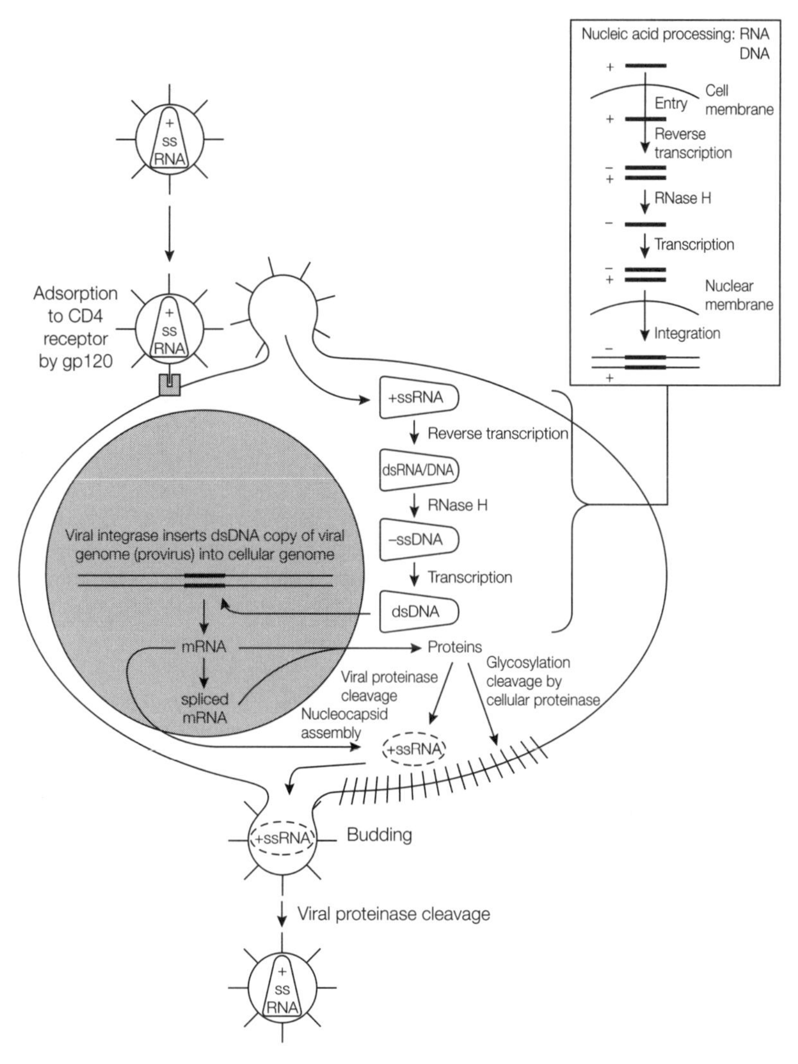

Fig. 5. HIV replication. From Harper, D., Molecular Virology, *2nd edn, © BIOS Scientific Publishers Limited, 1998.*

sequestered into the capsid, VP0 is cleaved into VP2 and VP4 and the mature virion is formed. Poliovirus exits the cell by lysis, releasing several hundreds of progeny virions.

(iii) **HIV**. The capsid of HIV assembles in the cytoplasm. Capsid assembly of the icosahedron follows the normal pattern, but the control mechanisms are obviously important here as the mature virion contains within its central core many components: proteins, two identical strands of RNA and the reverse transcriptase, integrase and protease complex of polymers. Budding of HIV is via the plasma membrane.

Many viruses (e.g. influenza, reovirus) have multisegmented genomes. Each segment, during infection, codes for at least one virus protein essential in replication. In order to be infectious the virion must, therefore, contain a copy of each segment of RNA. Quite how this is achieved remains a mystery!

J8 VIRUS INFECTION

Key Notes

Virus spread

Viruses can gain access to the host through the skin and mucous membranes, via the respiratory or gastrointestinal tracts or through sexual contact. Without spread from host to host, viruses will cease to replicate and become extinct. A knowledge of how viruses spread and their routes of entry and exit from the host has allowed a study of the epidemiology of infections, which in turn has helped control the spread.

Clinical results of infection

The outcome of virus infection is dependent on a number of factors (e.g. age and immune status of the host). Infections are either localized at or near the site of entry, or systemic, where the virus spreads from its point of entry to involve one or more target organs. The outcome of a viral infection follows one of several patterns: (a) inapparent infection; (b) disease syndrome, virus eradication and recovery; (c) latency; (d) carrier status; (e) neoplasia; (f) death.

Related topics

Viruses and the immune system (J9)
Virus vaccines (J10)
Antiviral chemotherapy (J11)

Virus spread

In order to persist and evolve in nature, viruses need a large population of susceptible hosts and an efficient means of spread between these hosts. The normal route is termed **horizontal** spread. The most common route of entry and exit of viruses is the **respiratory route**. Following their inhalation, viruses usually infect and replicate in the epithelial cells of the upper and lower respiratory tract (e.g. rhinoviruses, coronaviruses, influenza, para-influenza and respiratory syncytial virus). The viruses produced in these airways exit their host via sneezing and coughing. Many viruses which enter and exit via this route are not 'respiratory viruses' (e.g. chicken pox, measles, German measles). In these cases the virus leaves the respiratory tract to set up infection in other target organs. The **oral–gastrointestinal** route is used mainly by those viruses responsible for gut infections (rotavirus, Norwalk virus and the enteroviruses, including polio and coxsackie viruses). Vast numbers of virus particles can be excreted in fecal material (e.g. in the order of 10^{12} particles g^{-1}), facilitating the easy spread of these viruses in conditions of poor sanitation. Thus, the drinking of fecally contaminated water and consumption of contaminated shellfish or other food prepared by unhygienic food handlers are ways in which these viruses are spread.

Whilst the skin normally provides an impenetrable barrier to virus invasion, infectious viruses can enter following **trauma to the skin**. This may be from the bite of an animal vector (e.g. rabies via an infected canine, yellow fever and dengue via an infected mosquito). HIV, hepatitis B and hepatitis C may be transmitted by the injection of blood or blood products either in the form of a blood transfusion, a needle-stick injury, or by intravenous drug abuse.

Sexual transmission of viruses is an important route for the spread of HSV, the papilloma viruses, hepatitis B and HIV.

Viruses may also be transmitted **vertically** – that is, from mother to offspring via the placenta, during childbirth, or in breast milk. Examples are rubella virus (German measles) and cytomegalovirus (CMV), acquired by the mother during pregnancy and transmitted to the developing embryo, often leading to severe congenital abnormalities and/or spontaneous abortion. Some viral infections, for example, HSV infections, if acquired *in utero* or during birth, can present as an acute disease syndrome in the neonate. In the case of HIV and hepatitis B transmission, the neonate may be born with an asymptomatic infection, the virus persisting in a **carrier state** and developing into disease much later.

Clinical results of infection

The outcome of a viral infection is dependent on a number of factors including **age**, **immune status** and **physiological well-being** of the host. Thus, HSV infection is usually fatal in the neonate but not in the older child. Epstein–Barr virus (EBV) causes a very mild febrile illness in young children but infectious mononucleosis (glandular fever) in teenagers. CMV in a healthy individual may cause a mild febrile illness, but in immunosuppressed individuals can lead to fatal pneumonia. Measles rarely causes severe complications in healthy well nourished children, but kills around 900 000 children per year in 'developing' countries where malnutrition is a problem. Upon infection viruses either remain **localized** or become systemic (*Fig. 1*). The outcome of a virus infection is considered below.

Inapparent (asymptomatic) infection

Many virus infections are **sub-clinical**, there being no apparent outward symptoms of disease. This is virtually always true in the immune host where recovery from a previous infection or vaccination protects the host from virus growth following reinfection by the wild-type virus. However, several viruses (e.g. respiratory and enteroviruses) may not produce clinical symptoms in some non-immune individuals. Thus, polio virus, in 80% of infected individuals, replicates in the epithelial cells of the gastrointestinal tract, is excreted in the feces, but causes no symptoms.

Disease syndrome, virus eradication and recovery

This is the pattern following most viral infections in otherwise healthy individuals – clinical symptoms of various severity (i.e. a **disease syndrome**) followed by virus eradication by the immune system, recovery and often life-long immunity. This is true of most childhood infections, for example, measles, mumps and German measles, and most respiratory diseases, of which there are a high number of viruses responsible. A vast spectrum of other viruses also follows this pattern, including those of hepatitis A virus (infectious hepatitis), rotavirus (gut infections) and coxsackie virus (myocarditis, pericarditis, conjunctivitis).

Figure 2 shows the possible routes of infection by respiratory viruses. One important respiratory pathogen of infants (**respiratory syncytial virus**, RSV) causes severe necrosis of the bronchiolar epithelium, which sloughs off, blocking the small airways. This leads to obstruction of air flows and respiratory disease. Children recovering from acute RSV bronchitis are often left with a weakened and vulnerable respiratory system, predisposing them to a lifetime of chronic lung disease. Following respiratory infection with measles, the virus replicates in local lymph nodes that drain from the infected tissue. The virus thus enters the blood (**a primary viremia**) where it grows on epithelial surfaces

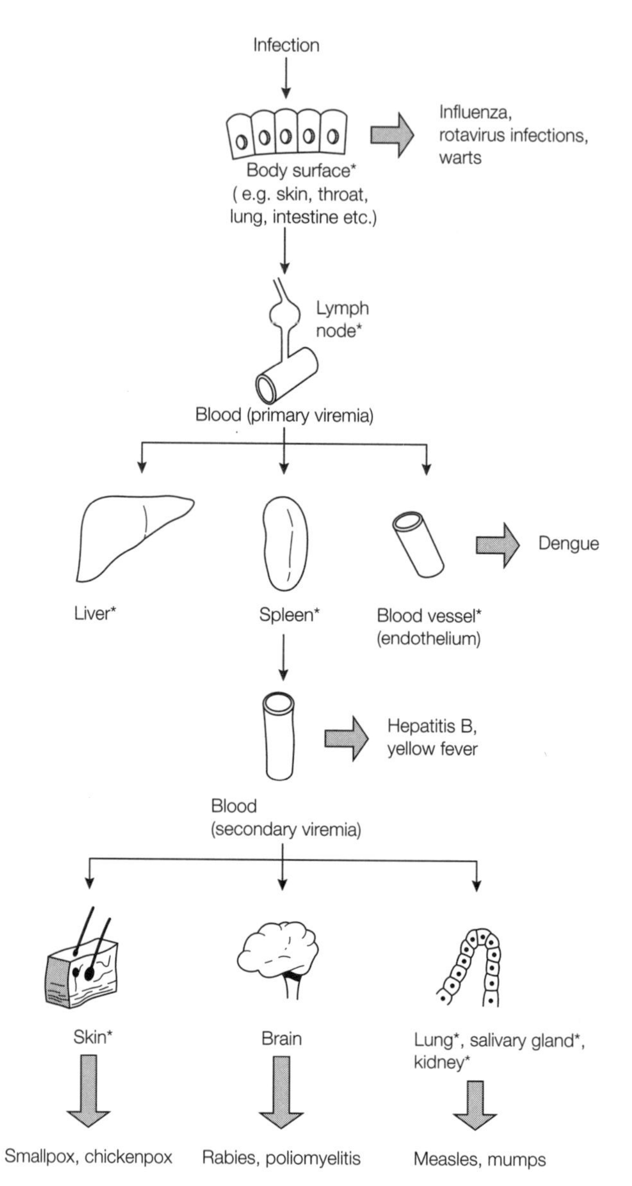

*Fig. 1. Virus spread within the host. Different viruses have different modes of spread within the host. Some viruses remain localized at the site of entry, whereas others may spread to involve other tissues. Routes of infection are shown, together with examples of possible clinical outcomes. *Possible sites of replication; ⇒, sites of shedding. Redrawn from Phillips and Murray (eds),* The Biology of Disease, *1995, with permission from Blackwell Science Ltd.*

before entering the blood again (**a secondary viremia**). At this point the patient is highly infectious but does not have the distinctive measles rash, which appears about 14 days post-infection (see *Fig. 3* for course of events).

Latency

A restricted range of viruses, most being in the **herpesvirus family** (HSV, varicella zoster, Epstein–Barr virus and CMV) are not eradicated from the body

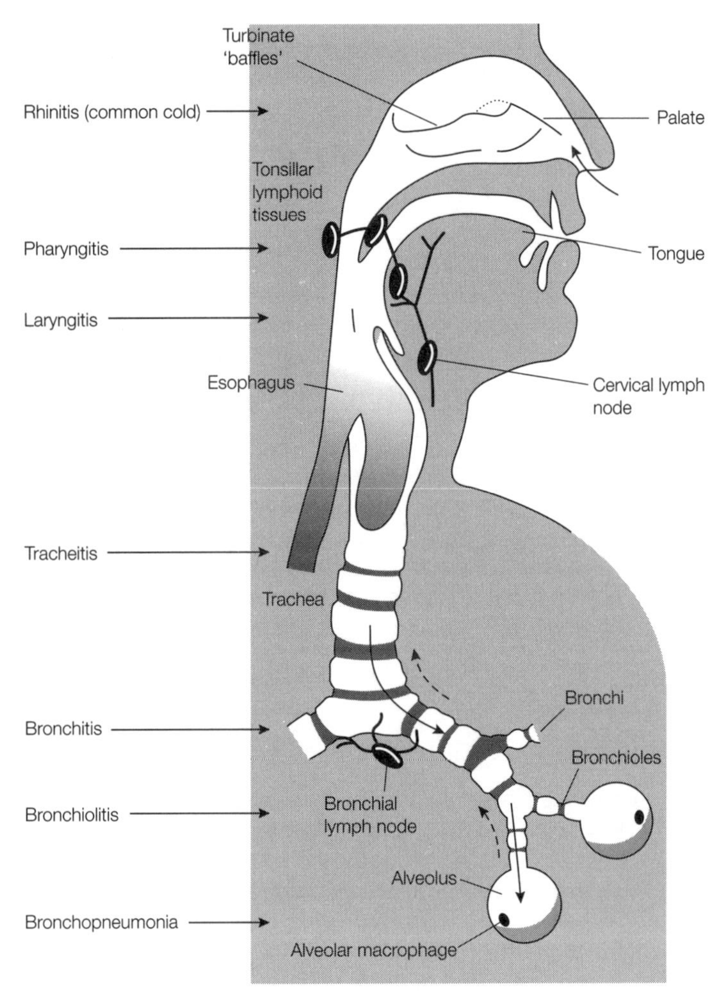

Fig. 2. Routes of infection in the respiratory tract. Virus infections can produce a variety of respiratory disorders, depending on the area of respiratory tract infected. Dotted arrow, mucociliary flow; solid arrow, air flow. Redrawn from Phillips and Murray (eds), The Biology of Disease, *1995, with permission from Blackwell Science Ltd.*

following recovery, but instead become **latent** within the host. Virus replication may be initiated some time later (**reactivation**) and cause clinical symptoms which are either similar or different to those observed in primary infection.

HSV can reactivate many times during the life of an individual and produce the typical painful cold sore lesions on the mouth or genitals. The virus lays dormant in the trigeminal or sacral ganglia between periods of reactivation. It is not understood what happens to reactivate the virus at the cellular level, although a number of stimuli including menstruation, exposure to UV light and stress are responsible for initiating it.

A very different clinical syndrome may result from the reactivation of varicella zoster virus. Thus, the primary infection produces chicken pox, whereas reactivation is associated with the development of **shingles**. Shingles is characterized by a localized area of extremely painful vesicles which clear after 1–2

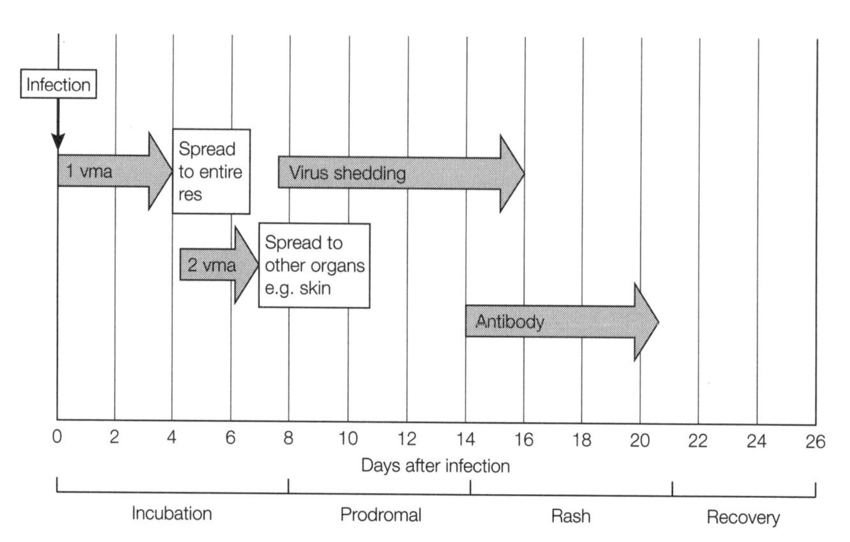

Fig. 3. The course of clinical measles. Measles is characterized by a primary viremia (1 vma) followed by a second viremia (2 vma). Virus shedding occurs, at which point the patient is highly infectious. A prodromal phase (prior to clinical symptoms) is followed by the typical symptoms and signs of measles, including a rash. The appearance of antibody is usually followed by complete recovery and life-long immunity. Res = reticulo endothelial system. Redrawn from Phillips and Murray (eds), The Biology of Disease, *1995, with permission from Blackwell Science Ltd.*

weeks. However, the patient may suffer from very severe **post-herpetic neuralgia** which can persist for months or years.

Carrier, chronic or persistent state

Although a rare event following viral infection, the virus carrier state is induced by **hepatitis B**, **hepatitis C** and **HIV**; 5–10% of individuals infected with hepatitis B virus, and 80% of those infected with hepatitis C virus will carry the infective particles in their blood for months or years. Estimates suggest that >300 million individuals worldwide carry these viruses in their blood and body fluids. These viruses have adapted to avoid the clearance mechanisms of the host immune system. Hepatitis B is transmitted in body fluids but is also passed vertically from mother to offspring. The exact details on hepatitis C routes of transmission, other than in blood and body fluids, are not known.

Unfortunatley, chronic carriers of these viruses have a greatly increased risk of developing **cirrhosis of the liver** and **hepatocellular carcinoma**. At present, in the UK, hepatitis C infection is the predominant condition that leads to the need for liver transplantation.

HIV, following a primary infection, replicates at a low level in **T4 lymphocytes** and other cells. The infected individual becomes **HIV antibody-positive** and excretes virus in a range of body fluids. The virus, in most individuals if untreated, will eventually progress to more rapid replication and cause the clinical syndrome known as **AIDS**. Approximately 10% of individuals infected with the virus to date, and after >20 years untreated viral infection have not progressed to AIDS. These individuals are important in our studies on HIV and its interaction with the host, **i.e.** why are they not progressing?

Hepatitis B and C and HIV are often referred to as persistent viruses.

Neoplastic growth

Introduction of genetic material (**viral oncogenes**) and the rearrangement or switching on of cellular genes (**cell oncogenes**) are events which can be mediated by viruses. In some situations this can contribute to the development of **neoplasia**. **Feline leukemia** virus causes a range of lymphoblastic leukemias in cats, indeed is the second highest cause of death in cats (the highest is road accidents!). Many viruses (e.g. **hepatitis B**, **hepatitis C**, **EBV**, **HSV** and **papilloma viruses)** have been implicated as being co-factors in the development of a range of malignancies.

Death

Whilst many viruses are fatal in distinct circumstances and others in a percentage of victims, some are always fatal. **Rabies**, **HIV** infection leading to **AIDS**, and a range of **neurological conditions** resulting from viral infections (e.g. subacute sclerosing panencephalitis) are examples.

While the incubation period of rabies is 30–90 days, after initial symptoms of the disease are manifest, patients die within 7–12 days. Following a bite, the virus may enter the peripheral nerves and then moves to the spinal cord and brain where it replicates. The virus leaves the cells and spreads to virtually all the tissues of the body including the salivary glands where it is shed in the saliva. The patient develops a variety of abnormalities including **hydrophobia** (aversion to water) **rigidity**, **photophobia** (aversion to light), focal or generalized convulsions and a variety of **autonomic disturbances**. Development of a flaccid paralysis and onset of coma precede death.

J9 Viruses and the immune system

Key Notes

The immune system	The whole armory of the immune system is harnessed to combat viral infections. Viruses thus react specifically with B cells and T cells. In addition, non-specific responses (e.g. interferon and natural killer cells), have a role in combating virus infection.
Virus antigens	During their replication, viruses code for a range of viral proteins (structural and non-structural) which are foreign to the host. Many induce protective antibody and cytotoxic T lymphocyte (CTL) responses. Others, while not protective, are important in diagnostic tests. The smallest part of a molecule with antigenicity is referred to as an epitope.
Antigen recognition	Antibodies secreted from B lymphocytes (assisted by T4 helper cells) specifically react with antigenic sites on virus particles and may neutralize virus infectivity. CTLs recognize antigens on and lyse infected cells. Antigen–antibody recognition is extremely specific and is determined by the detailed molecular structure of the antigen (epitope) inducing a unique response from a lymphocyte.
Primary response to viruses	Interferon and natural killer cells respond first in combating virus-infections. Specific CTL responses take 7–10 days and antibody slightly longer. The secondary response is almost immediate. The primary antibody response is usually an IgM response with smaller concentrations of IgG. The secondary response consists mainly of the more efficient IgG molecules.
Viruses and the immunocompromised	Individuals who are immunosuppressed suffer worse than most from viral infections where the clinical syndrome may be more severe than in an immunocompetent host. HIV causes immunosuppression. In addition, many individuals are born with genetic defects which leave them immunosuppressed; others are suppressed by medication.
Virus-induced immunopathology	The immune response mounted against an invading virus may itself be responsible for the damage and disease state which follows infection.
Evasion of the immune system	Many viruses have adapted to counteract the various host defense mechanisms.

Related topics

Virus proteins (J3)
Virus infection (J8)
Virus vaccines (J10)

The immune system

Virus infections are countered by the host **immune** system. The immune system consists of a wide range of morphologically and antigenically distinct white cells with different functional responses to viral invasion. Simplistically, the major type of white cells are **lymphocytes**, these being **T cells** or **B cells**. T cells (**cytotoxic (CD8)** or **helper (CD4)** cells) pass through the thymus during their maturation. B cells give rise to antibody-producing plasma cells. T4 helper cells help B cells react to highly specific antigenic stimuli (e.g. viral proteins). In addition to these specific responses the host mounts a series of non-specific responses, for example, production of **cytokines** (e.g. **interferon**), **natural killer cells, mucociliary responses** and **macrophages**. It is the outcome of the competition between the immune system and the progress of viral replication that determines the outcome of the infection. In **self-limiting** infections viruses are 'cleared' by the immune system. Good humoral immunity requires the induction of high levels of IgG antibodies.

Virus antigens

During the virus-replication cycle, the virus genome is expressed to give a number of proteins which are recognized by the host as being foreign. Many of these proteins induce a **protective response** in the host (important in **vaccine design**). In many viruses these proteins are laid down in the plasma membrane of the infected cells where they are recognized by both T cells and B cells (antibodies). It is usually the viral **structural proteins** (e.g. envelope and capsid proteins) that induce this protective antibody response.

Antigenic recognition

Antibodies, secreted by B cells, recognize whole viral protein antigens, usually in the fluid phase. The specific interaction between the antibody and the antigen has been likened to a 'hand-in-glove' reaction. Antibodies **neutralize** the infectivity of virus particles by agglutination, conformational change, etc. This, of course, reduces dramatically the number of host cells infected by virus. Cytotoxic T cells recognize virus antigens via a **cell-surface heterodimer** (the **T-cell receptor**) in association with the **major histocompatability complex** (MHC-class 1) which is present in almost all nucleated cells with the exception of neurons. Thus, cytotoxic T cells interact with the antigens present on the plasma membrane of infected cells, killing the cell before high concentrations of virus can be produced.

In order for antibody to be secreted by B cells, the specific antigen must be presented to T4 helper cells, this being achieved by macrophages and dendritic cells (**antigen-presenting cells**). In this instance the specific recognition of the antigenic site is made in association with the **MHC-class II** protein (present on selected cells in the body, e.g. macrophages and dendritic cells). Primary antigen recognition usually results in the formation of **memory B and T cells** which persist, making antigenic recognition much faster during re-infection.

Primary response to viruses

During a primary infection, the mass of antigen available to stimulate the immune response increases as the organism replicates. Initial responses are made by the host via **interferon** and **natural killer** (NK) cells. Interferons are host cell proteins with antiviral activity. It is α and β interferon that are recognized as having an antiviral effect. Once induced they have an effect against a wide range of viruses. Interferon acts in two stages: **induction** which results in the derepression of the interferon gene and the release of interferon which produces an **antiviral state** in other cells. NK cells exist as large granular lymphocytes in peripheral blood and mediate cell lysis. They are not antigen-specific and do not exhibit immunological memory. They appear to have an important role in the control of some virus infections.

The specific T-cell responses peak early (7–10 days) and decline within 3 weeks post-infection. The initial (primary) antibody response usually peaks later than the rise in cytotoxic T lymphocytes (CTL). Antibodies are often barely detectable during the acute stage of infection but increase dramatically 2–3 weeks post-infection. High levels of antibody may linger for several months. Indeed, where virus cannot be isolated from the host, a dramatic increase in **specific antibody** titer from the **acute** to the **convalescent** stage of infection is the main indicator of the cause of the infection. Upon reinfection the **secondary** response is virtually immediate in terms of CTL and antibody response. The mechanisms that combat viral infections are outlined in *Table 1*.

Table 1. Mechanisms to combat virus infection

Stage of infection	Immune response	Mechanism
Early infection (first line of defense)	Interferon, natural killer cells	Inhibits virus replication
	Soluble mucosal surface IgA antibody	Kills virus
Viremia (virus in the blood)	Antibodies, complement	Kills virus (neutralizes infectivity)
	Macrophages	Reduces spread
		Digests antibody complexes
Target organs	Antibody, complement	Lysis of infected cells
	Cytotoxic T cells, complement	Kills virus-infected cells, reducing virus replication
	Interferon	Inhibits virus replication

Viruses and the immunocompromised

Individuals with natural or artificially induced immunosuppression are at risk from a range of viral infections but particularly those which are responsible for latent infections (e.g. HSV, CMV, chicken pox virus) where the virus reappears to cause disease (**recrudescence**). Thus infants with **severe combined immunodeficiency (SCID)** develop recurrent infections early in life (e.g. rotavirus in the gut, which induces prolonged diarrhea). **Immune suppression** which follows transplant surgery may lead to, for example, generalized shingles (reoccurrence of chicken pox), CMV pneumonia or genital warts. Individuals with little or no antibody production (**hypogammaglobulinemia**) often excrete viruses for many years (e.g. poliovaccines from their gut).

Some viruses induce immunosuppression. Measles is known to slightly suppress T cell responses but the significant culprit is the **human immunodeficiency virus** (HIV). This virus infects T4 helper cells, predominantly by attachment to the **CD4 receptor** where it persists for several months or years before, usually, progressing to high concentrations of virus which seriously deplete the T4 cell population, thus resulting in serious immunosuppression and susceptibility to a wide range of pathogens (AIDS).

Virus-induced immunopathology

The immune response to viruses can result in damage to the host, either by the formation of immune complexes or by direct damage to infected cells.

Complexes can form in body fluids or on cell surfaces. **Chronic immune complex glomerulonephritis** can occur in mice infected neonatally with lymphocytic choriomeningitis virus. In adult mice, direct damage to infected and non-infected brain cells by a T-cell-dependent mechanism is responsible for most of the fatal tissue damage. This is also thought to be true for liver damage in chronic active hepatitis in man.

Potentially fatal hemorrhagic fever can result when an individual previously infected with Dengue virus is later infected with a different strain of Dengue, the pathology resulting from the immune response to the second strain. Viruses may also evoke autoimmunity, probably via **molecular mimicry** (production of an antigen which shares conserved sequences with a host cell protein). As a result, antibodies or T cells are produced which also react with host proteins.

Evasion of the immune system

During coevolution with their hosts, viruses have adopted several measures to counteract the various host defense mechanisms. These include inhibition of peptide processing, resistance to serum inhibitors, resistance to or poor inducers of interferon, inhibition of phagocytes, suppression of immune responses, low capacity to evoke an immune response, alteration of lymphocyte traffic, effects on cellular modulators, depression of complement activity, resistance to the immune response and antigenic variation. Thus, the high mutation rate of many RNA viruses, e.g. HIV, creates mutants that are no longer recognized by antibody molecules. Influenza virus undergoes constant change in the environment (so-called antigenic 'drift' and 'shift' of its hemagglutinin molecule) leading to the appearance of different strains of virus, capable of replication in previously immune hosts. Most of the herpesviruses escape the immune system whilst latent in their host and many during replication (e.g. Epstein-Barr virus, EBV) may specifically inhibit the intracellular transport mechanism by which viral peptide fragments are presented to the cell-mediated immune system. Adenovirus, poxvirus and cytomegalovirus interfere with peptide presentation by downregulating the expression of the MHC class I protein. Vaccinia virus, reovirus, adenoviruses and EBV are examples of viruses capable of inhibiting the action of interferon, usually by interfering with the action of a cellular protein kinase enzyme. Both herpes simplex virus (HSV) and EBV structural proteins interfere with the complement cascade mechanism. HSV, in addition, has a structural protein, present on infected cell surfaces, which has Fc-receptor activity, thus preventing complement fixation or opsonization by phagocytes.

The ability of viruses to evolve more rapidly than their host means that virus replication and its subsequent disease will continue to threaten the health of the world's population. The evolutionary process has also led to the emergence of new viruses capable of replicating in different species, including humans.

J10 VIRUS VACCINES

Key Notes

Vaccination

Vaccination is the use of a vaccine to stimulate the immune response to protect against challenge by wild-type virus. A good vaccine is designed so as to mimic the host response to infection seen with the wild-type infection and to protect against disease but to have minimal side effects. Vaccines are either live (attenuated) virus or inactivated (killed) virus. Sub-virion protein or DNA vaccines are being developed at present, although most are at an experimental stage.

Live (attenuated) vaccines

Produced by continued passage in tissue culture or by genetic manipulation (e.g. gene deletion). These vaccines are effective in stimulating the full range of immune responses. They are the preferred type of vaccine. Examples are measles, mumps and German measles vaccines.

Inactivated (dead) vaccines

Produced by chemical inactivation of wild-type virulent virus, they are generally less effective than live vaccines but do give significant protection against virus challenge. Examples are influenza and hepatitis A vaccines.

'Sub-virion' vaccines

These are mainly experimental (apart from hepatitis B surface antigen vaccine) and are composed of 'subunits' of the virion. They are used when other vaccines are ineffective or technologically not possible to manufacture.

DNA vaccines

These are totally experimental vaccines and consist of injecting a host with a plasmid containing DNA which encodes antigenic portions of the virus.

Related topics

Virus proteins (J3)
Virus infection (J8)
Viruses and the immune system (J9)

Vaccination

The early experiments of **Jenner** with cowpox were the basis upon which many of today's vaccination (*vacca* is Latin for cow) programs were derived. Jenner introduced the mild **cowpox** virus into some small boys whom he later challenged with **smallpox** virus. The boys were protected against the disease. Cowpox shares a number of antigens with smallpox but does not cause disease in humans. Hence, cowpox replication had induced an immune response which, upon subsequent challenge with smallpox, was sufficient to prevent the smallpox virus undergoing any significant infection. A hybrid cowpox virus, of unknown origin, vaccinia virus, was used in a vaccination program which in 1977 succeeded in eradicating smallpox from the world. Vaccines are basically of two types, live (**attenuated**) or inactivated (**dead**) vaccines. More recently, vaccines which contain selected virus proteins (e.g. hepatitis B vaccine) have been developed. Many, however, are only at the experimental stage of development. Even more recently have been experiments using DNA as a source of vaccination (see later). Vaccines are administered by a number of routes – **intramuscular**, **intradermal**, **subcutaneously**, **intranasal** or **oral**.

Live (attenuated) vaccines

It is generally accepted that live vaccines are the preferred vaccines and examples are shown in *Table 1*. Initially, they were derived by continued passage of virulent wild-type virus in tissue culture, this process selecting for **avirulent viruses** which will replicate *in vivo* but not cause disease. Such viruses could, however, revert back to wild-type phenotype and indeed in some cases during the early years of their use did. This rather empirical approach has now been replaced by the genetic 'manufacture' of viruses with deleted or mutagenized genes. The risk of reversion with these vaccines is minimal. However, many (e.g. Sabin polio vaccine) are the original mutants created by tissue culture passage over 40 years ago.

Live vaccines have distinct advantages over killed. They require small amounts of input virus (which, however, must replicate) they can induce local immunity (i.e. mucosal IgA), they may be given by the natural route of infection (e.g. Sabin polio vaccine which is given orally) and they are usually cheaper. The drawbacks to their use include reversion to virulence, ineffectiveness if not kept live (i.e. refrigerated or freeze-dried), ineffectiveness in individuals with, for example, infections of the gut, where the polio virus vaccine will not grow, contamination with adventitious agents and limited use in immune-suppressed patients. Recent experiments have shown that, for poliovirus type 3, reversion to a virulent form involves only two amino acid changes.

The World Health Organization has an ambitious vaccination program which seeks to get vaccines to most children of the world and to eradicate many viruses in the next decade. Polio has been eradicated from virtually all 'developed' countries, and by using National Immunization Days many millions of people are being made immune to polio and other viruses. It is hoped to eradicate polio globally by 2005, at which point the vaccination program will probably cease.

Table 1. Currently available live attenuated viral vaccines

Vaccine	Vaccine type	Uses
Oral polio	Attenuated trivalent	Routine childhood immunization; mass campaigns
Measles	Attenuated (Schwarz, Moraten, others)	Routine childhood immunization; mass campaigns
Rubella	Attenuated (RA 27/3)	Routine childhood immunization; adolescent girls; susceptible women of childbearing age
Mumps	Attenuated (Urabe or Jeryl Lynn)	Routine childhood immunization
Measles, mumps, rubella (MMR)	Attenuated	Routine childhood immunization (1 or 2 doses)
Varicella	Attenuated (Oka)	Routine childhood immunization (USA); vaccination of susceptible people
Yellow fever	Attenuated (17D)	Routine immunization or mass vaccination in endemic areas; vaccination of travellers to endemic areas

Inactivated (dead) vaccines

The early polio vaccine (**Salk vaccine**) still used in many countries typifies the approach to dead vaccination; other examples are shown in *Table 2*. High titers of wild-type virus are grown in tissue culture and inactivated chemically by the use of, for example, β-propiolactone or formaldehyde. The killed virus is administered parenterally (subcutaneously or intradermally) in order to stimulate an immune response. With no subsequent virus replication (as seen with live vaccines) the immune response follows that of a 'primary response' to an inert antigen. The necessary high levels of IgG are therefore stimulated only by multiple injections.

Killed vaccines have the advantage of non-reversion to virulence, are not 'inactivated' in the tropics, can be administered to immune-suppressed patients and are usually not affected by 'interference' from other pathogens (important in many developed countries). They are, however, expensive, do not stimulate local immunity, are considered dangerous to manufacture (prior to inactivation) and require multiple doses. The recently derived **hepatitis A vaccine** is a dead vaccine, primarily because it was technically impossible to develop an attenuated virus capable of liver growth but no disease.

It is considered that the use of dead vaccines alone, which often allow some wild-type virus replication at local sites (e.g. polio) will not be sufficient to totally eradicate viruses from the world.

'Sub-virion' vaccines

Attempts to produce these vaccines, often referred to as **subunit** vaccines have been prompted by a number of factors. Many viruses do not grow in tissue culture (e.g. hepatitis B and C); some are considered by some scientists to be too dangerous for use as live or killed vaccines (e.g. HIV); the vaccine may become latent in the body and become reactivated (e.g. HSV); the vaccine may be poorly effective and have side effects (e.g. influenza virus vaccines).

Table 2. Currently available inactivated viral vaccines

Vaccine	Vaccine type	Uses
Inactivated polio	Killed whole virus	Immunization of immunocompromised people; universal childhood immunization (some developed countries)
Hepatitis B	Purified Hepatitis B surface antigen produced by recombinant yeast, mammalian cells or from plasma	Routine childhood or adolescent immunization: immunization of high-risk adults; post-exposure immunization
Influenza	Killed whole or split virus	Vaccination of high-risk individuals or the elderly
Rabies	Killed whole virus	Post-exposure vaccination; pre-exposure; veterinarians/travelers
Hepatitis A	Killed whole virus	Pre-exposure vaccination: high-risk people and travelers
Japanese B encephalitis	Killed whole virus	Pre-exposure vaccination of travelers to endemic areas; routine or mass vaccination in endemic areas
Tick-borne encephalitis	Killed whole virus	Pre-exposure vaccination of travelers to endemic areas; (?)routine or mass vaccination in endemic areas

The basis of the approach is to develop a vaccine product which, when injected into the host, will induce protective immunity. The virus protein of choice for such a vaccine is usually a capsid or envelope protein, often the protein which binds to the cell receptor (e.g. gp120 of HIV, hemagglutinin of influenza). These proteins can be extracted from the virion by chemical treatment (e.g. influenza hemagglutinin), concentrated from the plasma of infected patients and inactivated (e.g. hepatitis B surface antigen), engineered by recombinant DNA technology (e.g. hepatitis B surface antigen) or synthesized as a peptide (still experimental, but examples include foot and mouth disease virus).

These vaccines may suffer from some of the drawbacks of dead whole virion vaccines in as much as they do not replicate in the host. They are not good inducers of immunity and presentation of virus antigens in such a way as to optimally stimulate the immune system is a very important area of research. Antigens may be injected with adjuvants (immune stimulating chemicals, e.g. alum) or as chimeric proteins linked to e.g. host interleukins. Another approach under study at present is to introduce the selected viral gene into a virus vector (e.g. vaccinia virus). The vaccination virus is then used to vaccinate the host, the virus replicating and expressing a number of antigens (including the cloned gene product) which in turn stimulate immunity.

The only subunit vaccine in general use at present is the hepatitis B surface antigen which has been genetically engineered into and expressed by yeast cells. The vaccine is used routinely to induce protection against hepatitis B in members of the medical profession.

DNA vaccines

These appear to be an exciting innovative approach to vaccination. DNA-encoding antigenic portions of viruses (e.g. protective antigens) are inserted into a plasmid which can be injected into the host as 'naked' DNA free of protein or nucleoprotein complexes. Host cells take up the DNA and may express the virally encoded proteins. Such vaccines stimulate excellent cytotoxic T-cell responses as the plasmid DNA provides internal adjuvant action through its 'immunostimulating sequences', i.e. unmethylated CpG. Experimental influenza vaccines in animals have proved effective in challenge experiments and clinical trials are underway with e.g. HIV.

J11 ANTIVIRAL CHEMOTHERAPY

Key Notes

Historical perspective

Viruses utilize the whole machinery of the cell for their replication and hence most antiviral compounds are also toxic to normal cells. However, recent knowledge of viral gene products has allowed a more selective approach to drug design.

Drug targeting

Our knowledge of the virus growth cycle and those various stages of replication which require virus-specific proteins has allowed drug targeting by pharmaceutical companies.

Drug design

Determining the three-dimensional structure of, for example, viral enzymes has allowed drugs to be designed which react specifically with, and inhibit parts of, viral molecules.

Effective concentration 50 and toxicity

Determining the antiviral effectiveness of a compound is achieved usually in tissue culture by assaying its ability to reduce viral infectivity and is expressed as the EC_{50}. Toxicity is assessed and expressed as selective toxicity.

Clinical trials

Phases I–IV of clinical trials assess the pharmacokinetics, pharmacology, metabolism and clinical efficacy of the compound.

Modes of action

Most compounds are nucleoside analogs, others inhibit, for example, proteases, and egress of virus from the cell. Aciclovir, probably the most successful compound to date, is specifically phosphorylated by viral enzymes before being selectively incorporated into the growing virus DNA chain, acting as a chain terminator.

Drug resistance

Viruses can mutate to form drug-resistant strains, which is a common problem related to antiviral chemotherapy. Combinatorial approaches and strict adherence to drug-taking regimes can help to prevent this.

Related topics Virus replication (J7) Virus infection (J8)

Historical perspective

As obligate intracellular parasites with very restricted genetic-coding capacity viruses rely heavily on utilizing the metabolic machinery of the cell for their replication. It was therefore considered by many that the concept of **selective toxicity** was an unattainable goal and that interference with viral replication would always bring unacceptable damage to the host. Viruses by definition are resistant to the action of antibiotics.

In the last decade a series of discoveries based on our increasing knowledge of the viral replication cycle and the specific gene products encoded by viruses have resulted in a number of successful antiviral agents being prescribed.

However, some still have a level of toxicity which, while acceptable for some diseases (e.g. AIDS), would not be tolerated for less severe diseases.

Chemotherapeutic agents fall into three broad groups. **Virucides** directly inactivate viruses (e.g. detergents, solvents), **antivirals** strive to inhibit viral multiplication but not host-cell metabolism and **immunomodulating agents** attempt to enhance the immune response against viruses (e.g. administration of interleukins). In this chapter **antivirals** will be discussed.

Drug targeting

Viral replication requires the virus to pass through a number of stages which are common to all viruses: **attachment and penetration**, **uncoating of the nucleic acid**, **transcription and translation**, **replication of nucleic acids** and **release of mature progeny**. This is demonstrated in diagrammatic fashion for the growth cycle of HIV in *Figure 1*.

To date, most antivirals are directed at inhibiting one of the steps shown in *Figure 1*, although the most common target is interference with nucleic acid metabolism by using many **nucleoside analogs**. Several compounds that interfere with influenza effect the disassembly of particles by blocking a virus protein complex that acts as an ion channel. Many viruses produce a virus-specific protease to process their proteins and this has been targeted in the search to develop an inhibitor of HIV. A series of inhibitors of picornaviruses act by directly binding to the virion capsid, blocking the interaction between the virion and the receptor on the cell surface that facilitates its entry and disassembly. A recent anti-influenza drug targets the virus neuraminidase and inhibits egress of virus from the cell.

Drug design

In the early days, most new antiviral compounds were discovered in an **empirical** fashion. Chemists would produce a wide range of, for example, nucleoside analogs, originally in many cases as anti-cancer drugs, which were tested in tissue culture to determine their antiviral activities. However, with our knowledge of the molecular basis of virus replication, the availability of the entire **nucleotide sequences** of virus genomes and the **three-dimensional protein structure** derived from **X-ray diffraction analysis**, compounds can be designed to interact with specific targets involved in the replicative cycle of viruses. In practice the most useful targets are **virus-induced enzymes** which often have properties different to those of the counterpart enzymes induced by the host cell (e.g. **thymidine kinase**, **DNA polymerase**, **reverse transcriptase** and **protease)**. When the enzyme can be crystalized and its three-dimensional structure determined the synthesis of appropriate molecules to interact with particular sites on that enzyme can be determined. However, future antiviral compounds should lack toxicity and should be able to be produced cheaply from available precursors. The chance of a wide range of new products appearing is therefore remote.

Effective concentration 50 and toxicity

Antiviral compounds are usually assessed at an early stage for their ability to interfere with viral growth in tissue culture. The virus is assayed by, for example, $TCID_{50}$ or plaque production with and without the drug. The percentage of reduction in infectivity is plotted against $\log_{10}$ of the drug concentration. The concentration of compound that reduces virus titer by 50% is measured and expressed as the effective dose 50 concentration (ED_{50}) (*Fig. 2)*. Tissue culture can also be used to assess the toxicity of compounds, although this is often tested in animals and humans. Animals play a vital role in the study

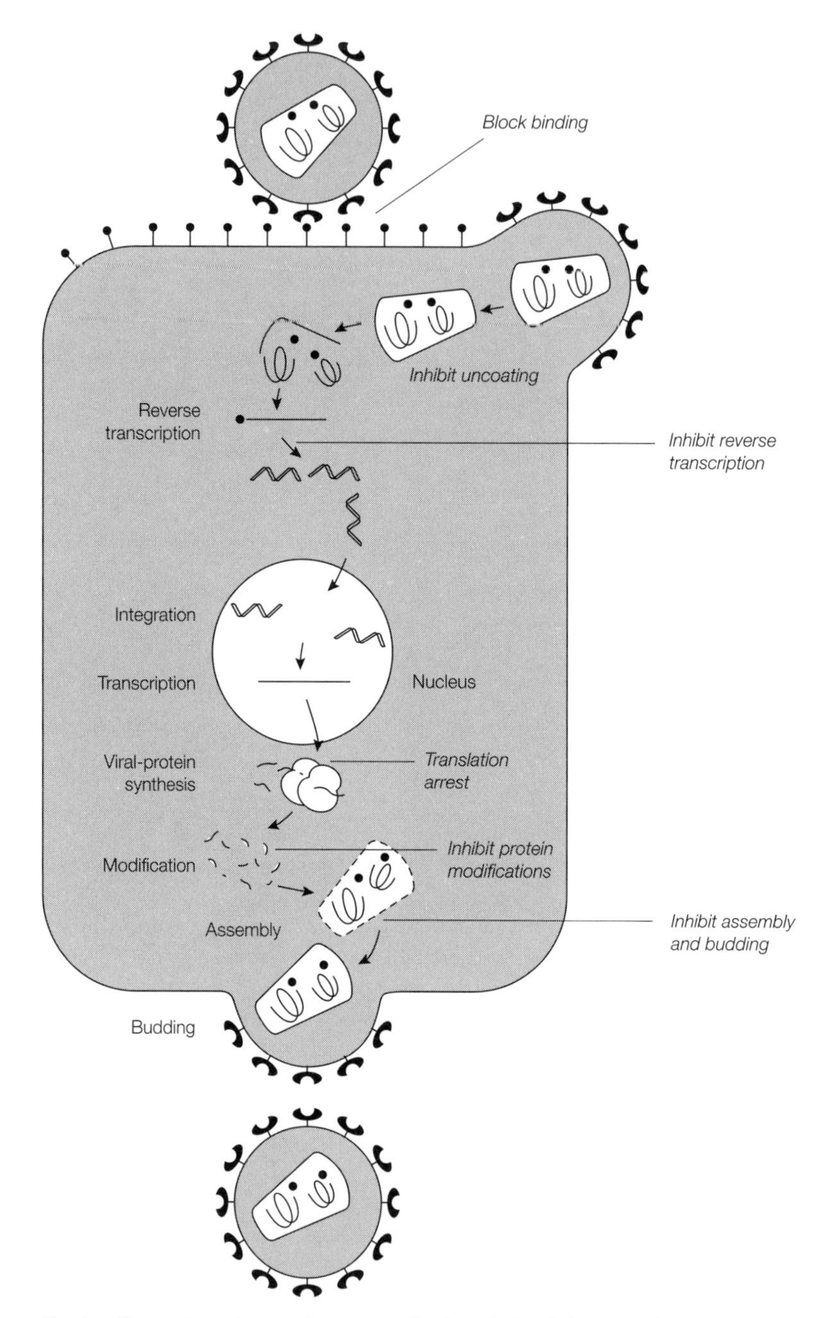

Fig. 1. The various stages of a virus replication. Antiviral drugs can be used to inhibit any of these steps. Redrawn from Yarchoan, R. et al., Scientific American, *vol. 259, no. 4, p. 90, 1988.*

of toxicity and a number of statutory tests are carried out to determine the risks and side effects before new compounds enter clinical trials. The ratio of the 50% toxic concentration to the 50% inhibitor concentration is termed the **selective index**. If this is close to unity the compound is toxic. A high selective toxicity suggests a useful compound. A **benefit/risk** ratio is also considered when using

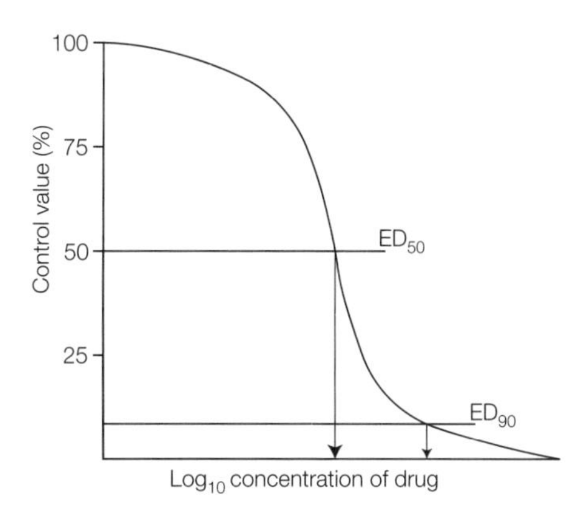

Fig. 2. Determination of effective dose (ED_{50}, ED_{90}) for an antiviral drug in tissue culture.

a compound, that is, side effects will be tolerated more if the risk from the disease is high (e.g. AIDS).

Clinical trials

Phase I involves administering the drug to healthy human volunteers where studies on the pharmacokinetics, pharmacology and metabolism of the compound are monitored.

In **Phase II** the compound is administered to diseased patients, and data similar to those above are obtained as the metabolism may be different in the diseased patient. Usually <100 patients are used for these trials.

In **Phase III** the drug is usually tested for its clinical efficacy by comparing with **placebos** or existing drugs. The main aim is to determine the **benefit/risk ratio** for the therapeutic course, this requiring 100–1000 patients.

Phase IV studies are usually conducted following marketing approval and increased experience of treating patients, providing more information on safety and efficacy.

Unfortunately, most potential antiviral compounds, although good inhibitors of viral growth in tissue culture, fail to pass successfully through all of the phases of clinical studies.

Modes of action

Table 1 highlights the modes of action of selected antiviral compounds. Perhaps the most successful antiviral compound is the nucleoside analog **acyclovir** (ACV – now named **aciclovir**) with more than 40 million patients treated. The drug inhibits the replication of HSV, and has been administered prophylactically for over 10 years to individuals, with no ill effects, to suppress recurrences of genital herpes. The compound is related to the natural nucleoside guanosine. To become active the compound must be converted into the triphosphate form by three phosphorylation steps (i.e. ACV mono-, di- and then triphosphate). These steps are not carried out in normal uninfected cells and hence ACV is a poor substrate for cellular enzymes. In contrast, the **HSV thymidine kinase** can convert ACV to ACV-monophosphate (ACV-MP). Cellular enzymes convert ACV-MP to the triphosphate (ACV-TP) where it enters the nucleoside pool and competes with guanosine triphosphate as a substrate for **HSV DNApolymerase**. Cellular DNA polymerases are much less sensitive to inhibition. As the ACV residue is linked to the growing chain of viral DNA it forms a **chain terminator**

Table 1. Mechanism of action of antiviral drugs

Drug	Mechanism(s) of action	Virus
Aciclovir	Nucleoside analog, inhibits nucleic acid synthesis: active form produced by viral thymidine kinase	HSV, VZV
Ribavirin	Nucleoside analog, inhibits nucleic acid synthesis, possibly by inhibiting viral RNA polymerase	RSV, LFV Hep C
Amantadine Rimantadine	Inhibit virus coating, maturation and egress from cell	Influenza
Lamivudine	Reverse transcriptase inhibitor – inhibits replication of virus	Hep B
Relenza	Inhibits neuraminidase and blocks release of virus	Influenza
Ganciclovir	Nucleoside analog, blocks nucleic acid synthesis by inhibiting viral thymidine kinase and other enzymes	CMV
Azidothymidine Dideoxyinosine Dideoxycytidine	Nucleoside analogs, inhibit nucleic acid synthesis by inhibition of reverse transcriptase	HIV
Indinavir	Anti-protease inhibitor, prevents cleavage of virus proteins	HIV
Foscarnet	Blocks protein synthesis by inhibition of RNA and DNA	CMV, HSV
Idoxuridine	Nucleoside analog, inhibits DNA synthesis	HSV
Vidarabine	Nucleoside analog, blocks DNA synthesis by inhibiting DNA polymerases	
Inteferon	Renders normal cells 'immune' to infection by interfering with virus transcription	Hep C

HSV, herpes simplex; VZV, Varicella–Zoster; RSV, respiratory syncytial; CMV, cytomegalovirus; HIV, human immunodeficiency virus.

as there is no 3′-OH group on the ACV sugar moiety to link the next residue to the growing chain of virus DNA.

Interferon is a natural human product (a cytokine) which acts on the surface of normal cells to render them immune to virus replication. The compound used to treat e.g. hepatitis C infections, is, however, toxic, the side effects mimicking those of an influenza infection. A new version, polyethylene glycol interferon, is more effective and less toxic and is administered subcutaneously once weekly for one year. Effective anti-HIV treatments are those involving combinations of drugs, e.g. two nucleoside analogs plus an anti-protease inhibitor. Such regimes have been highly effective in reducing virus loads and raising CD4 counts in HIV-infected individuals.

Drug resistance

Viruses adapt to become resistant to anti-viral drugs with examples in all areas of chemotherapy. Mutants resistant to Azidothymidine occur with a very high frequency, whereas combinatorial approaches appear to reduce the risk of viral resistance. Most important are measures which ensure that patients adhere to the regime of drug-taking which in some cases is highly complex and involves a vast number of tablets.

J12 Plant viruses

Key Notes

Historical aspects
Plant viruses were discovered over a century ago and have featured greatly in contributing to our knowledge of virus structure (e.g. tobacco mosaic virus, turnip yellow mosaic virus and tomato bushy stunt virus).

Plant viruses
Plant viruses are diverse in size, shape and biochemistry and are present in many virus families which include animal viruses (e.g. rhabdoviridae).

Disease and pathology
Viruses are singly responsible for grave economic loss estimated at being over $70 billion worldwide. They cause necrosis, wilting, mosaic formation and other damage, which reduces yields and value of crops, etc.

Transmission, infection and systemic spread
Plant viruses are transmitted mainly by invertebrate animals (e.g. aphids, leaf hoppers) or through infected seeds or 'manually' by contaminated implements. They gain entry by penetrating cuticles of plant cells and need to spread systemically to cause disease (via plasmodesmata).

Control of plant virus disease
Infected plants are virtually impossible to 'cure'. Control is by use of naturally resistant plant varieties or more recently genetically manufactured resistant varieties and by eradication of the transmission vector.

Viroids
These are 'virus-like' infectious agents composed solely of RNA with a complex tertiary structure (e.g. potato spindle tuber viroid).

Related topics
Virus structure (J1)
Virus taxonomy (J2)
Virus proteins (J3)
Virus nucleic acids (J4)
Virus replication (J7)

Historical aspects

Tobacco mosaic virus (TMV) has figured predominantly in early studies on virus structure and replication. The virus, which causes mosaicing of the leaves of the tobacco plant, was first described as being a 'contaguim virus fluidum' in 1898. In 1935 the virus was crystalized and shown to be a 'globular protein'. TMV was the first virus to be observed under the electron microscope, and the first to demonstrate the intrinsic infectivity of extracted RNA and to be assembled *in vitro* from purified preparations of viral RNA and coat protein molecules.

It was studies on turnip yellow mosaic virus and tomato bushy stunt virus that revealed the morphological details of icosahedral viruses.

Plant viruses

There are over 1000 plant viruses which have been classified by the ICTV. The virus usually receives its name by a combination of the host plant and the type of disease produced (e.g. **tobacco mosaic**, **turnip yellow mosaic**, **turnip bushy stunt**, **cauliflower mosaic**, **tomato spotted wilt**).

Plant viruses are diverse in their morphology, nucleic acid composition and replication patterns. It is beyond the scope of this book to detail all viruses but *Figure 1* highlights the various morphologies and groupings of plant viruses.

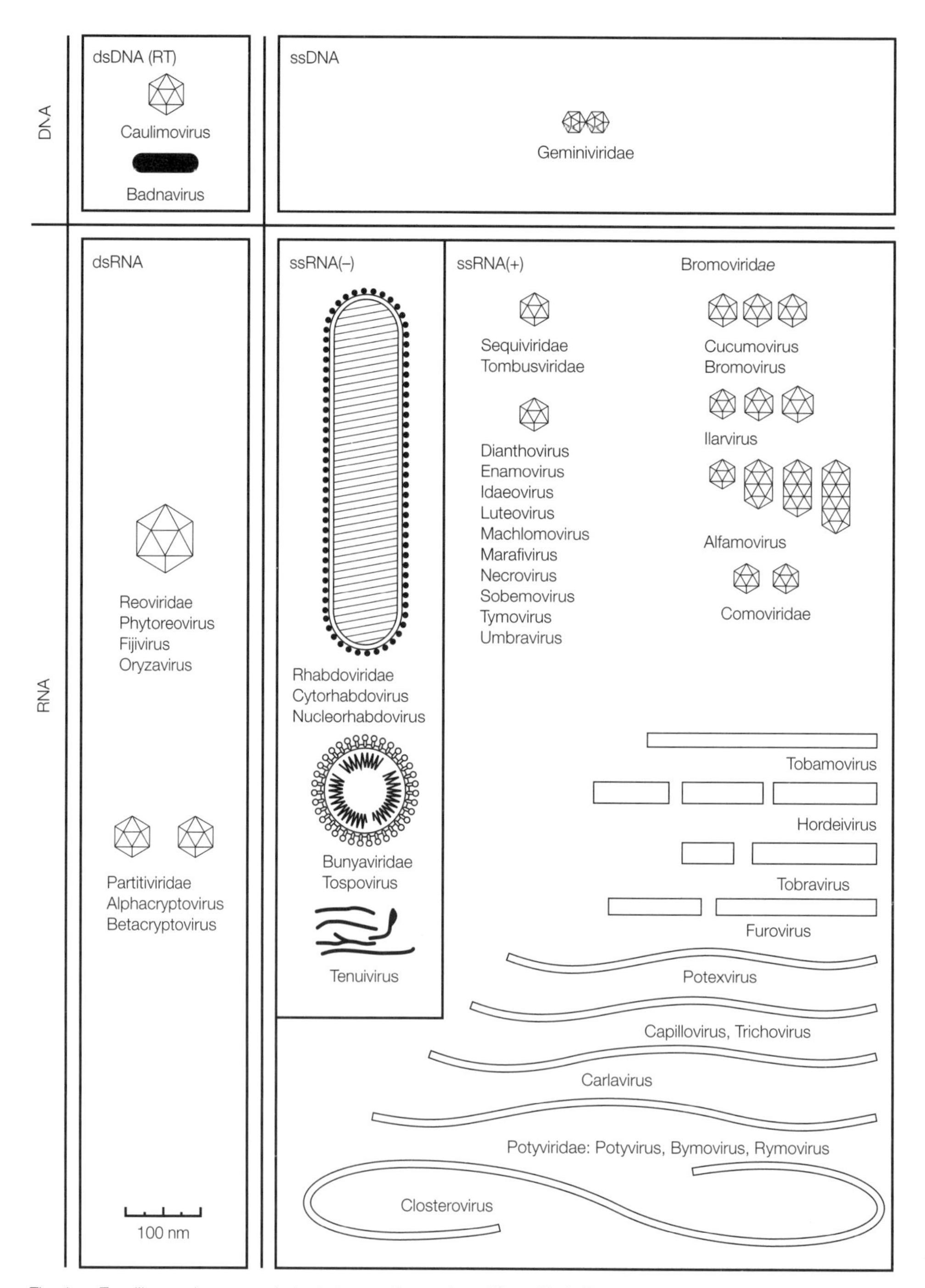

Fig. 1. Families and genera of plant viruses. Reproduced from Sixth Report of the International Committee on Taxonomy of Viruses, *Springer-Verlag.*

They may be **dsDNA** (Badnavirus, e.g. rice tungrobacilliform virus), **ssDNA** (geminivirus, e.g. maize streak virus), **dsRNA** (reoviruses, e.g. clover wound tumor virus), **RNA viruses with ambi sense RNA** (rhabdoviridae, e.g. potato yellow dwarf virus), **positive-sense ssRNA viruses** with monopartite genomes (e.g. Tobamoviridae tobacco mosaic virus), **bipartite ssRNA** viruses (e.g. Comoviridae cowpamosaic virus) and **tripartite ssRNA** viruses (e.g. Cucumoviridae cucumber mosaic virus).

Disease and pathology

The impact of viruses globally on the world's plants is enormous and estimates of $70 billion have been quoted as being the monetary loss per year as a result of virus infections. It appears that crops of all types can be affected. One disease, **tristenza**, has wiped out or made non-productive more than 50 million citrus trees worldwide. Swollen shoot disease has destroyed more than 200 million cocoa trees in parts of western Africa. Rice plants, barley, vegetable and field crops all succumb to infections. Many perennial plants are chronically infected with virus. The morphological outcome of a plant infection can be seen as mosaic formation, yellowing, molting or other color disfiguration, stunting, wilting and necrosis. These conditions result from cell and tissue damage which will significantly reduce the yields and commercial value of crops. Thus, photosynthesis, respiration, nutrient availability and hormonal regulation of growth can all be affected. How plant viruses cause such changes remains for the main part a mystery. Plant viruses code for few proteins, but as is the case in many bacteriophage proteins, they are multifunctional and many of their interactions with host proteins, although not part of the virus-replication process, result in pathology.

Transmission, infection and systemic spread

Most plant viruses depend on some kind of 'agent' for their dissemination in the natural setting. Invertebrate animals are the most important of these agents, examples being **aphids**, **leaf hoppers**, **mealybugs**, **whiteflies**, **thrips**, **mites** and **soil nemotodes**. Some viruses are passed directly to progeny plants through infected pollen or seeds and others are present in tubers, bulbs and cuttings. A few viruses can be transmitted as a result of contact with **contaminated implements** (e.g. hoes) or by direct contact between neighboring plants. The stem and leaf surfaces of plants can in many ways be compared to the human skin. As such they do not have receptors for virus attachment and have a protective function. For plant viruses to be infective they must penetrate this barrier and gain access to the metabolic machinery of the plant cell. It is thus through wounds that viruses enter plant tissue, either via a vector or mechanically by manual damage. In the laboratory, infection is achieved by rubbing the leaf with a cloth soaked in a virus suspension, using a fine mesh abrasive method to create the necessary wounds. In effect this procedure mimics the plaque assay described for animal viruses, but is not as accurate! (*Fig. 2*). On entry the virus particle is uncoated and goes through a replication cycle similar to that of animal viruses, this cycle varying from virus to virus. Many plant viruses code for a **movement protein** in addition to enzymes and structural proteins. *Figure 3* represents a diagram of the cycle of TMV. Plant viruses rarely cause significant damage and disease unless they become **systematically** distributed throughout the plant. Failure for this to happen explains why some plants are resistant to particular virus infections. Movement is facilitated by a virus protein which is involved in the transport of virus or virus nucleic acid through the fine pores (**plasmodesmata**) in the cell walls that interconnect plant cells. Movement also occurs

Fig. 2. Focal assay of tobacco mosaic virus on the leaf of a plant. Reproduced from Dimmock and Primrose, Introduction to Modern Virology, *4th edn, 1994, with permission from Blackwell Science Ltd.*

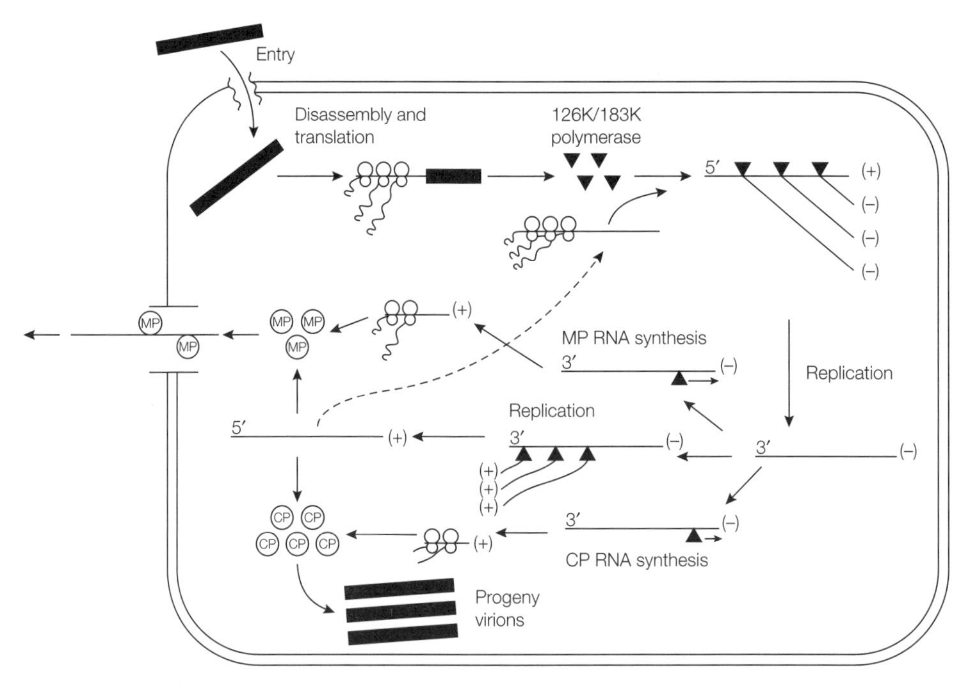

Fig. 3. Diagram of stages of TMV infection. All the events shown are presumed to occur in the cytoplasm of infected cells. MP, movement protein; CP, coat protein. Redrawn from Fields et al. *(eds),* Fundamental Virology, *3rd edn, 1996, with permission from Lippincott-Raven Publishers.*

through the companion and sieve cells of the phloem, this being facilitated by the viral coat protein.

Control of plant virus disease

Once infected, it is almost impossible to 'clear up' a virus infection in a plant by use of antiviral agents. Likewise plants do not mount immune responses. Until recently it was a reliance on **horticultural practice** (use of virus-free seeds, eradicating vectors, choosing the time of planting, etc.) that reduced the extent of virus infection. More recently **genetic engineering** has allowed the construction of plants which show **natural resistance** to virus infections. Other scientific measures made use of the fact that previous infections with a non-pathogenic virus appeared to protect the plant from an infection by a more pathogenic strain (but not because of immunity!). This phenomenon has been extended by transforming the plant with the coat protein gene of TMV and then challenging the plant with infectious TMV. The resulting **transgenic** plant was protected from challenge by the wild-type TMV. The phenomenon is referred to as **pathogen-derived resistance** to virus diseases. How and why this phenomenon works is not clearly understood, although in practice the method is receiving a great deal of attention from plant breeders and molecular biologists.

Viroids

Viroids are small, **unencapsidated ssRNA molecules** – the smallest known pathogens of plants. There are some 25 viroids which vary in nucleotide sequence. The first to be examined in detail was **potato spindle tuber viroid** (PSTVd) responsible for significant loss to the potato industry.

The RNA is a covalently closed circle and ranges in size from 246 to 357 nucleotides in length. The RNA has a complex secondary and tertiary structure which gives it a rod-like shape and **resistance to nucleases**. The RNA does not have a characteristic open reading frame and so does not act as mRNA. How this piece of RNA causes disease is largely unknown. It replicates in plant cells with the aid of host-cell enzymes (e.g. RNA polymerase II).

Viroids are spread by plant propagation (e.g. cuttings and tubers) through seeds and by manual mishandling with contaminated implements.

J13 Prions and transmissible spongiform encephalopathies

Key Notes

The agent

Prions (proteinaceous infectious particles) are not conventional infectious agents. They have no nucleic acid but consist entirely of protein. The infectious form (PrP^{sc}) arises from a modified cellular protein (PrP^{c}). Prions induce fatal transmissible spongiform encephalopathies (TSEs).

Pathogenesis of TSEs

TSEs are fatal chronic degenerative diseases with a very specific underlying pathology denoted by the laying down of protein deposits as plaques or fibrils (known as amyloids) in the kidneys, spleen, liver and significantly the brain. Post-mortem examination of brain tissue reveals a spongy appearance, which reflects the formation of holes in the tissue due to the cytotoxic effects of such deposits. TSEs have a long incubation period and are invariably fatal.

Molecular nature of prions

Evidence that prions are composed entirely of protein comes from experiments designed to selectively inactivate proteins or nucleic acids. Prions are resistant to heat (135°C for 18 minutes), ultraviolet light, ionizing radiation, DNAse and RNAse and Zn^{2+} catalyzed hydrolysis, treatments that selectively destroy DNA and/or RNA. However, they are sensitive to urea, SDS, phenol and other protein-denaturing chemicals. Evidence suggests they are proteins of approx. 254 amino acids in length. They 'replicate' by converting the cellular PrP (PrP^{c}) to the infectious (PrP^{sc}) form.

Animal TSEs

Scrapie, the most extensively studied, causes a neurological disease in sheep. Probably transmitted orally (e.g. by placentas) sheep have different genetic susceptibilities. Bovine spongiform encephalopathy (BSE) appeared in 1986 as a result of feeding cattle with scrapie-contaminated food-stuff. BSE ('mad cow' disease) appears to have infected humans.

Human TSEs

The neurological condition of Kuru resulted from the cannibalistic ritual of the Fore people of New Guinea. The disease, orally transmitted, has an incubation period of up to 30 years. Creutzfeldt-Jakob disease is sporadic, iatrogenic and familial and affects 1 in 10^6 individuals worldwide. New variant CJD (vCJD) appears to be different and is probably caused by the BSE agent.

The agent

Prions are not viruses and appear to be a new class of infectious agents that lead to chronic progressive infections of the nervous system, inducing common pathological effects, the results of which, after a long incubation period of perhaps several years, are invariably fatal. The clinical syndromes for which they are responsible are known collectively as **transmissible spongiform encephalopathies** (TSEs). Examples, in animals, include **scrapie** in sheep, **transmissible mink encephalopathies** (TME**), feline spongiform encephalopathy** (FSE) and **bovine spongiform encephalopathy** (BSE or 'mad cow' disease). There are four forms of human TSE – **Creutzfeldt-Jakob disease** (CJD), **fatal familial insomnia** (FFI – an inherited disease), **Gerstmann-Strausster-Scheinker disease** (GSS) and **Kuru**.

There has been much speculation as to the molecular nature of the infectious agents responsible for these disease conditions, but in 1972 Stanley Prusiner penned the term **prion (proteinaceous infectious particle)**. His hypothesis was that these agents are totally free of nucleic acid and consist solely of protein, a hypothesis for which much supportive evidence has accumulated. In 1997 Prusiner was awarded the Nobel Prize for his studies.

Pathogenesis of TSEs

Whilst the various conditions induced by prions have subtle differences there are many features in common. All have long incubation periods, with the agent replicating in a number of tissues, e.g. spleen, liver, before appearing in high concentrations in the brain and CNS towards the terminal stages of the disease. The agents are responsible for severe degeneration of the brain and spinal cord and, once clinical signs appear, the condition is invariably fatal. Pathologically the disease is characterized by the appearance of abnormal protein deposits **(amyloids)**, in various tissues, e.g. kidney, spleen, liver and brain. Amyloids arise from the accumulation of various proteins, which take the form of **plaques** or **fibrils**. Amyloidosis is also a characteristic feature of e.g. Alzheimer's disease, although this condition is not transmissible. The protein deposits are cytotoxic and are responsible for the resultant pathology, namely the sponge-like appearance of the brain where small holes can be visualized by microscopy in thin sections of brain tissue taken at post-mortem. It is this appearance which gave the spongiform encephalopathies their name. There is no conventional immune response to the agent, although the immune system plays an important part in the development of the disease before the agent gets into the CNS.

A diagnosis of TSE can only be made by demonstrating the presence of prion proteins in tissue taken at post-mortem. This is usually by immunohistochemical staining of the prion protein or by transfer of the agent to a permissive experimental animal.

Molecular nature of prions

Prions differ from all other forms of infectious agent in several respects. Primarily they lack any nucleic acid and, therefore, have no genomes to code for their progeny and yet they do 'reproduce' themselves.

Evidence for the proteinaceous nucleic acid-free nature of the agent has come from a number of experiments in which the chemical and physical nature of the infectious agents have been examined. They are resistant to heat inactivation, indeed high temperature autoclaving at 135°C for 18 minutes does not eliminate infectivity. Further evidence suggests that heating at 600°C (dry heat) does not totally kill a suspension of the agent. They are resistant to both ultraviolet light and ionizing radiation, treatments which normally damage microbial genomes (nucleic acid). Resistance is also shown to DNAse and RNAse treatment and to Zn^{2+} catalyzed hydrolysis. They are, however, sensitive to urea, SDS, phenol and other protein-denaturing chemicals. These characteristics are indicative of an agent which is of a proteinaceous character, lacking nucleic acid, rather than a virus, i.e. a novel agent or a so-called **prion (PrP)**.

The only known and demonstrated agent is a modified protein coded for by a cellular gene. The cellular isoform, **PrP^c^**, undergoes conformational change to the so-called infectious or scrapie form **PrP^sc^**. **Pri^sc^** is the term used for the infectious form of any TSE, although some authors use specific terms e.g. PrP^{cjd}. The two proteins have different confirmations. PrP^{sc} can propagate itself by inducing normal prion protein molecules to adopt the abnormal conformation. The PrP^{c} structure is that of a **30% helix** with no β sheet whereas on converson the PrP^{sc} adopts the **30% helix, 45% β sheet** configuration (*Fig. 1*) PrP^{c} exists in all mammals and birds examined to date and the protein is found anchored to external surfaces of cells by a glyolipid moiety. Its exact function is unknown.

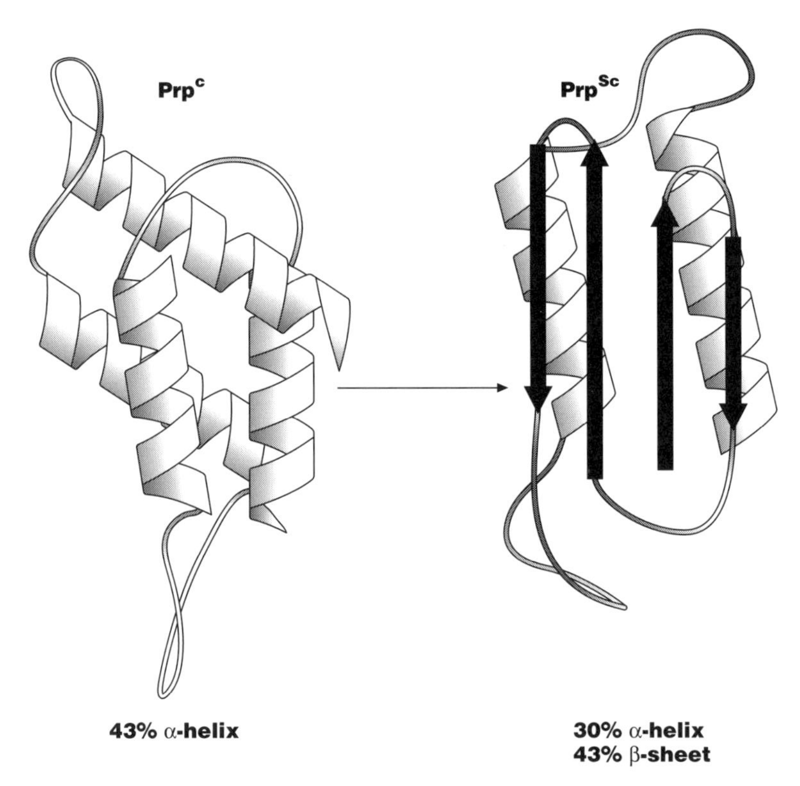

Fig. 1. Conformational changes in PrP. Redrawn from Principles of Molecular Virology, *A. Cann, Academic Press, London, 2001.*

However, transgenic 'knock-out' mice which lack the PrP^{c} gene appear to be normal. Infection of these mice with PrP^{sc} does not lead to disease, which is good evidence that PrP^{c} is necessary in the pathogenesis of TSEs. It also suggests that PrP^{sc} is itself not capable of replication.

As with conventional microbes there are different strains of prion. In humans there are two common versions of the prion protein that differ in a single amino acid (**valine or methionine at coding position 129**) but there are also rare mutant forms, many of which are associated with inherited susceptibility to prion disease. Sheep too have several different forms linked to susceptibility, whereas cattle have two forms which do not appear to be associated with susceptibility to BSE.

The infectious agents causing FSE and disease in exotic ruminants in zoos were found to be indistinguishable from the BSE strain, and although classical CJD is distinct, the transmissible agent that causes **new variant CJD** (vCJD) is the same as that which causes BSE.

There are a number of questions to be answered as to the cross-species barriers that may or may not exist in the infectivity of these agents. It appears that BSE arose by feeding cattle with scrapie-contaminated food-stuffs i.e. cross-species infectivity. Some strains of BSE can be propagated by several animal species, each of which have their own but different normal prion protein. The propagated prion (PrP^{sc}) is identical in conformation regardless of the host species, demonstrating that its final structure is independent of the host.

Certainly the BSE prion appears to have infected humans and resulted in vCJD. To what extent this has happened nationwide we still do not know.

Animal TSEs

Scrapie has been recognized as a distinct infection in sheep for over 250 years. A major investigation into its etiology followed the vaccination of sheep for louping-ill virus with formalin-treated extracts of ovine lymphoid tissue, unknowingly contaminated with scrapie prions. Two years later, more that 1500 sheep developed scrapie from this vaccine. The scrapie agent has been extensively studied and experimentally transmitted to a range of laboratory animals, e.g. **mice and hamsters**. Infected sheep show severe and progressive neurological symptoms, such as abnormal gait. The name owes itself to the fact that sheep with the disorder repeatedly scrape themselves against fences and posts. The natural mode of transmission between sheep is unclear although it is readily communicable in the flocks. The placenta has been implicated as a source of prions, which could account for horizontal spread within flocks. In Iceland scrapie-infected flocks of sheep were destroyed and the pastures left vacant for several years. However, reintroduction of sheep from flocks known to be free of scrapie for many years eventually resulted in scrapie! Sheep have also been infected by feed-stuff contaminated with BSE, to which they are susceptible.

Bovine spongiform encephalopathy (BSE, 'mad cow' disease) appeared in Great Britain in 1986 as a previously unknown disease. Affected cattle showed altered behavior and a staggering gait. Post-mortem revealed protease-resistant PrP in the brains of the cattle and the typical spongiform pathology. To date >190 000 cattle have been infected with this agent which, it is thought, initiated from the common source of **contaminated meat and bone meal** (MBM) given to cattle as a nutritional supplement. MBM was initially prepared by rendering the offal of sheep and cattle using a process that involved steam treatment and hydrocarbon solvent extraction. In the late 1970s, however, the solvent procedure was eliminated from the process, resulting in high concentrations of fat in the MBM; it is postulated that this high fat content protected scrapie prions in the sheep offal from being completely inactivated by the steam. Thus the initial MBM contained only sheep prions, and the similarity between bovine and sheep PrP was probably an important factor in initiating the BSE epidemic. Bovine PrP differs from sheep PrP at 7 or 8 residues. As the BSE epidemic expanded, infected bovine offal began to be rendered into MBM that contained bovine prions, this being fed to cattle!

Since **1988** the practice of using dietary protein supplements for domestic animals derived from rendered sheep or cattle offal has been forbidden in the UK. Statistics argue that this food ban has been effective in getting the epidemic under control (*Fig. 2*).

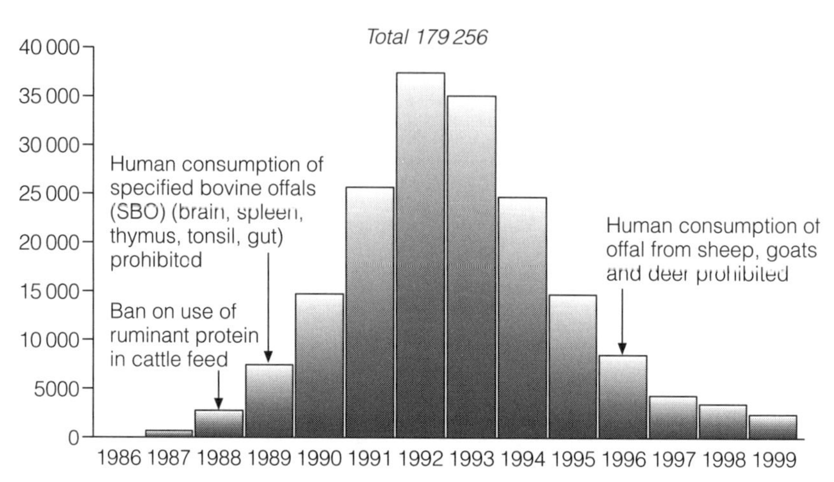

Fig. 2. Reported incidence of BSE in the UK, 1986–1999. Redrawn from Principles of Molecular Virology, *A. Cann, Academic Press, London, 2001.*

Brain extracts from BSE cattle have transmitted disease to mice, cattle, sheep and pigs after intracerebral inoculation. Disease has also followed in mink, domestic cats, pumas and cheetahs after oral consumption of BSE prions in food-stuffs.

Evidence is accumulating that the BSE prion has been transferred to humans and is responsible for vCJD. The oral route is suspected. The source of the infected material is unknown, although BSE-contaminated cattle products are the probable source. In **1989** the human consumption of specified bovine offals (brain, spleen, thymus tonsil, gut) was prohibited in the UK. In **1996** this ban was extended to sheep. Furthermore, the brain and CNS are now removed from all cattle at abattoirs prior to the distribution of meat.

Human TSEs

Kuru was the first human TSE to undergo extensive investigation. The first cases were recorded in the 1950s and occurred in male and female adolescents and adult women members of the **Fore people** in the highlands of New Guinea. The Fore people practised **ritual cannibalism** as a rite of mourning for their dead, the women and children but not the adult men taking part in the ceremony. Those with the disease demonstrated progressive loss of voluntary neuronal control, followed by death less than one year after the onset of symptoms. The incubation period for Kuru can be up to 30 years although normally it is shorter. At one point Kuru was the leading cause of death of women in the tribes. As the cannibalistic ritual ceased so did transmission of the disease. However, Kuru demonstrates that a human form of TSE can be acquired via the oral route.

Patients with **Creutzfeldt-Jakob disease** present with a progressive sub-acute or chronic decline in cognitive or motor function. CJD is rare, with an annual worldwide incidence of approximately **one per million population**. The disease is acquired **sporadically** (transmission route largely unknown), is **iatrogenic** (accidental transmission) or **familial** (10% of cases). Thus CJD has been transmitted by corneal transplantation, contaminated EEG electrode implantation, surgical operations using contaminated instruments, human pituitary growth hormone, human pituitary gonadotrophin hormone and dura mater grafts.

Recently accumulated evidence suggests that caucasian patients who are homozygons for **Met** or **Val** at **codon 129** of the PrP gene are more susceptible to sporadic CJD.

In April 1996 a new variant of CJD (**vCJD**) was described in the UK. The disease has features which distinguish it from other forms of CJD. These include an **early age of onset** (average 27 years as opposed to 65 for CJD), **prolonged period of illness** (average 13 months as opposed to 3 months for CJD) **a psychiatric presentation as opposed to neurological symptoms and the absence of the typical EEG appearances of CJD**. In terms of pathology vCJD bears a close resemblance to that of Kuru, particularly with regard to the type of amyloid plaque formation.

The official **UK Spongiform Encephalopathy Advisory Committee** have stated that 'although there is no direct evidence of a link, on current data and in the absence of any credible alternative, the most likely explanation at present is that these cases are linked to exposure to BSE. Indeed, one cluster of cases has previously been tentatively linked to blood on a butcher's knife.

As more cases of vCJD appear in the UK, epidemiologists will be in a position to predict the maximum number of deaths that may follow over the next two decades. However, one worry which at the moment of writing is not substantiated, is that the vCJD agent may be transmitted between humans.

The vCJD scenario has raised a number of political, sociological and scientific questions which will continue to be debated into the future.

FURTHER READING

There are many comprehensive textbooks of microbiology, but no one book that can satisfy all needs. Different readers subjectively prefer different textbooks and hence we do not feel that it would be particularly helpful to recommend one book over another. Rather we have listed some of the leading books which we know from experience have served their student readers well.

General reading

Madigan, M.T., Martinko, J.M. and Parker, J. (1997) *Brock Biology of Microorganisms*, 8th Edn. Prentice-Hall Inc., Upper Saddle River, NJ.

Prescott, L., Harley, J.P. and Klein, D.A. (1996) *Microbiology*, 3rd Edn. Wm C. Brown Communications Inc., Dubuque, IA.

Singleton, P. and Sainsbury, D. (1987) *Dictionary of Microbiology and Molecular Biology*, 2nd Edn. John Wiley & Sons, New York.

Collier (1997) *Topley & Wilson's Principles of Bacteriology, Virology and Immunity*, 9th Edn. Arnold Publishing, London.

Tortora, G.J., Funke, B.R. and Case, C.L. (1998) *Microbiology: An Introduction*, 6th Edn. The Benjamin-Cummings Publishing Co., Redwood City, CA.

More advanced reading

The following books are recommended for readers who wish to study a particular area in more depth.

Section A

Roberts, D. McL., Sharp, P., Alderson, G. and Collins, M. (Eds) (1996) *Evolution of Microbial Life: Society for General Microbiology 54th Symposium*. Cambridge University Press, Cambridge.

Section B

Dawes, I.W. and Sutherland, I.W. (1992) *Microbial Physiology*, 2nd Edn. Blackwell Scientific Publications, Oxford.

Hames, B.D., Hooper, N.M. and Houghton, J.D. (1997) *Instant Notes in Biochemistry*. BIOS Scientific Publishers Limited, Oxford.

Moat, A.G. and Foster, J.W. (1995) *Microbial Physiology*, 3rd Edn. Wiley-Liss, New York.

Neidhardt, F.C., Ingraham, J.L. and Schaechter, M. (1990) *Physiology of the Bacterial Cell: A Molecular Approach*. Sinauer Associates, New York.

Rhodes, P.M. and Stanbury, P.F. (1997) *Applied Microbial Physiology: A Practical Approach*. Oxford University Press, Oxford.

Schlegel, H.G. (1993) *General Microbiology*, 7th Edn. Cambridge University Press, Cambridge.

Section C

Alberts, B., Bray, D., Lewis., Raff, M., Roberts, K. and Watson, J.D. (1994) *Molecular Biology of the Cell*, 3rd Edn. Garland Publishing, New York.

Beebee, T.J.C. and Burke, J. (1992) *Gene Structure and Transcription: In Focus*. IRL Press, Oxford.

Calladine, C.R. and Drew, H.R. (1997) *Understanding DNA: The Molecule and How It Works*, 2nd Edn. Academic Press, Orlando, FL.

Creighton, T.E. (1993) *Protein Structures and Molecular Properties*, 2nd Edn. W.H. Freeman and Co., New York.

Kornberg, A. and Baker, T. (1991) *DNA Replication*, 2nd Edn. W.H. Freeman and Co., New York.

Lewin, B. (1997) *Genes VI*. Oxford University Press, Oxford.

Lodish, H., Baltimore, D., Berk, A., Zipursky, S.L., Matsudaira, P. and Darnell, J. (1995) *Molecular Cell Biology*, 3rd Edn. Scientific American Books, New York.

Turner, P.C., McLennan, A.G., Bates, A.D. and White, M.R.H. (1997) *Instant Notes in Molecular Biology*. BIOS Scientific Publishers Limited, Oxford.

Smith, C.A. and Wood, E.J. (1991) *Biological Molecules*. Chapman & Hall, London.

Section D

Barrow, G.I. and Feltham, R.K.A. (1993) *Cowan and Steel's Manual for the Identification of Medical Bacteria*, 3rd Edn. Cambridge University Press, Cambridge.

Cappucino, T.G. and Sherman, N. (1996) *Microbiology: A Laboratory Manual*, 4th Edn. The Benjamin-Cummings Publishing Co., Redwood City, CA.

Chan, E.C.S., Pelczar, M.J. and Krieg, N.R. (1993) *Laboratory Exercises in Microbiology*, 6th Edn. McGraw-Hill.

Evans, W.H. and Graham, T.M. (1989) *Membrane Structure and Function*. IRL Press, Oxford.

Isaac, S. and Jennings, D. (1995) *Microbial Culture*. BIOS Scientific Publishers Limited, Oxford.

Logan, N.A. (1994) *Bacterial Systematics*. Blackwell Scientific Publications, Oxford.

Penn, C. (1991) *Handling Laboratory Microorganisms*. Open University Press, Milton Keynes.

Section E

Brown, T.A. (1992) *Genetics: A Molecular Approach*, 2nd Edn. Chapman & Hall, London.

Dale, J. (1998) *Molecular Genetics of Bacteria*, 3rd Edn. Wiley-Interscience, New York.

Freidberg, E.C., Walker, G.C. and Siede, W. (1995) *DNA Repair and Mutagenesis*. American Society for Microbiology, Washington, DC.

Hardy, K. (1986) *Bacterial Plasmids*, 2nd Edn. Van Nostrand Reinhold, New York.

Maloy, S.R., Cronan, J.E. and Freifelder, D. (1994) *Microbial Genetics*, 2nd Edn. Jones & Bartlett Publishers Inc., Portola Valley, CA.

Section F

Janeway, C.A. and Travers, P. (1997) *Immunobiology: The Immune System in Health and Disease*, 3rd Edn. Garland Publishing, New York.

Mims, C., Dimmock, N., Nash, A. and Stephen, J. (1995) *Mims' Pathogenesis of Infectious Disease*, 4th Edn. Academic Press, London.

Russell, A.D. and Chopra, I. (1996) *Understanding Antibacterial Action and Resistance*, 2nd Edn. Ellis Horwood, Hemel Hempstead.

Salyers, A. and Whitt, D.D. (1994) *Bacterial Pathogenesis: A Molecular Approach*. American Society for Microbiology, Washington, DC.

Schaechter, M., Medoff, G. and Eisenstein, B.I. (1993) *Mechanisms of Microbial Disease*. Williams & Wilkins, Baltimore, MD.

Section G

http://phylogeny.arizona.edu/tree/eukaryotes/eukaryotes.html

Hall, J.L. and Hawes, C.R. (1991) *Electron Microscopy of Plant Cells*. Academic Press, London.

Sadava, D. (1993) *Cell Biology: Organelle Structure and Function*. Jones & Bartlett, London.
Smith, C.A. and Wood, E.J. (1996) *Cell Biology*. Chapman & Hall, London.

Section H

http://phylogeny.arizona.edu/tree/eukaryotes/fungi/fungi.html
Alexopoulos, C.J., Mims, C.W. and Blackwell, M. (1996) *Introductory Mycology*, 4th Edn. John Wiley & Sons, Chichester.
Carlile, M.J. and Watkinson, S.C. (1994) *The Fungi*. Academic Press, London.
Deacon, J.W. (1998) *Modern Mycology*. Blackwell Science Ltd, Oxford.
Ingold, C.T. and Hudson, H.J. (1993) *The Biology of Fungi*. Chapman & Hall, London.
Isaacs, S. (1992) *Fungal-Plant Interactions*. Chapman & Hall.
Jennings, D.H. and Lysek, G. (1996) *Fungal Biology*. BIOS Scientific Publishers Limited, Oxford.
Manners, J.G. (1993) *Principles of Plant Pathology*. Cambridge University Press, Cambridge.

Section I

http://phylogeny.arizona.edu/tree/eukaryotes/green-plants.html
Bold, H.C. and Wynne, M.J. (1985) *Introduction to the Algae*. Prentice-Hall, New Jersey.
South, G. and Whittick, A. (1987) *Introduction to Phycology*. Blackwell Scientific Publications, Oxford.
van den Hoek, C. and Mann, D.G. (1996) *Algae, an Introduction to Phycology*. Cambridge University Press, Cambridge.

Section J

Anderson, O.R. (1988) *Comparative Protozoology: Ecology, Physiology, Life History*. Springer-Verlag, New York.
Biagini, G.A., Rutter, A.J., Finlay, B.J. and Lloyd, D. (1998) Lipids and lipid metabolism in the microaerobic free-living diplomonad *Hexamita* sp. *Eur. J. Protistol.* **34**(2), 148–152.
Cox, F.E.G. (ed.) (1993) *Modern Parasitology: A Textbook of Parasitology*. Blackwell Scientific Publications, Oxford.
Gordon, G.LR. and Phillips, M.W. (1998) The role of anaerobic gut fungi in ruminants. *Nutr. Res. Rev.* **11**(1), 133–168.
Green, R.F. and Noakes, D.L.G. (1995) A little bit of sex is as good as a lot. *J. Theoret. Biol.* **174**(1) 87–96.
Ridley, R.G. (1997) Plasmodium: Drug discovery and development – an industrial perspective. *Exper. Parasitol.* **87**(3), 293–304.
Rosenthal, P.J. (1998) Proteases of malaria parasites: New targets for chemotherapy. *Emerg. Infect. Dis.* **4**(1), 49–57.
Stevens, J.R., Tibayrenc, M. (1996) *Trypanosoma Brucei* evolution linkage and the clonality debate. *Parasitol.* **34**(2), 148–152.

INDEX